Sustainability science: A multidisciplinary approach

Edited by Hiroshi Komiyama, Kazuhiko Takeuchi, Hideaki Shiroyama and Takashi Mino

United Nations University Press

TOKYO · NEW YORK · PARIS

SYNERGY

Indian Edition, 2012

ISBN: 978-93-82059-01-1

First published in 2011 by the United Nations University Press,
The United Nations University, 53-70 Jingumae 5-chome,
Shibuya-ku, Tokyo 150-8925
For copyright reasons, this edition is for sale in South Asia only.

Indian Edition Published by:

Synergy Books India

(A Unit of M/s Bookwell Publications)
24/4800 Ansari Road, Darya Ganj,
New Delhi-110002
Phone: 011-23268786, 23257264
Fax: 011-23281315
Website: www.synergybooksindia.com
E-Mail: synergybooksindia@gmail.com
bookwell@vsnl.net

Printed at : DK FINE ART PRESS (P). LTD. Delhi-92

Contents

Figures

Tables

Plate

Box

Contributors

Megan Bailey is a doctoral student in the Fisheries Economics Research Unit at the University of British Columbia's Fisheries Centre, Canada. Her research focuses on applying game theory to internationally shared resources in an effort to encourage cooperation among fishing nations, in the hope of achieving sustainable fisheries.

Kensuke Fukushi is Associate Professor of the Integrated Research System for Sustainability Science at The University of Tokyo, Japan. He has an adjunct appointment at the Department of Urban Engineering as well as the Graduate Program in Sustainability Science at the same university.

Harumoto Gunji is an Associate Professor in the College of Education and the Institute for Global Change Adaptation Science, Ibaraki University, Japan.

Mitsutsugu Hamamoto is Professor of Environmental Economics in the Faculty of Economics, Dokkyo University, Japan.

Keishiro Hara is an Associate Professor in the Research Institute for Sustainability Science (RISS) at Osaka University, Japan. His research interests include urban environmental management, sustainable resources management and sustainability science methodologies. He received his PhD in Environmental Studies from The University of Tokyo in 2004.

Hideyuki Hirakawa is an Associate Professor in the Center for the Study of Communication-Design (CSCD) at Osaka University, Japan. He has published articles and books on science and technology governance, especially focusing on public engagement.

Takeru Hirota graduated from the Division of Electrical, Electronic and Information Engineering, Graduate School of Engineering, Osaka University, Japan. Currently he works at JSOL Corporation.

Gakushi Ishimura is Assistant Professor at the Center for Sustainability Science, Hokkaido University, Japan. His research focuses on the sustainable management of natural resources by applying economic and quantitative analysis. He holds a PhD in resource management from the University of British Columbia, Canada.

Yuya Kajikawa is a Project Lecturer in the Innovation Policy Research Center, Institute of Engineering Innovation, Graduate School of Engineering, The University of Tokyo, Japan.

Hironori Kato is an Associate Professor in the Department of Civil Engineering, The University of Tokyo, Japan. His main research concerns are transportation planning and transportation policy. He has published papers on public transit planning, project evaluation and travel demand modelling.

Michinori Kimura is a Researcher in the Division of Sustainable Energy and Environmental Engineering, Graduate School of Engineering, Osaka University, Japan.

Hiroshi Komiyama became Chairman of the Mitsubishi Research Institute, Inc. and President Emeritus at The University of Tokyo, Japan, in April 2009, after completing a four-year presidency at The University of Tokyo.

Kouji Kozaki is Associate Professor in the Division of Information and Quantum Sciences, the Institute of Scientific and Industrial Research (ISIR), Osaka University, Japan.

Yasuhiko Kudo is a postdoctoral fellow at the Center for Sustainability Science (CENSUS), Hokkaido University, Japan. He holds a PhD degree in agricultural economics. His specialized field is agricultural economics and regional economics. He is currently studying sustainable community improvement, mainly in agriculture.

Terukazu Kumazawa is a postdoctoral fellow in the Ritsumeikan Global Innovation Research Organization (R-GIRO), Ritsumeikan University, Japan.

Nobuo Kurata is Professor of Applied Ethics and Philosophy in the Graduate School of Letters at Hokkaido University, Japan. He has published articles on environmental ethics and applied ethics. He has also contributed to some participatory technology assessments.

Hirotaka Matsuda is a Project Lecturer in the Transdisciplinary Initiative for Global Sustainability (TIGS) / Integrated Research System for Sustainability Science (IR3S), The University of Tokyo, Japan.

Takanori Matsui is an Assistant Professor in the Division of Sustainable Energy and Environmental Engineering, Graduate School of Engineering, Osaka University, Japan. His primary research topics are in sustainability design and risk.

Makiko Matsuo is a Project Researcher in the Graduate School of Public Policy and the Graduate Schools for Law and Politics, The University of Tokyo, Japan.

Masahiro Matsuura is Associate Professor of Ocean Policy at the Graduate School of Public Policy at The University of Tokyo, Japan. He has published many books on negotiation and consensus-building, including *Localizing Public Dispute Resolution in Japan* (VDM-Verlag, 2005).

Takashi Mino is a Professor in the Division of Environmental Studies and Senior Advisor to the Graduate Program in Sustainability Science, Graduate School of Frontier Sciences, The University of Tokyo, Japan. He originally majored in wastewater engineering and his current interests concern sustainability education.

Riichiro Mizoguchi is a Professor in the Division of Information and Quantum Sciences, the Institute of Scientific and Industrial Research (ISIR), Osaka University, Japan.

Akihisa Mori is an Associate Professor in the Graduate School of Global Environmental Studies, Kyoto University, Japan. Through his research on development and environmental policies in East Asia, he has written and edited several books, including *Environmental Aid* (Yuhikaku, 2009) and *Environmental Policy in China* (Kyoto University Press, 2008).

Mitsuhiro Nakagawa is a Professor in the Department of Regional and Environmental Science, College of Agriculture, Ibaraki University, Japan. His major research concerns are the prediction of global food supply and demand, sustainable rural development and environmental education.

Motoharu Onuki is a Project Associate Professor of the Graduate Program in Sustainability Science (GPSS), Graduate School of Frontier Sciences, The University of Tokyo, Japan. He also works for the Asian Program for Incubation of Environmental Leaders (APIEL) at GPSS, an educational programme to foster environmental leadership.

Mitsuru Osaki is a Professor in the Research Faculty of Agriculture and Vice Director of the Center for Sustainability Science (CENSUS) at Hokkaido University, Japan. He has carried out many collaborative researches and projects on sustainability in the areas of tropical peat management, rehabilitation of tropical forest, rhizosphere management, biochar, super high-yielding crop varieties and bioenergy.

Hisashi Otsuji belongs to the Department of Science Education, College of Education, Ibaraki University, Japan. His research is in the new field of the cultural studies of science education, and his interest focuses on identifying invisible cultural influences on science education, in particular Buddhism.

Osamu Saito is an Assistant Professor at Waseda Institute for Advanced Study, Waseda University, Tokyo, Japan. His work focuses on ecosystem services management, with a particular interest in the interlinkages between ecological, human and social systems. His experience includes socio-ecological studies on the ecosystem services provided by the agricultural rural landscape (*satoyama*) in both Japan and other Asian countries.

Jin Sato is Associate Professor in the Institute of Advanced Studies on Asia at The University of Tokyo, Japan. He has a PhD in International Studies from The University of Tokyo, and was a postdoctoral fellow at Yale University, USA, in 1998/99. He is currently a visiting fellow at Princeton University, USA, conducting research on democracy and resource governance in Asia.

Takamitsu Sawa is President of Shiga University, Chairman of the Council of Transportation and member of the Central Environmental Council of the Japanese government. He was President of the Society of Environmental and Policy Studies in Japan from 1995 to 2006. He was awarded a Purple Ribbon Medal in 2007.

Yoshiyuki Shimoda is a Professor in the Department of Sustainable Energy and Environmental Engineering, Osaka University, Japan. He specializes in urban energy system engineering, end-use energy demand forecasting, building physics and urban climatology.

Hideaki Shiroyama is a Professor of Public Administration in the Graduate School of Law and Politics and Graduate School of Public Policy, The University of Tokyo, Japan.

Akimasa Sumi is Professor and Executive Director of the Transdisciplinary Initiative for Global Sustainability (TIGS), Integrated Research System for Sustainability Science (IR3S), at The University of Tokyo. Japan. Before that, he was the Director of the Center for Climate System Research (CCSR), The University of Tokyo, and has been conducting research on climate change and climate modelling.

Makio Takemura is the president of Toyo University, Japan, and a Professor in the Faculty of Humanities of the university. He was also the representative of the Transdisciplinary Initiative for Eco-Philosophy at Toyo University, which is one of the cooperating institutions of the Integrated Research System for Sustainability Science (IR3S) at The University of Tokyo, Japan. His speciality is Buddhist studies, especially Japanese Buddhism. He holds a PhD in Literature from The University of Tokyo. His most recent book is *Introduction to Buddhism as Philosophy* (Kodansha, 2009).

Kazuhiko Takeuchi is a Professor in the Graduate School of Agricultural and Life Sciences and the deputy executive director of the Integrated Research System for Sustainability Science (IR3S) at The University of Tokyo, Japan. He is also a Vice-Rector of the United Nations University and Director of the United Nations University Institute of Sustainability and Peace.

Makoto Tamura is an Associate Professor in the Institute for Global Change Adaptation Science (ICAS),

Ibaraki University, Japan. He received his MA and PhD degrees from the Graduate School of Arts and Sciences, The University of Tokyo, Japan. His research interests are impact assessment and countermeasures for climate change, and the interrelationship between economic activity and environment.

Noriyuki Tanaka is Professor and Vice Director at the Center for Sustainability Science (CENSUS), Hokkaido University, Japan, and also chair of the Japan Science and Technology Agency (JST) strategic coordinated training programme for Sustainability Leaders and Sustainability "Meisters" (StraSS). He also teaches at the Division of Sustainable System Development, Graduate School of Environmental Science, Hokkaido University.

Nobuyuki Tsuji is employed as an Associate Professor (fixed term) by the Sustainability Governance Project, Center for Sustainability Science, Hokkaido University, Japan. He has a PhD degree in mathematical ecology. He is currently studying the mathematical analysis of regional sustainability.

Takahide Uegaki is an educational coordinator of the Graduate Program on Sustainability Science at Ibaraki University, Japan. He received his PhD degree from the Tokyo University of Agriculture and Technology. His studies are mainly based on environmental philosophy and his recent interest is the social theory of the three unsustainabilities.

Michinori Uwasu is currently Assistant Professor in the Sustainability Design Center, Osaka University, Japan. When the section was written, he was Assistant Professor in the Research Institute for Sustainability Science, Osaka University, Japan.

Helmut Yabar is Associate Professor at the Graduate School of Life and Environmental Sciences, University of Tsukuba, Japan. His research interests include integrated resource and waste management, sustainability indicator development and the linkages between environmental policy and technology innovation.

Masaru Yarime is Associate Professor of the Graduate Program in Sustainability Science (GPSS) in the Graduate School of Frontier Sciences, The University of Tokyo, Japan. His research interests include corporate strategy, public policy and institutional design for sustainability innovation, university–industry collaboration, and the structural analysis of knowledge creation, diffusion and utilization.

Hiroyuki Yoshikawa is Director-General of the Center for Research and Development Strategy at the Japan Science and Technology Agency. Previously he was President of The University of Tokyo and President of the International Council of Science. He was also President of the National Institute of Advanced Industrial Science and Technology in Japan.

Preface

This book forms part of a series on sustainability science. Sustainability science is a newly emerging academic field that seeks to understand the dynamic linkages between global, social and human systems, and to provide a holistic perspective on the concerns and issues between and within these systems. It is a problem-oriented discipline encompassing visions and methods for examining and repairing these systems and linkages.

The Integrated Research System for Sustainability Science (IR3S) was launched in 2005 at The University of Tokyo with the aim of serving as a global research and educational platform for sustainability scientists. In 2006 IR3S expanded, becoming a university network including Kyoto University, Osaka University, Hokkaido University and Ibaraki University. In addition, Tohoku University, the National Institute for Environmental Studies, Toyo University, Chiba University, Waseda University, Ritsumeikan University and the United Nations University joined as associate members. Since the establishment of the IR3S network, member universities have launched sustainability science programmes at their institutions and collaborated on related research projects. The results of these projects have been published in prestigious research journals and presented at various academic, governmental and social meetings.

The *Sustainability Science* book series is based on the results of IR3S members' joint research activities over the past five years. The series provides directions on sustainability for society. These books are expected to be of interest to graduate students, educators teaching sustainability-related courses and those keen to start up similar programmes, active members of NGOs, government officials and people working in industry. We hope this series of books will provide readers with useful information on sustainability issues and present them with novel ways of thinking and solutions to the complex problems faced by people throughout the world.

Integrated Research System for Sustainability Science

1

Introduction

1-1

Sustainability science: Building a new academic discipline

Hiroshi Komiyama and Kazuhiko Takeuchi

1-1-1 Introduction

In scientific and academic circles worldwide, the opportunity to develop the emerging discipline of sustainability science has never been greater. This new science has its origins in the concept of sustainable development proposed by the World Commission on Environment and Development (WCED), also known as the Brundtland Commission (WCED, 1987). Defining sustainable development as "development that meets the needs of the present without compromising the ability of future generations to meet their own needs", the WCED garnered global support for its argument that development must ensure the coexistence of the economy, society and the environment. Today, sustainability is recognized the world over as a key issue facing twenty-first-century society.

However, it has also been remarked that the idea of sustainable development increasingly appears to be linked to political agendas, raising concerns about the solidity of its analytical basis; and the scientific and technological underpinnings of the concept remain unclear to many (Cohen et al., 1998). During the 1990s, the International Council for Science (ICSU) initiated studies of science and technology for sustainable development. There were, with growing frequency, calls for a science of sustainability that would be predicated on recognition of the fundamental links between science and technology, and between economics and society, while remaining free from political bias of the sort seen, for example, when North–South issues are raised in debates over sustainable development (Clark and Dickson, 2003; ICSU, 2002; Kates et al., 2001).

As a result, efforts to build a new sustainability science have accelerated in academia, particularly in Europe and North America. Spearheading the development of a broadly comprehensive discipline, one that extends beyond problem-specific research, is the Forum on Science and Innovation for Sustainable Development, administered by Professor Bill Clark of the John F. Kennedy School of Government at Harvard University and his colleagues under the auspices of the American Association for the Advancement of Science (AAAS), one of the world's preeminent scientific organizations. Researchers from Europe and Japan as well as the United States participate in this forum, thus enhancing collaborative efforts towards a science of sustainability on a global scale (Clark, 2007).

In Europe, several institutions have emerged that address specific issues of sustainability science. The activities of the Tyndall Centre for Climate Change Research, a network of universities based at the University of East Anglia in the United Kingdom, and the Potsdam Institute for Climate Impact Research (PIK) in Germany, whose work has a significant influence on the European Union's climate policy, have attracted international attention. The Stockholm Resilience Centre (SRC) in Sweden has become a global hub for research on the resilience of ecosystems, a concept key to the study of ecological and societal adaptability to climate change. And the Interuniversity Research Centre on Sustainable Development (CIRPS), a network based at Sapienza University of Rome, Italy, has done significant research on energy sustainability, particularly on the development of new transport-related energy policies.

At The University of Tokyo, the need for a new academic discipline of sustainability science has grown increasingly evident during 10 years of collaborative research and education initiatives through the Alliance for Global Sustainability (AGS) with the Massachusetts Institute of Technology (MIT), the Swiss Federal Institute of Technology (ETH), and Chalmers University of Technology in Sweden. The university has been fortunate to receive support from the Special Coordination Funds for Promoting Science and Technology (SCF) of Japan's Ministry of Education, Culture, Sports, Science and Technology (MEXT) for the purpose of building a sustainability science network in Japan and working towards sustainability from a global perspective, particularly in Asia. In August 2005, The University of Tokyo inaugurated the Integrated Research System for Sustainability Science (IR3S) and invited universities throughout Japan to participate, thus launching a full-scale effort to set up a nationwide research network. In April 2006, IR3S began an active programme of research and educational activity addressing sustainability issues on a global scale, but with a special focus on Asia.

The series of which this volume is a part is a compilation of the fruits of research, education and other socially contributory activities by IR3S since 2005. Hence most of the sections in this series are written by Japanese researchers at IR3S member universities and institutions, and the examples cited herein are found primarily in Asia, in keeping with IR3S's own focus since its inception. However, the methodologies employed, particularly as they pertain to the development of sustainability science, are of course applicable to any region of the world. Moreover, the issues addressed here are relevant to problems of sustainability anywhere, in developed and developing countries alike, not just in Japan and Asia.

1-1-2 Organization and activities of IR3S

The five participating universities of IR3S have all established centres for sustainability-related research and education:

- the Transdisciplinary Initiative for Global Sustainability (TIGS) at The University of Tokyo,
- the Kyoto Sustainability Initiative (KSI) at Kyoto University,
- the Research Institute for Sustainability Science (RISS) at Osaka University,
- the Sustainability Governance Project (SGP) at Hokkaido University, and
- the Institute for Global Change Application Science (ICAS) at Ibaraki University.

Each centre has its particular area of focus: TIGS on the development of sustainability strategies through knowledge structuring; KSI on socioeconomic reform and technological strategies; RISS on the design of a resource-circulating society based on ecologically sound industrial technology; SGP on the creation of sustainable biomass production zones and regional governance; and ICAS on climate change adaptation strategies appropriate for the Asia Pacific region.

To promote collaborative research among its participating institutions, IR3S initiated three flagship projects: "Sustainable Countermeasures for Global Warming", "Development of an Asian Resource-Circulating Society" and "The Conceptual Framework of Global Sustainability: Appropriate Reform of the Socioeconomic System and the Role of Science and Technology". The results of these joint research projects, embodying as they do the objective of furthering sustainability science that is the mission of IR3S, are reported in detail elsewhere in this series.

Under the aegis of TIGS, IR3S is also working with several cooperating institutions on specific research problems outside the purview of the three flagship projects. These institutions and their research topics are:

- Toyo University (philosophy of coexistence)
- the National Institute for Environmental Studies (long-term scenarios for environmental policy)
- Tohoku University (environmental risk)
- Chiba University (food and health)
- Waseda University (politics and journalism)
- Ritsumeikan University (strategic innovation).

In 2009, the final year of SCF support of IR3S, these institutions were joined by United Nations University, headquartered in Tokyo, which will cooperate in the development of an international sustainability research meta-network. The cooperating institutions not only complement the research activities of the IR3S participating universities but may be said to represent the formation of a full-fledged "Team Japan" amply equipped to further the goals of sustainability science. With the ending of the SCF funding period, the plans call for the establishment of a Sustainability Science Consortium independent of The University of Tokyo-based IR3S with the addition of more cooperating institutions so as to form a truly well-rounded "Team Japan".

In parallel with its joint research projects, IR3S has given equal weight to sustainability science education programmes. To foster a new discipline such as sustainability science, education programmes must be implemented that will train a new generation of professionals in the field. Requiring as it does an academic framework that spans multiple disciplines, ranging from the sciences to the humanities, sustainability science needs specialists who are well versed in their fields but at the same time have a broad grasp of how their area of specialization fits into the larger picture of sustainability-related issues. The University of Tokyo has established a Graduate Program in Sustainability Science on its Kashiwa campus at the Institute of Environmental Studies, Graduate School of Frontier Sciences. To nurture an international outlook in its graduates and to facilitate the enrolment of students from abroad – particularly from the developing nations of Asia – courses in this programme are conducted in English.

The other IR3S participating universities have also established sustainability science education programmes reflective of their varying strengths and specialties. The five participating universities inaugurated the Sustainability Science Collaborative Education Program, a master's degree program, with the shared aim of nurturing graduates who fully understand the diverse, international, interdisciplinary nature of the sustainability concept and are capable of putting this understanding into practice through public activity. Through this joint education programme, participating universities permit students from one another's campuses to attend classes and field training programmes at each university, thus expanding their range of options in acquiring sustainability-related knowledge and experience. The joint programme also affords students from the various partner campuses greater exposure to a diversity of courses conducted in English, of which there is a dearth at Japanese universities. Graduates of the programme are awarded joint course completion certificates from IR3S. The participating universities are also investigating the implementation of a dual/joint degree programme in sustainability science.

1-1-3 The concept of sustainability science

Through discussion among its participating institutions, IR3S has sought to clarify the concept of sustainability science, generally described as a discipline that points the way towards a sustainable society. In addition to addressing such issues as intergenerational equity in the context of sustainable development, the problem of sustainability is approached at three levels of "system" – global, social and human – as defined below. All three systems are crucial to the coexistence of human beings and the environment, and it is the authors' view that the current crisis of sustainability can be analysed in terms of the breakdown of these systems and the linkages among them.

The global system comprises the entire planetary base for human survival: the geosphere, atmosphere, hydrosphere and biosphere. The Earth sustains human life by providing natural resources, energy and a supportive ecosystem. The global system can exhibit great fluctuations in the Earth's climate and crust – the subject of the earth sciences – that profoundly affect human activity and survival. Conversely, the rapid expansion of human activity has also become a significant factor in fluctuations in the global system. Global warming and the destruction of the ozone layer are two salient examples of this human-induced change.

The social system consists of the political, economic, industrial and other structures created by human beings that provide the societal base for a fulfilling human existence. "Fulfilment" is often assumed to depend on economic growth and technological advancement, but this development also contains the seeds of such social problems as environmental pollution and the growing inequality between rich and poor. These problems, of which environmental issues are representative, transcend the confines of the social system in their impact, extending to the global system. Another social problem, the declining birth rate in developed countries (particularly in Asia), may be said to raise questions about the sustainability of the family, a fundamental unit of the social system. Issues such as these challenge us to re-examine notions of what constitutes a wealthy or fulfilled society.

The human system is the sum total of factors affecting the survival of individual human beings; it is, of course, intimately connected to the social system. The healthy functioning of the human system requires the establishment of lifestyles and values that enable people to live healthily, safely and securely – that is, not merely to survive but to experience a fulfilled life. In reality, however, human beings are adversely affected physically and emotionally by diseases, mental illness and inequities in the social system. An increase in such problems puts pressure on the social system. As this stress increases and the environment deteriorates, the human system itself becomes less healthy. Emblematic of this trend are the problems of extreme poverty – hunger, disease, lack of shelter, exclusion – which are especially prevalent in developing countries and are targeted by the UN Millennium Project under the framework of quantified Millennium Development Goals (UN Millennium Project, 2005). Disparities in values, as reflected in religious tensions, are also among the problems that threaten the sustainability of the human system. In the extreme, the weakening of sustainability and the concomitant impact on the health of the human system are manifested in increasing conflict and war.

What types of problems occur on a global scale as a result of the highly interactive relationship among these three systems, and what visions or scenarios do the solutions to these problems demand? Some examples are depicted in Figure 1.1.1.

A representative problem arising from the interaction between the global and social systems is global warming, which demands the development of a low-carbon society that embraces systemic and technological reforms leading to significantly reduced emissions of the gases that contribute to global warming. An example of a problem arising from the interaction between the social and human systems is the generation of waste. Here, what is required is the construction of a resource-circulating society, i.e. one capable of sustainable production and consumption (Sotherton et al., 2004). This demands the implementation of reduce–reuse–recycle (3R) policies, the development of manufacturing processes predicated on resource recirculation, and the cultivation of resource-conserving lifestyles. Finally, the interactive relationship between global and human systems involves particularly serious problems that directly affect human survival. Examples include the spread of infectious diseases and other health risks associated with global warming, the effect on human health of increased ultraviolet exposure owing to destruction of the

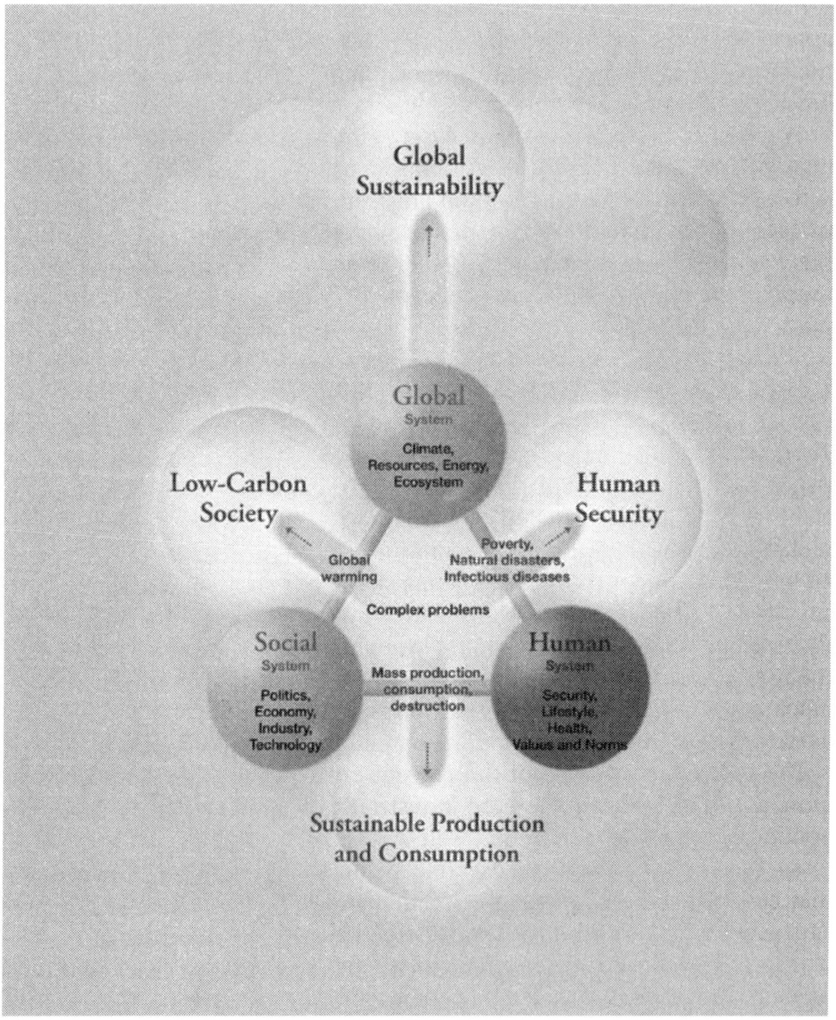

Figure 1.1.1 Addressing sustainability science through the lens of three systems, and the linkages among them.
Source: Komiyama and Takeuchi (2006).
Note: Please see page 467 for a colour version of this figure.

ozone layer (McCarthy et al., 2001; McMichael et al., 2003), and forced evacuations and loss of habitat caused by rising sea levels (Nicholls, 2004). Because these problems threaten human security and safety, it is essential to solve them if society is to achieve human well-being and sustainability. Necessary measures include the mitigation of infectious diseases and refugee relief.

Sustainability science must therefore adopt a comprehensive, holistic approach to the identification of problems and perspectives relevant to the sustainability of these global, social and human systems. The emerging discipline needs to be a dynamic and evolving field of enquiry that provides visions and scenario analysis pointing the way to global sustainability (Swart et al., 2004). The ultimate purpose of sustainability science is to contribute to the preservation and improvement of the sustainability of these three systems. Although sustainability science has its origins in the concept of sustainable development, the authors propose that it is, in reality, a far more multifaceted concept.

1-1-4 Structuring knowledge for sustainability science

Two obstacles that impede efforts to deal with the sustainability-related issues outlined above are the complexity of the problems and the specialization of the scholarship that seeks to address them. First, the sustainability crisis is caused by a multitude of factors, the complexity of global environmental problems being a classic example. Hence it is no easy task to gain an overarching view of such problems, let alone solve them. Second, the disciplines that examine these complex issues have themselves grown increasingly compartmentalized in recent years, so that much research is conducted from a highly restricted perspective with regard to both phenomena identification and problem-solving.

The fundamental cause of the current crisis in sustainability is the industrialization that followed the industrial revolution and the rapid economic growth it fostered. One result was the burgeoning consumption of fossil fuels and other nonrenewable resources, a level of consumption that has led some to call the twentieth century the "century of explosive expansion". Pollution, which first emerged as a severe problem in particular localities, developed into the global issue recognized today. As environmental and other problems become global in scale, their causes and effects grow increasingly complex; pollution generated in one part of the world, for example, may do its worst damage in an entirely different region. This complexity hampers both the effort to identify problems and the search for solutions.

For scholars, knowledge-structuring is an essential first step in the acquisition of a comprehensive view of sustainability issues. The problems that sustainability science confronts are not only complex but also interconnected. If scholars are to find solutions to them, they must first clarify the relationships among them, i.e. engage in problem-structuring. Next, they must assemble a platform of knowledge that not only affords an overview of the entire web of problems but also, by systematically

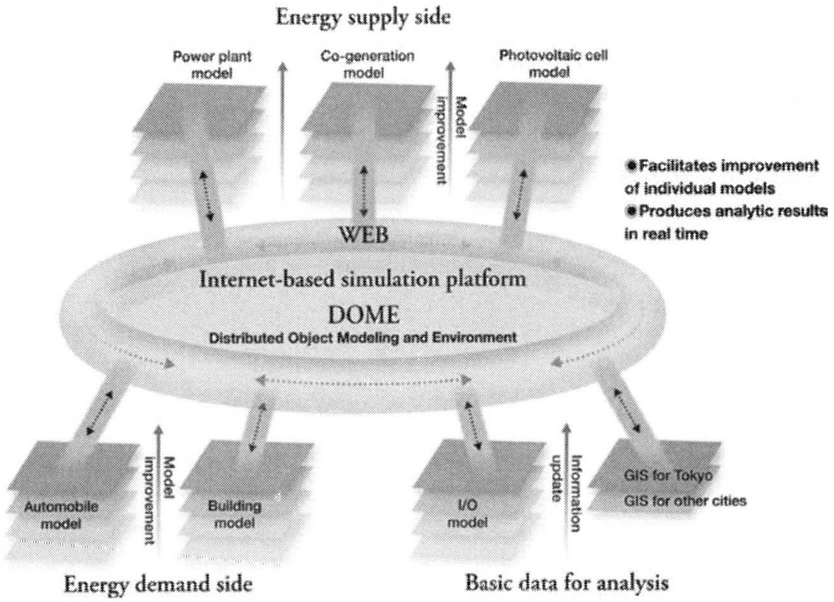

Figure 1.1.2 The Tokyo Half Project (THP): A system chart demonstrating the potential to reduce CO_2 emissions from Tokyo.
Source: Komiyama and Takeuchi (2006).

organizing disparate fields of enquiry, enables them to replace the current piecemeal approach with one that can devise and apply integrative solutions to these problems.

Structuring knowledge in this manner will stimulate existing disciplines and mechanisms, contribute to the development of scenarios for a better future, and point the way to new inventions (Komiyama et al., 2004). Knowledge-structuring is thus of critical value in identifying problems and responding to the needs of academia and industry. But nowhere is its effect more likely to be felt than in the field of sustainability science.

One example of knowledge-structuring at work is the Tokyo Half Project (Figure 1.1.2), a collaborative international research project by the AGS (Krains et al., 2001). The purpose of the project is to construct a model for computing total carbon dioxide (CO_2) emissions from Tokyo by quantifying the emission processes of the variety of CO_2 sources in the metropolis and compiling data on emissions from these sources on a common platform. This model can be used to evaluate the effect of efforts to reduce emissions from individual sources on total CO_2 emissions from Tokyo. It also serves to identify which measures would be required of consumers and industry to reduce CO_2 emissions by half, a target often cited in the post-Kyoto Protocol debate.

The field of sustainability science demands just this type of knowledge-structuring if scholars are to gain a comprehensive overview of this new discipline. The information technology revolution provides the means for integrating the exponentially growing knowledge base, and the proliferation of research utilizing such technology is anticipated. It is the authors' belief that this research can help resolve one of the fundamental dilemmas of contemporary scholarship: the inability of overly specialized disciplines to offer comprehensive solutions to the conditions that threaten the sustainability of global, social and human systems.

1-1-5 A transdisciplinary approach

Precisely because sustainability science includes global, social and human systems in its purview, and because the problems it addresses involve disparate elements (from science and technology to politics and economics, to human lifestyles and behaviour), the new discipline must necessarily embrace the social and natural sciences (Mihelcic et al., 2003). But as the body of academic and scientific research continues to grow, and as the disciplines engaged in research continue to compartmentalize, it becomes almost impossible for the individual researcher or research group to access and utilize this vast accumulation of data. Therefore, a framework needs to be constructed within which individual disciplines can provide quantifiable criteria and indicators related to sustainability. By integrating these criteria, scholars can structure their knowledge, their methods and their grasp of the issues they confront. This is the first step they must take if they are to progress from identifying problems to solving them.

It is important to note that, although these criteria and indicators must conform to scientific standards of objectivity, they must not be expected to yield a singular solution to any given problem. Indeed, a diversity of solutions should be sought in accordance with the specific environmental and cultural conditions of each nation or region. Any attempt to impose uniform solutions to global environmental problems will threaten the diversity of the Earth's regions and cultures in the same way that economic globalization does now. Destroying this diversity will, in turn, prevent the realization of a society that is truly sustainable in the sense that it fosters human fulfilment, not merely survival. If the process of structuring sustainability-related scholarship and its knowledge base yields different structuring models for different regions and nations, then structuring itself can be a driving force for greater diversity.

One problem unique to sustainability science lies in the process of shifting from the stage of phenomena identification and analysis to that of problem-solving. For sustainability science, this process necessarily differs from the conventional transition from basic to applied research, because solutions to problems may have to be sought before those problems have been sufficiently analysed or even identified. Global warming is the prime example of this dilemma. Future scenarios proposed by various models of global warming remain unverifiable, yet the search for solutions cannot wait.

The principle that must be applied here is the precautionary approach. But acceptance and implementation of this approach require a framework for obtaining the agreement of all sectors of society, and that is where interaction between scientists and the public is of the essence. What is demanded of sustainability science is not only the development of scientifically sound models for proposing future scenarios and evaluating the effects of different countermeasures and solutions, but also effective management of the process by which these forecasts and evaluations are accepted by society so as to generate the social reforms necessary to ensure global sustainability.

If sustainability science is to contribute practical solutions to the problems society faces, cooperation among researchers, industry and the general public is imperative. Only when society at large is inspired to act on the basis of their research and conclusions can sustainability scientists lay the foundations for the construction of a sustainable society.

Public acceptance of various approaches, both preventive and adaptive, to the solution of problems of sustainability requires public understanding of scientific findings, as well as of the uncertainty of future forecasts, as a basis for the adoption of technological and economic measures to combat these problems. The development of consensus is crucial to this process, and consensus can be achieved only by promoting dialogue between researchers and the public. Dialogue and consensus are the means by which a transdisciplinary science of sustainability can serve as a fulcrum for effecting the social change required for true sustainability.

The authors would like to emphasize the key role of education in this process. Sustainability science must nurture a generation of leaders who are capable of appreciating the significance of changes in global, social and human systems that occur over the extremely long term, and who choose the path of sustainability in implementing policies on the basis of this understanding. It is particularly crucial that concern with sustainability issues and a desire to act on them be instilled in the generation that comes of age in the mid-twenty-first century, when limits on energy and other resources – and the global environment in general – are predicted to reach a crisis point.

1-1-6 Sustainability science and the creation of a sustainable society in the twenty-first century

As the authors have pointed out, sustainability science requires the construction of a transdisciplinary academic framework that brings the natural sciences, social sciences and humanities together, structures academic knowledge and the issues it must address, and defines standards and indicators for sustainability. Based on this understanding of its mission, IR3S today sees its objective as the building of a sustainable society in the twenty-first century that combines the characteristics of a low-carbon society, a resource-circulating society and a society in harmony with nature (see Figure 1.1.3).

According to the Vision 2050 goals previously proposed by Komiyama and Kraines (2008), a sustainable society can be achieved by the year 2050 by tripling the efficiency of energy use, constructing a resource-circulating system and doubling the amount of renewable energy. Meanwhile, a research team led by the National Institute for Environmental Studies has suggested that Japan's carbon dioxide output could be reduced by 70 per cent by 2050 using a mix of existing advanced technologies. IR3S is now working with the National Institute to develop a year-2050

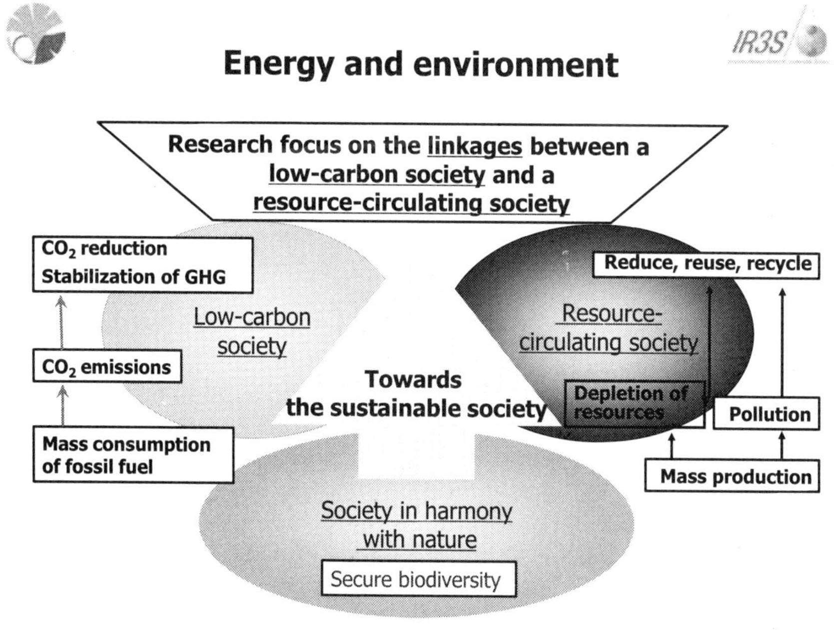

Figure 1.1.3 Integrating three scenarios of society for the achievement of a sustainable society.

global vision that combines scenarios for a low-carbon society in which CO_2 emissions are reduced by half, a resource-circulating society that recycles natural resources, and a society in harmony with nature that preserves ecological systems and biodiversity.

The integration of these three scenarios was adopted as a basic policy in “Becoming a Leading Environmental Nation Strategy in the 21st Century – Japan’s Strategy for a Sustainable Society”, approved in June 2007 by the Japanese Cabinet, and has been reflected in subsequent reevaluations of Japan’s environmental policy. This integration is an essential step towards strengthening ties among what have so far tended to be isolated policy-making processes, and hence towards fostering a more effective overall policy on sustainability. Efforts are also needed in the international arena to achieve a similar synergy among such initiatives as the United Nations Framework Convention on Climate Change (UNFCCC), the 3R Initiative and the Biodiversity Treaty. If synergy can be achieved not only at the policy debate level but also among professionals working towards these scenarios of a low-carbon society, a resource-circulating society and a society in harmony with nature, it should accelerate the development of a vision of a sustainable society for the twenty-first century that embodies optimum outcomes for all three scenarios.

With the aim of encouraging synergy of this sort in international policy, IR3S is currently pursuing joint research with the Institute for Global Environmental Strategies on biofuel use strategies for sustainable development in the Asia Pacific region. The project is supported by the Global Environment Research Fund of Japan’s Ministry of the Environment. The purpose of this research is to present scenarios for appropriate biofuel use based on socioeconomic analysis and lifecycle assessments addressing concerns that increased biofuel production may compete with food production or cause ecological damage, as well as forecasting how the development of second-generation biofuel technologies might affect these conditions. This research also entails developing proposals for regional policy packages in China, India, Indonesia and elsewhere, as well as for cooperation on interregional policy.

When devising policies for biofuel production and use, it is important to study the impact on the various actors involved. In developing countries such as Indonesia, for example, bioenergy projects often take the form of expanding the plantation-style cultivation of energy crops. However, such projects carry the risk of adversely affecting the lives of local residents. In such regions it is more desirable to promote small-scale bioenergy use through local initiatives that will improve, not threaten, the welfare of those communities. In this sense, it is necessary to assess the sustainability of bioenergy policies on the micro as well as the macro level.

Asia's rapid economic growth has fuelled concerns that increasing energy demand in the region and an accompanying spike in CO_2 emissions will further exacerbate climate change. The Energy Sustainability Forum, established by IR3S in cooperation with Showa Shell Sekiyu Corporation, studies the future of energy in Asia, including Japan. The "Triple 50" scenario advanced by Professor Shigefumi Nishio et al. of the Institute of Industrial Science at The University of Tokyo proposes raising energy efficiency to 50 per cent, lowering fossil fuel dependency to 50 per cent and increasing energy self-sufficiency to 50 per cent by the year 2030 in Japan (see Table 1.1.1). The implications in terms of future energy demand are shown in Figure 1.1.4. Professor Tetsuo Yuhara et al. have explained how

Table 1.1.1 The "Triple 50" scenario for Japan in 2030

Year	Energy self-sufficiency	Dependence on fossil fuel	Energy efficiency
2005	20%	80%	35%
2030	50%	50%	50%

Source: Yuhara (2008: 4).

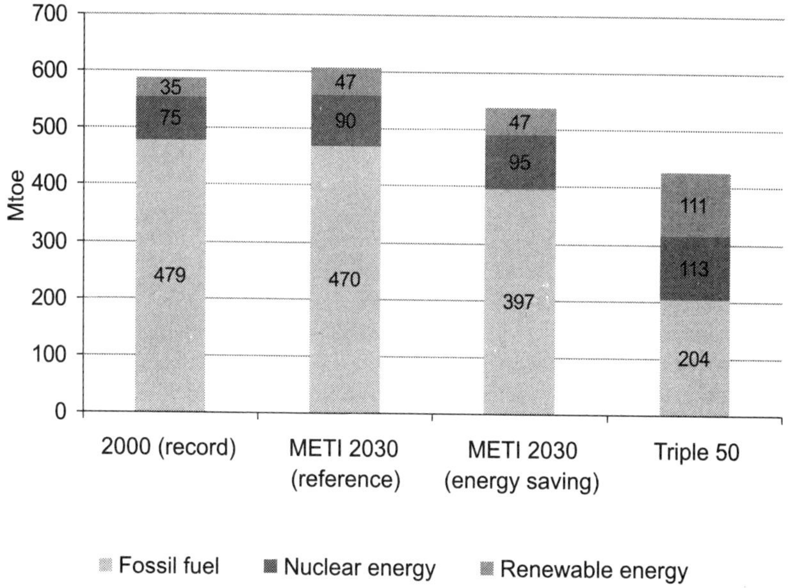

Figure 1.1.4 The "Triple 50" scenario for Japan: Forecasts of long-term energy demand in 2030 by the Agency for Natural Resources and Energy (Ministry of Economy, Trade and Industry) and Triple 50.
Source: Yuhara (2008: 4).
Note: Please see page 468 for a colour version of this figure.

Table 1.1.2 The "Triple 50" scenario for China: Lowering fossil fuel dependency to 50 per cent in 2050

Year	Fossil fuel	Nuclear energy	Renewable energy
2030	70%	10%	20%
2050	50%	20%	30%

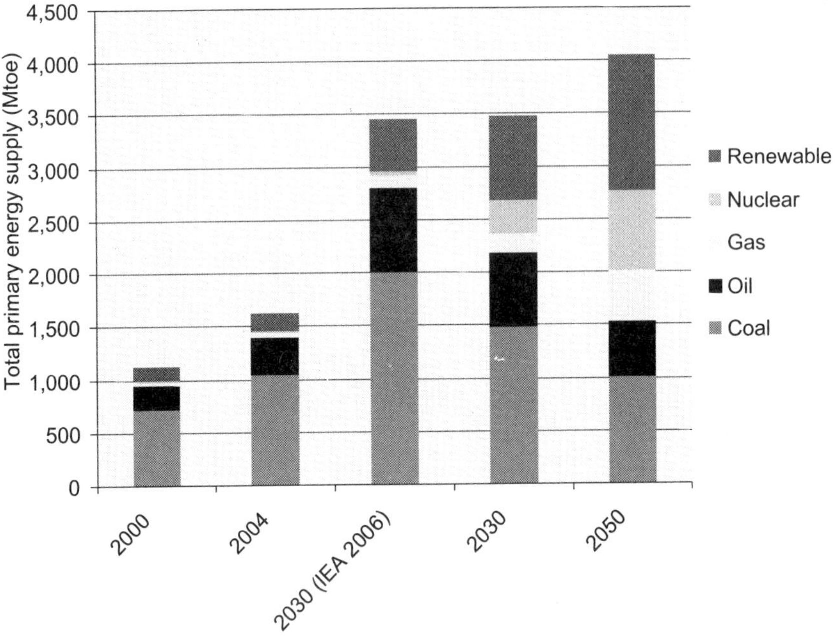

Figure 1.1.5 The "Triple 50" scenario for China: China's energy mix in 2000, 2030 and 2050.
Source: Yuhara (2008: 12).
Notes: 1. Given the current energy situation in China, the possibility of achieving the "Triple 50" scenario in China is not envisaged until 2050.
2. Please see page 469 for a colour version of this figure.

the Triple 50 scenario could be applied to China by the year 2050 (see Table 1.1.2 and Figure 1.1.5).

China faces numerous issues that require further study, including the safety of nuclear power, the introduction of green coal technology, and the feasibility of applying carbon capture and storage (CCS) technology. However, the IR3S believes in pursuing concurrent solutions for global and regional environmental problems and therefore hopes to continue working on the development of environmental and energy policies for China, India and other regions.

Other areas of research that IR3S feels a responsibility to pursue include environmental improvements and poverty eradication in developing countries. These nations are particularly vulnerable to the consequences of climate change, resource exhaustion and ecological degradation. Even as scholars seek solutions to these problems, they must take positive steps to introduce policies ameliorating their effects in developing countries. IR3S intends to work with institutions such as United Nations University to contribute to such efforts in the most impoverished nations of Asia and Africa.

1-1-7 Conclusion

A sustainable society is the ultimate means of achieving global sustainability; as a globally shared objective it requires a globally shared strategy. Yet, maintaining the vitality of human society also requires the preservation of natural and cultural diversity in the various regions of the world. If both of these goals are to be achieved without sacrificing one for the other, scholars and scientists must work together to devise a common strategy that enjoys the support of research groups worldwide but at the same time allows room for the development of solutions that recognize and enhance regional diversity.

This is, indeed, the heart of the mission for the global network of scholars currently building and expanding the field of sustainability science. Still in its infancy and limited in its impact on the world, sustainability science lacks the wherewithal to construct a global sustainability strategy if the effort emanates solely from universities and research institutes in one or two regions or the research networks they have formed. Academic and research entities around the world must join forces and invest their collective strength in the development of a truly comprehensive global policy shared by all. At the same time, they must commit themselves to protecting the diversity of their respective regions.

REFERENCES

Clark, W. C. (2007) "Sustainability Science: A Room of Its Own", *Proceedings of the National Academy of Sciences of the United States of America* 104: 1737–1738.

Clark, W. C. and N. M. Dickson (2003) "Sustainability Science: The Emerging Research Program", *Proceedings of the National Academy of Sciences of the United States of America* 100: 8059–8061.

Cohen, S., D. Demeritt, J. Robinson and D. Rothman (1998) "Climate Change and Sustainable Development: Towards Dialogue", *Global Environmental Change* 8: 341–371.

ICSU [International Council for Science] (2002) *Science and Technology for Sustainable Development.* World Summit on Sustainable Development Report 19.

Kates, R. W., W. C. Clark, R. Corell, J. M. Hall, C. C. Jaeger, I. Lowe, J. J. McCarthy, H. J. Schellnhuber, B. Bolin, N. M. Dickson, S. Faucheux, G. C. Gallopin, A. Grubler, B. Huntley, J. Jager, N. S. Jodha, R. E. Kasperson, A. Mabogunje, P. Matson, H. Mooney, B. Moore, T. O'Riordan and U. Svedin (2001) "Environment and Development: Sustainability Science", *Science* 292: 641–642.

Komiyama, H. and S. Kraines (2008) *Vision 2050: Road Map for a Sustainable Earth.* Tokyo/Berlin/Heidelberg/New York: Springer.

Komiyama, H. and K. Takeuchi (2006) "Sustainability Science: Building a New Discipline", *Sustainability Science* 1: 1–6.

Komiyama, H., Y. Yamaguchi and S. Noda (2004) "Structuring Knowledge on Nanomaterials Processing", *Chemical Engineering Science* 59: 5085–5090.

Krains, S. B., D. R. Wallace, Y. Iwafune, Y. Yoshida, T. Aramaki, K. Kato, K. Hanaki, H. Ishitani, T. Matsuo, H. Takahashi, K. Yamada, K. Yamaji, Y. Yanagisawa and H. Komiyama (2001) "An Integrated Computational Infrastructure for a Virtual Tokyo: Concepts and Examples", *Journal of Industrial Ecology* 5: 35–54.

McCarthy, J. J., O. F. Canziani, N. A. Leary, D. J. Dokken and K. S. White, eds (2001) *Climate Change 2001: Impacts, Adaptation, and Vulnerability*. Contribution of Working Group II to the Third Assessment Report of the Intergovernmental Panel on Climate Change. Cambridge: Cambridge University Press.

McMichael, A. J., D. H. Campbell-Lendrum, C. F. Corvalan, K. L. Ebi, A. Githeko, J. D. Scheraga and A. Woodward, eds (2003) *Climate Change and Human Health – Risks and Responses*. Geneva: World Health Organization.

Mihelcic, J. R., J. C. Crittenden, M. J. Small, D. R. Shonnard, D. R. Hokanson, Q. Zhang, H. Chen, S. A. Sorby, V. U. James, J. W. Sutherland and J. L. Schnoor (2003) "Sustainability Science and Engineering: The Emergence of a New Metadiscipline", *Environmental Science & Technology* 37: 5314–5324.

Nicholls, R. J. (2004) "Coastal Flooding and Wetland Loss in the 21st Century: Changes under SRES Climate and Socio-economic Scenarios", *Global Environmental Change* 14(1): 69–86.

Sotherton, D., H. Chappells, B. Van Vliet, eds (2004) *Sustainable Consumption: The Implications of Changing Infrastructures of Provision*. Cheltenham: Edward Elgar Publishing.

Swart, R. J., P. Raskin and J. Robinson (2004) "The Problem of the Future: Sustainability Science and Scenario Analysis", *Global Environmental Change* 14(2): 137–146.

UN Millennium Project (2005) *Investing in Development: A Practical Plan to Achieve the Millennium Development Goals. Overview*. New York: United Nations Development Programme.

WCED [World Commission on Environment and Development] (1987) *Our Common Future*. Oxford: Oxford University Press.

Yuhara, Tetsuo (2008) "Japan–China Strategic Partnership to Global Sustainability of Energy Issue – a Long Term Energy Vision and Pathway to 2050 with Innovative Energy Technologies", The University of Tokyo and Tsinghua University, Workshop on Sustainability in East Asia, 20 May. Available at: <http://www.adm.u-tokyo.ac.jp/res/res5/Prof.%20Yuhara&Ms.%20Kitamura_Energy%20Technology.pdf> (accessed 18 May 2010).

2

The connections between existing sciences and sustainability science

2-1

The structuring of knowledge

Yuya Kajikawa

2-1-1 Introduction

The structuring of knowledge has become a challenging issue because of the segmentation and specialization of our intellectual base due to a flood of information. Currently, there are more than 3,000 papers about sustainability and sustainable development, a quantity beyond our capacity to read so as to grasp the overall structure of sustainability science. And the number of papers continues to grow exponentially. But such concerns are nothing new, and indeed they were articulated in the 1960s by De Solla Price (1963). The increase in the amount of knowledge itself is not problematic, because knowledge is the driver that advances our society and civilization. But it is also a fact that we feel overwhelmed and frustrated by the lack of a comprehensive view. It is no exaggeration to say that, these days, we are drowning in a sea of information as we look for knowledge.

The growth of knowledge is inevitably accompanied by segmentation and specialization because the individual scientist feels compelled to focus on or specialize in only a few scientific sub-domains to keep up with the growth of those domains. Although specialization is an inevitable strategy for obtaining deeper scientific understanding, it is also common sense that the flood of information and consequent specialization make it difficult for scholars to obtain a comprehensive perspective not only on research domains other than their own, but also on their own specific research topics. Ziman (2001: 165) concisely describes this situation in an article discussing scientific education: "Research scientists are trained to produce specialized bricks of knowledge, but not to look at the whole building."

A similar sense of crisis is also expressed by other scholars. Yamaguchi and Komiyama (2001: 107) state: "In these decades, human beings have encountered fundamental difficulties due to the gap between the complexity of the problems encountered and the subdivision of our knowledge base. A person finds it difficult to grasp the whole of an issue because only a small part of the issue is native to his/her specific field." Börner et al. (2003: 180) observe that: "Traditional approaches struggle to keep up with the pace of information growth. In multidisciplinary fields of study it is especially difficult to maintain an overview of literature dynamics.... Researchers examining the domain from a particular discipline cannot possibly have an adequate understanding of the whole."

As a consequence, knowledge-structuring, which reorganizes existing knowledge in a clear and accessible manner so as to provide a comprehensive view of knowledge, is becoming an increasingly important field of expertise. Knowledge-structuring is especially important for sustainability science because of the latter's multidisciplinary characteristics, as will be explained below.

2-1-2 The structure of sustainability science

Sustainability is indisputably an important concept for society, the economy and the environment, but there is general agreement that sustainability is threatened by increasing population, resource extraction and climate change. It is now clear that future development is limited and constrained by the growing world population, the depletion of natural resources and our capacity for mitigating environmental change and adapting to it. Thus, for the further development of society, growth must be sought in a sustainable manner. Currently, emerging concerns about sustainability are apparent in a number of societal sectors, including the political and economic sectors, universities and the public at large. Reflecting its social importance, sustainability science is becoming a distinct scientific field (Kates et al., 2001; Komiyama and Takeuchi, 2006). However, the definition of sustainability and the scope of sustainability science are not often clear. This section will therefore review the definition of sustainability and the current status of sustainability science before discussing knowledge-structuring.

"Sustainability" literally means the ability to sustain, or a state that can be maintained at a certain level. The term has been used to express the state in which levels of harvest in agriculture, fishery and forestry are maintained within the capacity of the ecosystem, which is therefore recoverable. In that sense, sustainability means environmental sustainability – in other words, sustainability of the ecosystem's function to provide food, fish and other products and services. However, it is now used in a wider context. For example, the Brundtland Report by the World Commission on Environment and Development (WCED) related sustainability to development and defined sustainable development as "development that meets the needs of the present without compromising the ability of future generations to meet their own needs" (WCED, 1987: 43). The report, titled *Our Common Future*, also stated that global environmental problems resulted from both the South's enormous poverty and the North's unsustainable consumption and production.

The Brundtland Report broadened the definition of sustainability to encompass the entire range of human values (Ascher, 2007). This expanding definition is also seen in the so-called WEHAB targets (for Water, Energy, Health, Agriculture and Biodiversity) declared at the Johannesburg Summit of the United Nations World Summit on Sustainable Development (WSSD, 2002). In recent years, it has become common to represent sustainability by a set of concepts including social, human and environmental systems. In short, sustainability is achieved only when there is full reconciliation between (1) economic development, (2) meeting, on an equitable basis, growing and changing human needs and aspirations, and (3) conserving limited natural resources and the capacity of the environment to absorb the multiple stresses that are a consequence of human activities (Hay and Mimura, 2006).

Sustainability involves a wide range of issues, because the target of a threat to sustainability, and the root cause of that threat, will differ among stakeholders and depend on economic, environmental and social conditions that in turn differ from country to country and occur on different time scales; in many cases these are interlinked and have trade-off relationships. For this reason, sustainability science has or should have multidisciplinary or interdisciplinary characteristics, as many have repeatedly emphasized (Kates and Dasgupta, 2007; Komiyama and Takeuchi, 2006; National Research Council, 1999).

This characteristic of sustainability science is evidenced by the results of citation network analysis (Kajikawa et al., 2007) shown in Figure 2.1.1. Kajikawa et al. analysed the academic landscape of sustainability science through a citation network analysis of 29,391 papers that included the words "sustainability" or "sustainable" in their bibliographical records. A citation network is clustered so that there are many intra-cluster links and as few as possible inter-cluster links. The clustered network is visualized by using a large graph layout algorithm that locates groups of papers

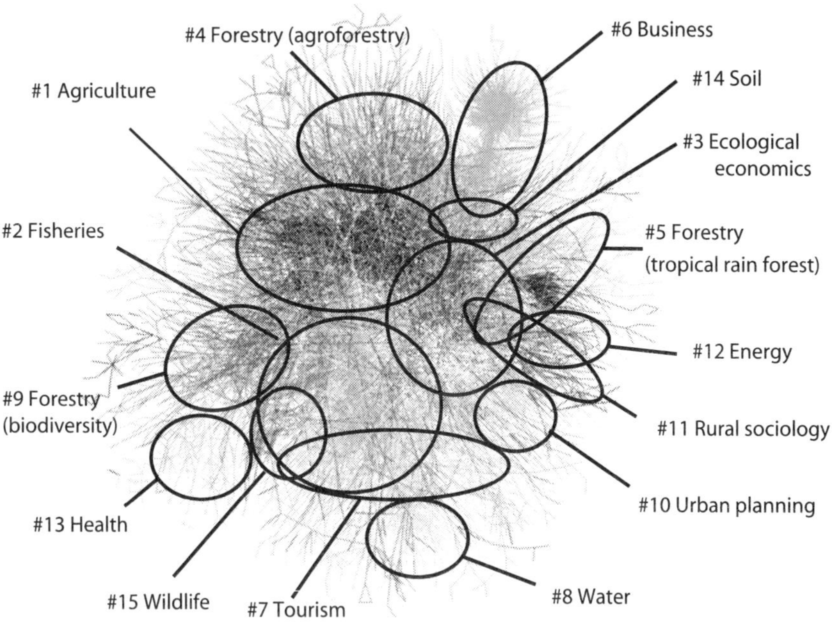

Figure 2.1.1 Visualization of the citation network of sustainability science.
Source: Kajikawa et al. (2007).
Note: Please see page 470 for a colour version of this figure.

citing each other in close positions. In Figure 2.1.1, only intra-cluster links are visualized to identify the position of each cluster.

Results show the existence of 15 main research domains, which are separated according to the target, i.e. what to sustain. These clusters are: Agriculture, Fisheries, Ecological Economics, Forestry (agroforestry), Forestry (tropical rain forest), Business, Tourism, Water, Forestry (biodiversity), Urban Planning, Rural Sociology, Energy, Health, Soil, and Wildlife. Some of these clusters focus on environmental issues but others are concerned with social development and human issues, which confirms the multidisciplinary character of such research.

Although the citation network clustering visualized in Figure 2.1.1 displays multidisciplinary characteristics to some extent, further light was shed on interdisciplinary characteristics through natural language processing (Kajikawa et al., 2007) and citation network analysis (Kajikawa and Mori, 2009). According to these results, commonly discussed topics in different citation clusters include climate change, welfare and livelihood. This indicates that one of the main research topics for sustainability science is the relationship between climate change and targets to be sustained, such as food production, fish catches, ecosystems in forestry, urban and rural development, health and the economy, with the ultimate goal of improving and sustaining the quality of human welfare. According to the results of this citation network analysis, interdisciplinary papers connecting disparate clusters focus on philosophical aspects of sustainability or relationships among individuals, society and policy. Interdisciplinary research should span the boundaries of disciplines in the design of total systems, visions and solutions to attain the goal of realizing a sustainable society. But such research is still scarce, and much more effort must be devoted to it.

2-1-3 The integration of knowledge in interdisciplinary fields

The necessity of interdisciplinary research and transdisciplinary expertise is repeatedly emphasized. But why? And how is it to be achieved?

The answer to the first question is that complex issues cannot be solved through segmented and specialized disciplines. Figure 2.1.2 is a schematic illustration of this discourse. As segmentation proceeds, it becomes increasingly difficult to develop solutions to a given problem. The essential cause of the problem is usually located somewhere along a long causal chain. A particular discipline may be adequate to the task of solving a specified part of the problem, but it lacks a comprehensive view, and a

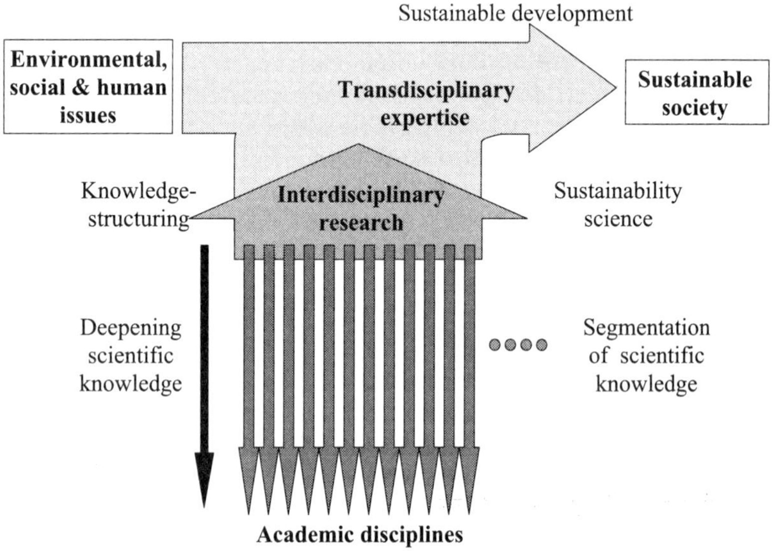

Figure 2.1.2 Integration of academic disciplines by knowledge-structuring to realize a sustainable society.

substantive, comprehensive solution is outside its scope. As specialization proceeds, knowledge tends to be produced within separate disciplines. But in some cases this knowledge is too narrow and too far removed from actual social needs. Alternatively, it may not be noticed or properly evaluated despite its relevance to a problem because specialists in other disciplines cannot track the updating of the knowledge within its own discipline. It is therefore essential that the fruitful outcomes of research in all scientific disciplines are utilized and that activities in these disciplines that are necessary to realize sustainability are encouraged. Whereas the development of discipline-based science has been the source of most scientific advances of the last century, it has also limited the capacity of science to address problems that span multiple disciplines (Perrings, 2007).

Therefore, knowledge-structuring is one of the vital roles of sustainability science. Currently, most publications related to sustainability publish purely monodisciplinary work in such fields as climate change research, agricultural research, forestry research and energy research (Kajikawa, 2008). But sustainability science should not merely be multidisciplinary, i.e. an agglomeration of monodisciplinary work.

One might think that an interdisciplinary field is itself a discipline. It is possible to regard an interdisciplinary research field as one that is currently in an intermediate state and will eventually become a discipline. For example, nanotechnology was an interdisciplinary field that required the integration of a variety of knowledge, including physics, chemistry and biology, but it can now be regarded as a distinct research field and thus a discipline. This may be true, but the direction in which an interdisciplinary field aims to develop differs from that of other disciplines, as shown in Figure 2.1.2. Other disciplines aim to develop a deeper understanding of the real or virtual world. Sustainability science, however, aims to propose solutions, to design society and, finally, to drive society in a sustainable direction by integrating salient and reliable knowledge produced in other disciplines. The structuring of knowledge is an essential part of such expertise. But what is knowledge-structuring? And how is it accomplished?

2-1-4 The structure of knowledge and knowledge-modelling

Knowledge-structuring is a set of processes to access, collect, analyse, assess, organize and finally represent knowledge based on the structure of knowledge. An essential component of knowledge-structuring is to elucidate a structure of knowledge so as to obtain a total view of knowledge that will facilitate communication among disciplines. Specialists in a given discipline share an epistemological view and knowledge framework of the issues they pursue. In interdisciplinary research, an essential, common framework is extracted from among those of different disciplines through knowledge-modelling. It is necessary for knowledge-modelling to elucidate the structure of knowledge.

Before discussing the structure of knowledge, it is necessary to consider what in fact scientific knowledge is, as has been discussed in Kajikawa (2008). The nature of knowledge has provoked great controversy in many disciplines. In a philosophical context, knowledge is defined as justified true belief (Dretske, 1981). According to this definition, most of the knowledge we believe to be knowledge is abandoned because it is seldom justified or true in the strict sense. This definition results in an infinite journey to seek a fundamental knowledge that is not justified by other knowledge. Instead of considering the definition of knowledge further, however, let us move on to look at the relationship between knowledge and modelling, because in scientific enquiry knowledge is acquired through modelling.

In their classic paper, Rosenblueth and Wiener describe the essence and role of the model in science (Rosenblueth and Wiener, 1945). They state that the intention and the result of a scientific activity are to obtain an understanding and control of some part of the universe. An inevitable step to obtaining these is the creation of a model, because no substantial part of reality is so simple that one can grasp and control it without abstraction. Modelling is a process that replaces the part of the universe under consideration with an abstract model that has a similar but simpler structure. In natural science, models are expressed by relationships among concepts and physical quantities – for example, a smaller raindrop falls more slowly than a larger one in air; a spear flies further when it is thrown with spin; Newton's equation of motion $f = ma$, and so forth (Yamaguchi and Komiyama, 2001). It is noteworthy that an abstract model including general features of the real world tends to be more highly evaluated than a mere collection of data. Similarly, structured knowledge is more valuable than a mere collection of discipline-specific knowledge, and knowledge-structuring needs knowledge on knowledge in order to integrate this knowledge.

One's perception of the real world is structured through modelling and then becomes knowledge. Similarly, knowledge is structured through knowledge-modelling. For knowledge-structuring, the structure of knowledge (i.e. the model of knowledge) must be stated explicitly in what is often called an ontology. Although ontology is discussed in detail in Section 2-3, a brief overview will be provided here.

Traditionally, ontology is a philosophical theory on the nature of existence. Researchers working in artificial intelligence have reincarnated this term as jargon for expressing a shared and common understanding of some domain that can be communicated between people and application systems (Gruber, 1995). According to this definition, ontology is a method of representing knowledge in a machine-readable manner for computer manipulation. Formally, the concept of ontology is defined as $O = (C, R, A, T)$, where O is an ontology, C is a set of concepts, R is a set of relations in C, A is an axiom to define O, and T is a top level in a hierarchy (Shamsfard and Barforoush, 2004). In short, ontology defines the basic terms and relations comprising the vocabulary of a topic area as well as the rules for combining terms and relations between terms with some rules called axioms. Ontology also defines the hierarchical structure of concepts, which is called taxonomy. This view of ontology is compatible with scientific knowledge. Indeed, a physics textbook written half a century ago was based on this view. In the preface of their book, Hix and Alley stated that "an excellent technique for utilizing this compilation of laws and effects in problem solution is to break the problem into input and output physical quantities" (Hix and Alley, 1958: 2). They called quantitative and qualitative relationships among concepts "laws" and "effects", respectively. Therefore, one can regard physical quantities as C and laws and effects as R. By combining a set of C and R, physical laws and effects enable the real world to be controlled by tracing the path from input to output passing through a series of laws and effects.

But the epistemological perspective on the structure of knowledge differs from discipline to discipline. A common framework of knowledge must therefore be built for sustainability science to integrate multidisciplinary knowledge and develop this interdisciplinary research field. What, then, is the structure of knowledge in sustainability science?

2-1-5 Towards knowledge-structuring in sustainability science

Currently, researchers are attempting to build a common modelling framework for sustainability science. For example, Ostrom (2007) proposed an analytical framework consisting of a resource system (e.g. fishery, lake, grazing area), resource units generated by that system (e.g. fish, water, fodder), the users of that system and the governance system, where all these components and their interactions are bound by other related ecosystems and constrained by social, economic and political settings. Turner et al. (2007) cited the importance of observing, monitoring and understanding system dynamics in a coupled human–environment system, spatially explicit modelling of the focal system, and assessment of system outcomes such as vulnerability, resilience or sustainability. In the context of vulnerability and resilience, an understanding of perturbation or

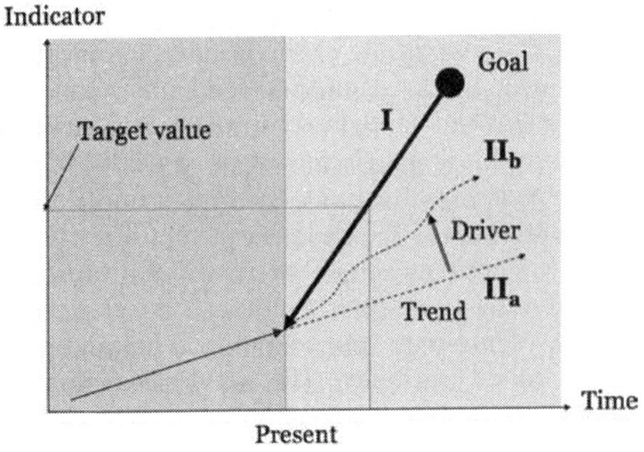

Figure 2.1.3 Research framework of sustainability science.
Source: Kajikawa (2008).
Notes: I: Backcasting; II_a: Forecasting; II_b: Realistic route to attaining goal by improving trend via external factors (e.g. human activities).

stresses/stressors to a system is crucial to understanding the hazards and risks of the system (Turner et al., 2003).

Reviewing the recent discourse, Kajikawa (2008) proposed that the following are key components of the framework of sustainability science: goal-setting, indicator-setting, indicator measurement, causal chain analysis, forecasting, backcasting and problem-solution chain analysis. Figure 2.1.3 is a schematic illustration of this framework.

Goals are broad, qualitative statements about objectives and visions. Examples of goals are reducing hunger, stabilizing climate and improving health. Indicators are quantitative measures selected to assess progress towards or away from a stated goal, and target values are quantitative values of indicators for attaining the goal at a specific time or within a certain timeframe. The historical trend of the value of an indicator is measured by a variety of methods, and is extrapolated to project the future. Trends are changes in the values of indicators over time, and driving forces are the processes that influence trends and the ability to meet agreed-upon targets that work as the principal drivers towards or away from sustainability goals and targets.

The above framework reveals basic characteristics of sustainability science. One is a broad time span with normative characteristics. It includes a time span from past to present, and from the present to the future. Whereas past data can be described, future data cannot be described but can only be predicted, and more importantly are determined as something that should be attained. The latter point indicates the normative nature of sustainability science. In short, goal-setting is a normative process based on visions and social and political processes rather than on scientific activity per se, but it should have some rational basis. Defining sustainability is ultimately a social choice about what to develop, what to sustain and for how long (Parris and Kates, 2003). Providing scientific data that have been reported in a variety of disciplines can stimulate constructive public discourse and help select prior targets. For that purpose, an infrastructure must be provided where individuals can feel that sustainability-related issues are their own issues. Essential to this infrastructure are micro-level data with detailed spatial-temporal resolution focusing on individuals and organizations, data on their current status and data predicted according to the actions they take.

Another characteristic of sustainability science is a problem-solving perspective. Komiyama and Takeuchi (2006) state that a challenge unique to sustainability science is the process of shifting from the stage of phenomena identification and analysis to that of problem-solving. Researchers try not only to understand certain phenomena through causal chain analysis but also to propose solutions through problem-solution chain analysis. Moreover, these solutions are not limited to technological ones, but include – indeed, emphasize – the social and political aspects of solutions. This effort must combine engineering, psychology, economics, institutional design, legal studies, political science and other social sciences. But it must also be kept in mind that these social, political, psychological and economic studies usually lack an elucidation of obstacles to solutions. If these are real solutions, why have they not been adopted and the problem already solved? It is clear that the deeper structure of problem-solution chains must be elucidated in order to recognize the essential issues that hamper the realization of solutions. This is not limited to sociopolitical solutions; obstacles to the technological solutions discussed in each domain are also seldom reported. Although proposing solutions from a different perspective from those of existing disciplines is a necessary role of sustainability science, it is also important to gather solutions proposed in those disciplines, to structure the current relationships among them and to elucidate obstacles and problem-solution chains.

It is now clear that interdisciplinary effort on knowledge-structuring is essential and that it should be addressed as a part of sustainability science, even if this still happens all too rarely. Engineering research is clearly needed to provide plausible solutions, but it has yet to play a prominent role in sustainability science. Research on research is also needed to collect and structure the problem-solution chains reported in fragmentary fashion in different research papers from different disciplines.

2-1-6 Conclusion

In this section, the circumstances that make knowledge-structuring a necessity were reviewed. The issue one faces is the segmentation and specialization of the intellectual base compared with the complexity of the problem that must be addressed, i.e. sustainability. The importance of knowledge-structuring is not limited to sustainability science, but it is particularly crucial to this field. The reason knowledge-structuring is necessary for sustainability science is the diffuse scope and interdisciplinarity of the field, as evidenced by the results of citation network analysis.

Knowledge-structuring is key to integrating the diverse knowledge of diverse disciplines. Knowledge-structuring is a process to collect, analyse, assess and organize existing knowledge. Knowledge becomes structured knowledge through this process, which is expected to bridge existing but diverse areas of knowledge in various academic disciplines.

To achieve a sustainable society, sustainability science must be a distinct discipline that is at the same time engaged in a transdisciplinary effort arching over existing disciplines, because a sustainable society will be achieved only when action accompanies knowledge. This proactive aspect of sustainability science will be discussed in Section 2-2.

Acknowledgements

This section is based on previous papers (Kajikawa et al., 2006, 2007; Kajikawa, 2008).

REFERENCES

Ascher, W. (2007) "Policy Sciences Contributions to Analysis to Promote Sustainability", *Sustainability Science* 2(2): 141–149.

Börner, K., C. M. Chen and K. W. Boyack (2003) "Visualizing Knowledge Domains", *Annual Review of Information Science and Technology* 37: 179–255.

De Solla Price, D. J. (1963) *Little Science, Big Science*. New York: Columbia University Press.

Dretske, F. (1981) *Knowledge and the Flow of Information*. Cambridge, MA: MIT.

Gruber, T. R. (1995) "Towards Principles for the Design of Ontologies Used for Knowledge Sharing", *International Journal of Human-Computer Studies* 43(5/6): 907–928.

Hay, J. and N. Mimura (2006) "Supporting Climate Change Vulnerability and Adaptation Assessments in the Asia-Pacific Region: An Example of Sustainability Science", *Sustainability Science* 1: 23–35.

Hix, C. F. and R. P. Alley (1958) *Physical Laws and Effects*. New York: John Wiley & Sons.

Kajikawa, Y. (2008) "Research Core and Framework of Sustainability Science", *Sustainability Science* 3: 215–239.

Kajikawa, Y. and J. Mori (2009) "Interdisciplinary Research Detection in Sustainability Science", unpublished paper presented at workshop, 12th International Conference on Scientometrics and Informetrics (ISSI 2009), Rio de Janeiro, 14–17 July.

Kajikawa, Y., K. Abe and S. Noda (2006) "Filling the Gap between Researchers Studying Different Materials and Different Methods: A Proposal for Structured Keywords", *Journal of Information Science* 32: 511–524.

Kajikawa, Y., J. Ohno, Y. Takeda, K. Matsushima and H. Komiyama (2007) "Creating an Academic Landscape of Sustainability Science: An Analysis of the Citation Network", *Sustainability Science* 2: 221–231.

Kates, R. W. and P. Dasgupta (2007) "African Poverty: A Grand Challenge for Sustainability Science", *Proceedings of the National Academy of Sciences* 104: 16747–16750.

Kates, R. W., W. C. Clark, R. Corell, J. M. Hall, C. C. Jaeger, I. Lowe, J. J. McCarthy, H. J. Schellnhuber, B. Bolin, N. M. Dickson, S. Faucheux, G. C. Gallopin, A. Grubler, B. Huntley, J. Jäger, N. S. Jodha, R. E. Kasperson, A. Mabogunje, P. Matson, H. Mooney, B. Moore III, T. O'Riordan and U. Svedin (2001) "Environment and Development: Sustainability Science", *Science* 292(5517): 641–642.

Komiyama, H. and K. Takeuchi (2006) "Sustainability Science: Building a New Discipline", *Sustainability Science* 1: 1–6.

National Research Council (1999) *Our Common Journey: A Transition towards Sustainability*. Washington, DC: National Academic Press.

Ostrom, E. (2007) "A Diagnostic Approach for Going Beyond Panaceas", *Proceedings of the National Academy of Sciences* 104(39): 15181–15187.

Parris, T. M. and R. W. Kates (2003) "Characterizing a Sustainability Transition: Goals, Targets, Trends, and Driving Forces", *Proceedings of the National Academy of Sciences* 100(14): 8068–8073.

Perrings, C. (2007) "Future Challenges", *Proceedings of the National Academy of Sciences* 104(39): 15179–15180.

Rosenblueth, A. and N. Wiener (1945) "Role of Models in Science", *Philosophy of Science* 12: 316–322.

Shamsfard, M. and A. A. Barforoush (2004) "Learning Ontologies from Natural Language Texts", *International Journal of Human-Computer Studies* 60(1): 17–63.

Turner, B. L. II, E. F. Lambin and A. Reenberg (2007) "The Emergence of Land Change Science for Global Environmental Change and Sustainability", *Proceedings of the National Academy of Sciences* 104(52): 20666–20671.

Turner, B. L. II, R. E. Kasperson, P. A. Matson, J. J. McCarthy, R. W. Corell, L. Christensen, N. Eckley, J. X. Kasperson, A. Luers, M. L. Martello, C. Polsky, A. Pulsipher and A. Schiller (2003) "A Framework for Vulnerability Analysis in Sustainability Science", *Proceedings of the National Academy of Sciences* 100(14): 8074–8079.

WCED [World Commission on Environment and Development] (1987) *Our Common Future*. Oxford: Oxford University Press.

WSSD [World Summit on Sustainable Development] (2002) *WEHAB Framework Papers*. Available at <http://www.un.org/jsummit/html/documents/wehab_papers.html> (accessed 25 May 2010).

Yamaguchi, Y. and H. Komiyama (2001) "Structuring Knowledge Project in Nanotechnology Materials Program Launched in Japan", *Journal of Nanoparticle Research* 3(2): 105–110.

Ziman, J. (2001) "Getting Scientists to Think about What They Are Doing", *Science and Engineering Ethics* 7(2): 165–176.

2-2

The structuring of action

Yuya Kajikawa and Hiroshi Komiyama

2-2-1 Introduction

Section 2-1 illustrated the multidisciplinary characteristics of sustainability science and cited the importance of developing an interdisciplinary intellectual base through the structuring of knowledge. Knowledge definitely plays a critical role in the advancement of sustainability science. But it must be noted that a sustainable society can be achieved only when knowledge is accompanied by action. Knowledge without action cannot change a situation, and action without knowledge leads to uncertain results.

In this context, the role of universities and research institutes has never been as significant as it is today. However, the traditional approaches of such institutions are not adequate. The traditional role of universities has been to generate and feed knowledge to society and policymakers, and to work with them through university–government, university–industry and other forms of collaboration. Although necessary, these activities are not sufficient if universities are to help achieve a sustainable society. Instead of merely being knowledge providers, universities must be more proactive in serving as a driving engine for social and political change by applying their knowledge in a manner useful to society and to policymakers (Figure 2.2.1). To fulfill this role, one of the challenges they face is the structuring of actions.

Action-structuring is a key concept in the design and integration of collective actions and is indispensable for transforming society in a sustainable

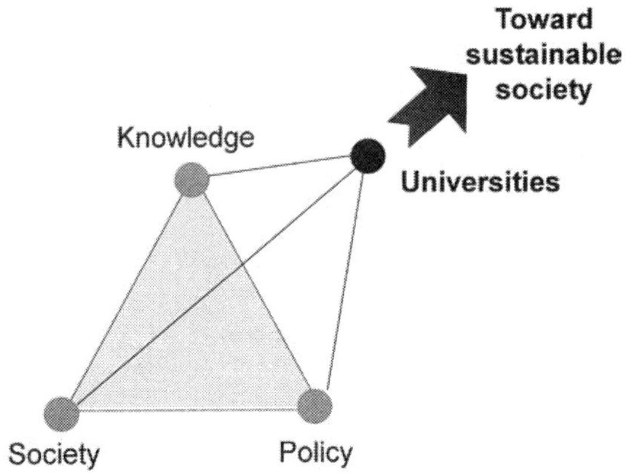

Figure 2.2.1 Driving engine for change.

direction with the goal of achieving a sustainable society. It is necessary because a single action has a limited impact on sustainability and related issues. As an example, consider a single eco-house with solar panels, a heat pump water heater system and high-performance air conditioners. Such an eco-house can reduce household energy consumption by 80 per cent compared with a conventional house, but it is clear that the impact of one eco-house on global climate change is negligible. Without proper orchestration or structuring, all of the activities by numerous individuals will have only a minuscule effect.

2-2-2 Action-structuring

Action-structuring has three phases: decomposition of actions into unit actions, integration of these into new actions, and promotion of collective actions (Figure 2.2.2).

The aim of decomposition of actions is to extract the unit actions that make up an action. By properly reverse-engineering an action, the individual actions (unit components) that were used for that specific action can be identified. Each large circle in Figure 2.2.2 is a category of action that includes a number of options for actions. Examples of these categories are reducing energy consumption and reducing water consumption. Each category may contain subcategories; for example, reducing energy consumption includes subcategories for lighting, machinery and heating. Each category can also be subdivided into a number of unit actions. By replicating this process with other actions, all the necessary components of knowledge that are currently available and implemented can be assembled. It is necessary to write down knowledge at the level of actions,

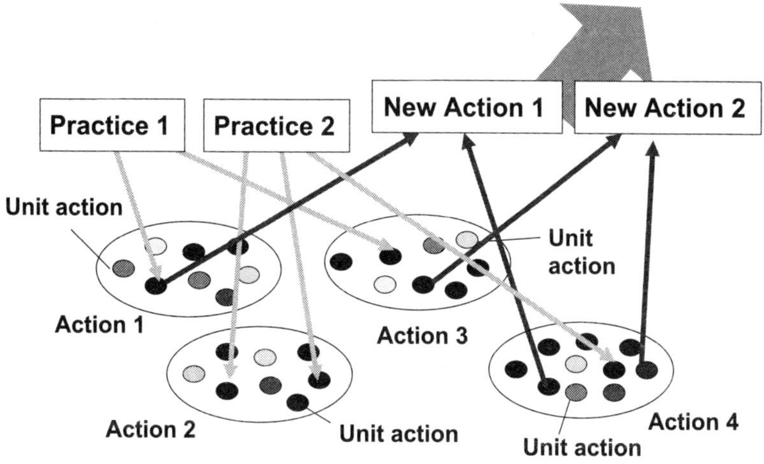

Figure 2.2.2 Action-structuring.

in other words, to write it in a manner that promotes action. Consider the case of actions to reduce energy consumption. Take the following statement as an example: "The quantum dot structure of a solar cell enhances the theoretical limit of light conversion efficiency." This statement definitely expresses some knowledge, but it does not promote action by anyone other than a few engineers. On the other hand, another statement of knowledge, such as "installing solar panels saves 90 per cent of the money we spend on energy consumption", encourages people to install solar panels and save energy. The point is to describe knowledge at the level of action for each individual.

Once decomposed, the unit components of action – for example, the installation of solar panels, heat pump water heaters, improved insulation or new air conditioners – can be integrated so as to enable people to take different actions suitable for their different situations. The process of integration and the resulting optimum mix of unit components will vary depending on the situation one faces, and they are determined by a number of constraints such as technology, economics, political feasibility and social acceptance. For example, triple-glazed windows work well and are suitable in boreal regions, but in warm regions it might be enough to adopt double-glazed windows if the balance of cost performance is considered.

Decomposition of actions is an approach to knowledge-structuring at the level of action. It can extract units of knowledge for actions, and new actions can be designed by combining a set of such units. Knowledge extracted through the decomposition of actions can be integrated to formulate other actions for other situations. These processes support the design of individual actions suited to specific situations. For action-structuring, however, it is necessary to parlay these individual actions into social action.

The last phase of action-structuring is the promotion of collective actions through the orchestration of various related actions. But how can those working in universities and research institutes promote collective actions and play the role of a driving engine more effectively? No single institution is capable of tackling the complex and intertwining issues we face, and therefore a number of research networks have to be established to integrate efforts in a synergetic manner. Building and operating a network of such networks is the key to promoting collective action. Here, the concept of a "network of networks" for collective action will be introduced.

2-2-3 A network of networks

A network of networks (NNs) is a network among existing networks (Figure 2.2.3). The NNs facilitates communication among universities, research institutes and other actors by spanning the boundaries of existing networks. The concept of creating an ecological "network of networks" to study global climate change and other broad-scale phenomena dates back to a 1991 workshop (Bledsoe and Barber, 1993), as cited by Peters et al. (2008). A network is expected to serve as a conduit of knowledge and a platform for collaboration. However, even through networking it is not always easy to secure sufficiently broad capabilities to achieve objectives because the members of a network are often limited to those who are already in the same circle. Therefore, one needs to go a step further: connect these networks and create an NNs to link otherwise mutually

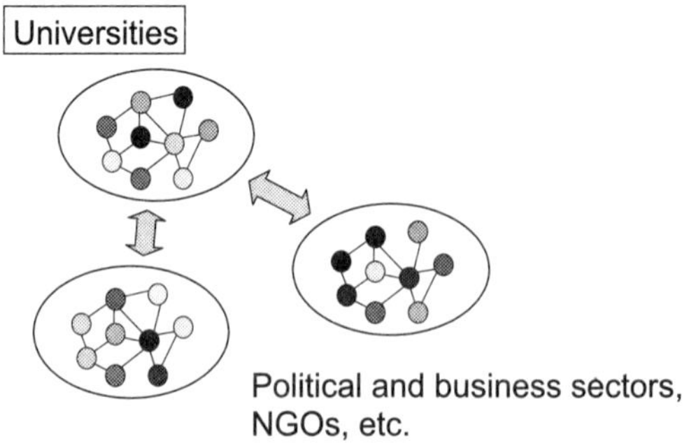

Figure 2.2.3 Network of networks.

isolated research institutes and sectors, thereby promoting a higher level of integration and securing a broader base of capabilities.

Collective action requires networks and flows of information between individuals and groups to oil the wheels of decision-making (Adger, 2003). These sets of networks are usefully described as an asset of an individual or a society and are increasingly termed "social capital". At its core, social capital theory provides an explanation for how individuals use their relationships with other actors in societies for their own and for the collective good. The NNs is expected to serve several functions in this regard.

One of the roles of the NNs is to promote knowledge diffusion and learning. This is especially important when conducting interdisciplinary research, but it is not limited to academic communities. An important role of sustainability science is the distribution of knowledge to society through communication among experts, decision-makers and the rest of society (Brewer, 2007; Sumi, 2007). And, more importantly, sustainability science must co-produce knowledge with society in order to strengthen the linkages between knowledge and action. Many studies reveal that the participation of diverse stakeholders in setting and implementing solutions is indispensable. As science and technology advance, knowledge tends to become centralized; in other words, essential information and knowledge tend to be monopolized by a particular group. Government actions are usually discussed among specialists, i.e. scientists, bureaucrats and politicians. Their decisions are then reported to the public through the media. When people are informed of results only after the fact, they are less likely to be convinced of their necessity, merits and origin. This tends to cause suspicion among citizens about these decisions, and this suspicion is an obstacle to the achievement of consensus in society. Consequently, outreach activities and information-sharing are crucial. Sustainability science has an important role to play in educating and promoting people who have multiple skills and perspectives to act on and solve problems collectively.

The second function of the NNs is mutual learning. Knowledge diffusion notwithstanding, information flow during the learning process is not unidirectional. Different stakeholders engage in joint practices where the acknowledgement and development of viable interdependencies are at stake (Bouwen and Taillieu, 2004). Learning about those interdependencies is a critical constitutive process. Through sharing problem perspectives and working with different kinds of knowledge and competencies, multiple actors or stakeholder parties co-construct a social learning process in an emerging community of practice (Wenger, 1998). Some can catalyse others' ideas and thoughts and vice versa. One can learn necessary knowledge from others and suggest what is unknown and what efforts are required to produce relevant knowledge. As discussed in Clark (2007), sustainability science is neither "basic" nor "applied" research, but rather an enterprise centred on "use-inspired basic research", where both the quest for fundamental understanding and considerations of use are important. Sustainability science and its objectives can be advanced through the establishment of an NNs among knowledge providers and users. The NNs will not only facilitate learning via the transfer of knowledge from one participant to another but also become the locus of novel knowledge creation through this learning process. The ability to learn about new issues and knowledge requires participation in them; hence a wide range of linkages is critical for knowledge diffusion, learning and innovation.

Another role of the NNs is in the area of fostering cognitive factors such as trust, legitimacy and visibility. As discussed by Coleman (1990), a closed network where members of the network are connected to each other can create strong social capital; it is a resource that helps the development of norms for acceptable behaviour and the diffusion of information about behaviour. Members of a closely knit network can trust each other to honour obligations, which diminishes the uncertainty of their exchanges and sense of worth and enhances their ability to cooperate in pursuit of their interests. Coordination is improved through repeated exchange among stable members of the group. Strong social capital of this sort can be expected to promote collective action against problems. It also consolidates the legitimacy and visibility of the community, which in turn attracts third parties that enhance members' competence and influence outside of the community.

These functions of the NNs provided by new linkages among existing networks intensify knowledge diffusion, learning, collaboration and the integration of diverse resources among these networks. A number of benefits can be expected to result from the diversity of information and the brokerage opportunities created by bridging separate cliques in a social network (Burt, 1992). Complex problems require an NNs for their solution because no single institution has all the necessary knowledge, skills, tools, connections and talents for such purposes. The NNs enables people to share knowledge, to learn mutually, to use complementary skills and pool resources, to breed innovation, to acquire visibility and legitimacy and, finally, to act collectively.

2-2-4 The functioning of a network of networks

There are several efforts under way to build networks of networks among leading universities, governments, industries and society. But the existence of an NNs does not guarantee that it will work well. What factors, then, are involved in the successful functioning of an NNs? Fadeeva (2004) proposes the following as key factors in the success of such collaborative undertakings: credible commitment, clarity of goals, clearly distributed responsibilities, involvement of relevant stakeholders, setting intermediate targets (to prevent targets from becoming diluted and to keep stakeholders motivated), monitoring progress towards achieving objectives, and establishing and using incentives and sanctions. Also essential are: illustrating a visionary scenario for attaining the goal, providing credible knowledge to realize the scenario, designing salient solutions, and having the leadership and legitimacy to drive a movement. These factors can be classified into the following categories: target-oriented activities, relational activities and content-based activities.

In the first category, target-oriented activities, key factors for success are clarity of goals, monitoring progress and setting intermediate targets. Target-oriented activities manage transitions in society by navigating participants towards a goal. Clarity of goals is especially important because sustainability is a vague and polysemic term. A given issue differs among different people, cultures and timeframes. Sustainability is a term with multiple meanings because it encompasses a variety of objectives, including environmental, social and human sustainability, as well as a variety of goal directions – equilibrium, growth, reduction. Sustainability may focus on multiple goals because different people have different aspirations in different time periods, over different time scales and in different contexts.

The second category of factors in the functioning of an NNs is relational activities. Many factors are included in this category: involvement of relevant stakeholders, clearly distributed responsibilities, incentives and sanctions, credible commitment, and legitimacy and leadership all affect the continuity and performance of the NNs. Socio-psychological factors must also be considered, such as will, strategy, intention, motivation, responsibility and ethics, as well as economic, political and social regulations and the institutions influencing them. The involvement of relevant actors at an early stage is important to legitimize a goal. Commitment, distributed responsibilities and incentives are the drivers that encourage participants to contribute to achievement of the goal in a distributed autonomous coordination system.

The components of the third category, content-based activities, are: illustrating a visionary scenario to attain the goal, providing credible knowledge to realize the scenario and designing salient solutions. These comprise a backcasting process, with feasibility assessment of the scenario supported by reliable knowledge. Backcasting is a normative approach to the realization of a goal by working backwards to show pathways between the goal and the present status. But it must also be noted that there are no panaceas and that solutions are valid under certain specific prerequisites. A core aspect of panaceas is the tendency to apply a single solution to many problems (Ostrom, 2007), but plausible solutions are dependent on the temporal and spatial scale. If one fails to be vigilant, one can become trapped by an inclination to apply panaceas regardless of the circumstances (Brock and Carpenter, 2007). Many papers have proposed a variety of solutions to the problems they address. When there is a solution, the problem is already solved or will be easily solved. When a problem remains, we can infer that solving the problem is hampered by certain obstacles, that is, the existence of other problems. Thus the structuring of knowledge about the problem-solution chain is crucial to finding the essential causes and bottlenecks of an issue and solving the root problem. Evaluation of the feasibility and unintended results of plausible solutions is a subtask of this problem-solution chain analysis.

2-2-5 Integration of knowledge through action

As already stated, a sustainable society can be achieved only when knowledge is accompanied by action. Action is essential and indispensable. But, in order to attain goals, one must also have a knowledge platform that provides reliable information on the actions of individuals and the predicted results of those actions. Action and knowledge are inextricably interrelated. In fact, knowledge is produced and updated through the interaction between action and knowledge and through mutual interaction between the real world and the abstract world and between issues and solutions. That is the driver for transforming society in a sustainable direction. Both action and knowledge are indispensable. If one acts but does not think, then one will make mistakes. Conversely, if one thinks but does not act, then one's knowledge is useless.

In a recent paper, Beers and Bots (2009) considered knowledge-sharing across scientific disciplines from three different aspects: knowledge-modelling, science webs and communities of practice. Here the last aspect will be discussed. (The first aspect is discussed in subsections 2-2-1 and 2-2-3; regarding science webs, an NNs is currently being implemented by The University of Tokyo.[1])

A community of practice as defined in Beers and Bots (2009) is a group that shares information, insights, experiences and tools related to an area of common interest. Community members frequently engage in open discussions to help each other solve problems. While exchanging ideas and experiences, they develop a shared terminology and a common set of "good practices". This knowledge is socially constructed, highly situated and, to an important extent, implicit. As such, it cannot be shared with other groups; knowledge transfer between two communities requires individuals who participate in both and are active as "knowledge brokers" and "boundary spanners". Thus, the building of an NNs requires a particular kind of expertise.

One of the roles of the NNs is to facilitate communication among research institutions by spanning the boundaries of existing networks, especially networks among universities. Launching a joint research and education programme is one way to make this NNs work, but it should not be limited to universities. Another, more important role is to increase the visibility of activities by participating universities and to influence political and business sectors, nongovernmental organizations (NGOs), and so on.

By forming an NNs, universities should be helping to move society towards sustainability. To move in that direction, the actions of a variety of stakeholders must be structured. There are numerous initiatives and efforts being promoted in the world towards the establishment of global sustainability. At the national level, Japan's former Prime Minister Yasuo Fukuda announced a vision for reducing 60–80 per cent of greenhouse gas (GHG) emissions by 2050. Even the United States, which moved slowly on these issues under the previous administration, is now getting serious and following suit under President Obama's leadership: the Green New Deal. At the regional or municipal level, the state of California, under the active leadership of Governor Arnold Schwarzenegger, has announced a plan to reduce GHGs by 80 per cent by 2050. Governor Shintaro Ishihara of Tokyo has publicized a vision of the "10-Year Project for a Carbon-Minus Tokyo". Boris Johnson, Mayor of London, announced the establishment of a Low Emission Zone, which came into effect on 4 February 2008 and covers almost all of Greater London. Most of these initiatives aim to achieve their goals by the year 2050. That may sound far in the future, but it is not. Considering the amount of work to be done, including the development of the necessary infrastructures, there is not much time. Hence a concurrent – not a sequential – approach is needed to shorten the process.

Universities need to be more proactive and show initiative and implementation ability through their actions as leaders in the promotion of global sustainability in the twenty-first century. Instead of merely being passive knowledge providers, they must take action to show that sustainability is possible without compromising their basic activities – research, education, campus life – and their outcomes. It is encouraging that many universities are actively undertaking sustainable campus projects. For example, The University of Tokyo has started the Todai Sustainable Campus Project (TSCP), with a short-term target of a 15 per cent reduction in

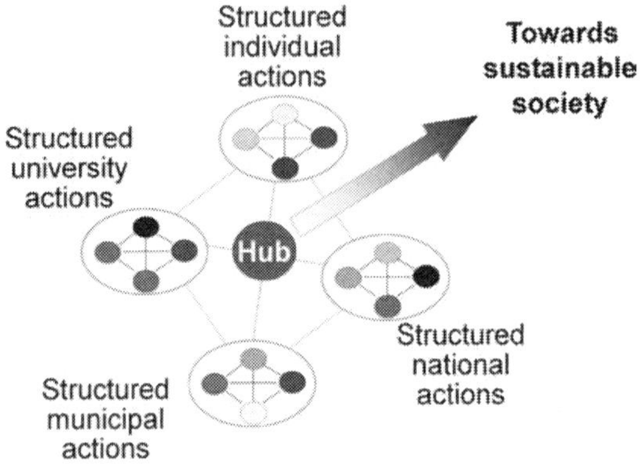

Figure 2.2.4 Action-structuring to realize sustainability.

CO_2 emissions by 2012 and a long-term target of a 50 per cent reduction by 2030. Harvard, Yale, the Massachusetts Institute of Technology and the University of British Columbia have similar programmes, as do numerous other universities. Through these social experiments, it is expected that valuable knowledge will be obtained in the quest to achieve sustainability.

The most important role of universities and research institutions in the twenty-first century is to serve as a hub for structuring the many actions being conducted at various levels by nations, municipalities, businesses, NGOs and individuals (Figure 2.2.4). For that purpose, some very powerful tools are now available: the structuring of action and the structuring of knowledge.

2-2-6 Conclusion

A sustainable society cannot be achieved solely through knowledge without action. But it is important to note that action without knowledge leads to uncertain results, and knowledge without action cannot change a situation. Knowledge and action have a reciprocal influence. Appropriate knowledge is necessary for relevant action, which in turn shapes necessary knowledge and integrates that knowledge. The traditional role of universities has been to generate and feed knowledge to society, but, in the context of sustainability science, universities must actively work with other actors, spanning existing boundaries through university–government, university–industry and other forms of collaboration with society. In this section, action-structuring was proposed as a key concept in this collaborative attempt to achieve sustainability.

Action-structuring has three phases: decomposition of actions, integration into new actions, and the promotion of collective actions. Through the decomposition and integration of actions, new actions suitable for particular conditions and constraints can be designed, but they must then be carried out by collective action. A network of networks can work as a device for promoting collective actions. The NNs is a network among existing networks that involves diverse stakeholders in setting issues, designing and implementing solutions, empowering them and maintaining reliability, visibility, confidence, legitimacy and trust in the solution.

For action-structuring through a network of networks, the following types of activity are indispensable. One is target-oriented activities, where the key factors for success are clarity of goals, monitoring progress and setting intermediate targets. Target-oriented activities manage transitions in society by navigating participants towards a goal. The second is relational activities involving relevant stakeholders with distributed responsibilities, incentives and sanctions, credible commitment, legitimacy and leadership. The third is content-based activities, whose components include illustrating visionary scenarios to attain a goal, providing credible knowledge to realize the scenario, and designing salient solutions. Finally, a knowledge platform must be provided through a science web and community of practice. Building a network of networks is the first step towards fulfilling these objectives and hence will contribute to the achievement of a sustainable society in the future.

Note

1. See <http://nns-u.org/>. The website includes a list of universities addressing sustainability and related issues; this list must continue to be expanded.

REFERENCES

Adger, W. N. (2003) "Social Capital, Collective Action, and Adaptation to Climate Change", *Economic Geography* 79(4): 387–404.

Beers, P. J. and P. W. G. Bots (2009) "Eliciting Conceptual Models to Support Interdisciplinary Research", *Journal of Information Science* 35: 259–278.

Bledsoe, C. and M. Barber (1993) "Ecological Network of Networks: Creating a Network to Study Ecological Effects of Global Climate Change", Report of a workshop sponsored by the Ecological Systems and Dynamics Task Group, US MAB Secretariat, US Department of State, Washington, DC.

Bouwen, R. and T. Taillieu (2004) "Multi-party Collaboration as Social Learning for Interdependence: Developing Relational Knowing for Sustainable Natural Resource Management", *Journal of Community & Applied Social Psychology* 14: 137–153.

Brewer, G. D. (2007) "Inventing the Future: Scenarios, Imagination, Mastery and Control", *Sustainability Science* 2(2): 159–177.

Brock, W. A. and S. R. Carpenter (2007) "Panaceas and Diversification of Environmental Policy", *Proceedings of the National Academy of Sciences* 104(39): 15206–15211.

Burt, R. (1992) *Structural Holes*. Cambridge, MA: Harvard University Press.

Clark, W. C. (2007) "Sustainability Science: A Room of Its Own", *Proceedings of the National Academy of Sciences* 104: 1737–1738.

Coleman, J. (1990) *Foundations of Social Theory*. Cambridge, MA: Harvard University Press.

Fadeeva, Z. (2004) "Promise of Sustainability Collaboration – Potential Fulfilled?", *Journal of Cleaner Production* 13: 165–174.

Ostrom, E. (2007) "A Diagnostic Approach for Going Beyond Panaceas", *Proceedings of the National Academy of Sciences* 104(39): 15181–15187.

Peters, D. P. C., P. M. Groffman, K. J. Nadelhoffer, N. B. Grimm, S. L. Collins, W. K. Michener and M. A. Huston (2008) "Living in an Increasingly Connected World: A Framework for Continental-scale Environmental Science", *Frontiers in Ecology and the Environment* 6(5): 229–237.

Sumi, A. (2007) "On Several Issues Regarding Efforts toward a Sustainable Society", *Sustainability Science* 2(1): 67–76.

Wenger, E. (1998) *Communities of Practice: Learning, Meaning and Identity*. Cambridge: Cambridge University Press.

2-3

The structuring of knowledge based on ontology engineering

Riichiro Mizoguchi, Kouji Kozaki, Osamu Saito, Terukazu Kumazawa and Takanori Matsui

2-3-1 Introduction

As one of the ultimate goals of research in any domain, knowledge-structuring has been carried out to date through the writing of papers and books as part of ordinary academic research activities. In this respect, what scholars are aiming at in the knowledge-structuring of sustainability science (hereafter referred to as SS) might seem to be nothing special. Nevertheless, SS researchers are particularly dedicated to the structuring of knowledge. There are two reasons for this:

1. SS is essentially an interdisciplinary field, covering so many domains that the conventional mode of research, conducted in a domain-specific manner, is not adequate for furthering SS.
2. SS is essentially oriented towards problem-solving. It demands intensive research not only to structure the findings that result from investigations of phenomena, but also to determine how those findings can contribute to solving sustainability-related problems. Furthermore, investigations need to be planned and executed from a problem-solving perspective from the outset, rather than through post hoc study of their application.

In short, SS must be developed as a new discipline by integrating and reorganizing existing fields through knowledge-structuring. It is an ambitious undertaking, but one worthy of the effort.

The purpose of this section is to discuss the knowledge-structuring of SS through ontological engineering, which is an advanced information technology. In a previous paper (Kumazawa et al., 2009), the authors discussed what they mean by knowledge-structuring in terms of reusability, versatility, reproducibility, extensibility, availability and interpretability. This section focuses more on concrete and technological aspects of knowledge-structuring, including a new version of a mapping tool developed for ontology exploration.

2-3-2 Reference model for knowledge-structuring in sustainability science

One of the major characteristics of SS research is a "problem-solving" orientation rather than the "understanding" orientation common to the conventional sciences. In addition to gaining an in-depth understanding of the critical phenomena we observe, researchers need to relate, formulate and reorganize their findings to contribute to the solutions required to make things sustainable. SS knowledge-structuring should be carried out with this requirement in mind. The idea of "knowledge-structuring" must be clearly articulated as a research goal. The authors addressed this topic in their previous paper (Kumazawa et al., 2009), in which they proposed a reference model for SS knowledge-structuring in order to roughly explain what they mean by knowledge-structuring.

Because of the complexity and diversity of sustainability issues, it is critical to identify and evaluate relationships among problems, causes, impacts, solutions and their interactions. Those relationships usually depend on the specific context of an individual case or problem. This is the source of a dilemma in SS knowledge-structuring, which must be generic enough to cope with many requirements at the same time. Problems and their solutions need to be explored within each problem's specific context. Therefore, SS knowledge-structuring must be carefully designed to fulfil the requirements of both generality and context-specificity. This is why the authors have developed a reference model. It enables them to make explicit the common aspects of possible forms of knowledge-structuring geared to specific purposes, as articulated by the following four questions:

- What is the underlying information structure of the problem?
- From which perspective are you looking at the problem?
- How can you describe and organize information on a problem within a specific context?
- How can contextualized and structured information facilitate essential problem-finding and -solving?

In this context, they developed a reference model for SS knowledge-structuring in which any specific tool and/or knowledge-structuring method that answers these four questions can be installed.

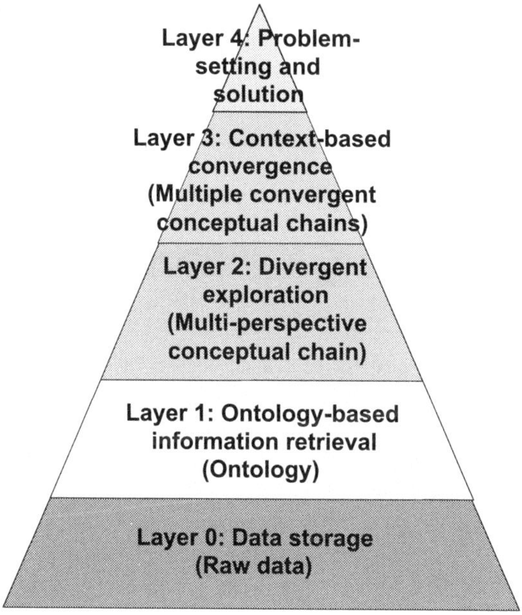

Figure 2.3.1 Layered structure of the reference model.

First of all, the reference model is not a solution to knowledge-structuring but a model to be referred to when discussing the knowledge-structuring of SS. It contributes to the evaluation and understanding of the differences and commonalities in knowledge-structuring tools and methods proposed in the future by providing a common framework in which to compare them. As shown in Figure 2.3.1, the model consists of five layers.

The bottom layer, Layer 0, is the data layer and stores raw data corresponding to the real world. Layer 1 is named the ontology layer; it stores the ontology for explaining and understanding the raw data at Layer 0, and describes the concepts and relationships related to SS that exist in the real world. Another function of the ontology is to provide a common vocabulary for promoting mutual understanding across domains. Typical tasks performed at Layer 1 include metadata generation for the virtual organization of the raw data and efficient retrieval of the raw data using the metadata.

Layer 2 is called "divergent exploration" because its main task is the divergent exploration of the conceptual world realized at Layer 1, which systematizes the concepts appearing in the SS world. Divergent exploration in an "ocean of concepts" uses divergent thinking across domains to guide researchers searching for interesting concepts and relationships that have been hidden in the conventional unstructured world. The ontology at Layer 1 must contribute to such exploration. Divergent exploration can be performed by obtaining what the authors call "multi-perspective conceptual chains" through the selection of arbitrary concepts according to the explorer's intention. Many ways of tracing conceptual chains are needed to handle the various aspects of SS. After generating these conceptual chains, the explorer moves on to a convergent thinking stage at Layer 3. The task of this layer is "context-based convergent thinking". At this layer, the explorer can set a specific context for a problem that he or she actually treats and obtain "multiple convergent conceptual chains" (Klein, 2004) in accordance with the given context. Examples of contexts include the social and environmental settings of a specific problem, implemented or planned countermeasures and policies for solving the problem and even trade-offs between the target problem and other problems, such as the trade-off between greenhouse gases (GHGs) and hydrofluorocarbons (HFCs), or between food security and biofuel production. At Layer 4, using all the information and knowledge obtained at the lower layers, the explorer will pursue essential problem-solving tasks such as problem-setting or even a new problem search, as well as information integration or abduction.

Whereas the bottom two layers are static, the top three layers are dynamic. The information in the top layers is dynamically generated as required by the tasks at those layers. This dynamism, one of the significant characteristics of the reference model, is achieved by combining the fine-grained conceptual organization of the ontology with an ontology exploration tool. This approach also helps resolve the generality vs. specificity dilemma. The authors believe that a static structure would be inadequate for dealing with the multi-perspective nature of SS. Another characteristic of the reference model is its layered structure, in which each layer is composed of a pair consisting of structured (or organized) information and a task. This reflects their understanding of SS as inherently problem- and use-inspired basic research.

2-3-3 Ontology-based knowledge-structuring

The authors applied the reference model to develop a knowledge-structuring system for SS. For Layer 0, they collected a comprehensive sample of literature and databases available on the World Wide Web. This work was conducted in parallel with the activities of the Research Institute for Sustainability Science (RISS) at Osaka University (Morioka et al., 2006) to develop a meta-database for SS, a conceptual map of a resource-circulating society and educational content of a core module for SS, entitled "Valuation Methods and Technical Aspects in Sustainability".

As a prototype tool at Layer 1, the authors constructed a trial SS ontology. For this, they first extracted the concepts for the SS ontology and the relationships between them from the meta-database of SS, the documents used as educational content and the database on the Environmental Information and Communication Network website.[1] Second, they discussed the architecture of the SS ontology and requirements for SS knowledge-structuring in monthly workshops that have been coordinated by RISS since 2006. On the basis of the information collected and the discussions in the workshops, a prototype version of the SS ontology was built as a required task at Layer 1. The authors conducted several of the kinds of research studies that are necessary for applying an ontology to sustainability-related domains, including targeting sustainable development indicators, risk communication and education (Brilhante et al., 2006; Friend, 1996; Macris and Georgakellos, 2006; Suzuki et al., 2005; Tiako, 2004).

Semantic web technology has been applied to develop systems for knowledge-structuring and data retrieval. For example, EKOSS (Expert Knowledge Ontology-based Semantic Search) is a knowledge-sharing platform based on semantic web technologies (Kraines et al., 2006). In order to fulfil the specifications of Layer 2, the authors also developed a conceptual mapping tool that enables a user to explore the SS ontology from that user's particular perspective and to generate a conceptual map accordingly. The following subsections explain this development process and its outcomes.

The underlying philosophy of ontology-based knowledge-structuring

One of the key ideas of the authors' approach to knowledge-structuring is the combination of static and dynamic structuring. By dynamic structuring they mean that the structured knowledge is not fixed in a predefined structure but is flexible enough to adapt to various requirements specified by users. The problem is how to realize such knowledge-structuring. A solution can be found in ontological engineering, in which fine-grained concepts are systematically represented to capture the "reality" of the target world. These structured concepts can be used for describing requirements.

Conventional methods of knowledge-structuring tend to model knowledge in a systematic manner based on a set of principles that capture the essential properties of the target domain. This is carried out for human consumption and is intrinsically static because it is done in papers and/or books. The authors' method of knowledge-structuring is different in that it is done on a computer, which can in principle adapt to any input requirements. To maximize this adaptivity, they exploit the flexibility of a computer program, which can change its behaviour. Such flexibility is critical to providing high adaptivity in a knowledge-structuring tool, giving domain experts the opportunity to derive unexpected causal chains from known factors.

2-3-4 Development of a sustainability science ontology

Overview of the ontology

The quality of the ontology the authors develop is the heart of their enterprise. First of all, it must be compliant with ontology theory. Some principles they have rigorously followed are: (a) forming a proper *is-a* hierarchy, (b) avoiding confusion between *is-a* and *part-of* relations, (c) maximal compliance with a reliable upper ontology, (d) use of proper property-inheritance and (e) identification of roles. These are content-independent quality requirements. Content-dependent quality measures include the following aspects of the ontology:

1. It covers a variety of domains within SS.
2. Each concept defined in the ontology has a rich slot description.
3. It organizes domain knowledge not in specific domains but in one world, to reflect the basic idea of knowledge-structuring across domains.
4. Its top-level categories reflect the essential properties of SS and are composed of the following five types:
   - Goal
   - Problem
   - Countermeasure
   - Assessment
   - Domain concept

The ontology is partially compliant with the YATO upper-level ontology developed by one of the authors (Mizoguchi, 2009).[2] By "partially" is meant that only *Domain concept* follows YATO. This is to make the particularity of SS explicit since, if the ontology were made completely YATO compliant, these categories would be hidden under the top categories of YATO. The authors believe that *Goal*, *Problem*, *Countermeasure* and *Assessment* are major top-level categories for SS which should not be hidden. *Goal*, which is further divided into *Structural goal* and *Situational goal*, represents the aims of SS research. The most significant is *Problem*, since this is what SS research is expected to solve and hence must be structured appropriately. *Problem* is further divided into *Problems of the earth system*, *Problems of the social system* and *Problems of the human system*, following the framework proposed by the Integrated Research System for Sustainability Science (IR3S). *Countermeasure* is divided into two subtypes, *Future-oriented countermeasure* and *Present countermeasure*. *Assessment* has three subclasses: *Assessment perspective*, *Assessment index* and *Assessment method*. *Domain concept* is equivalent to ordinary top-level categories in existing upper ontologies because it includes everything found in the world that SS tries to explain and address. Unlike existing ontologies of SS, this one has no domain-specific subdivision and is instead divided into *Substrate*, *Entity* and *Dependent entity* according to the Basic Formal Ontology (BFO) developed by Barry Smith (Grenon et al., 2004).[3] The ontology developed by the authors is a prototype that is now undergoing augmentation into a more complete one. Because their main goal here is to investigate the basic ideas of knowledge-structuring of SS and to develop a computational model for it, the completeness of the ontology is secondary. Needless to say, however, the ontology must follow a good design philosophy that will be used and shared among domain experts for a long time. Figure 2.3.2 is a snapshot of the prototype version of this ontology.

Development of the sustainability science ontology

Constituents of ontology

Although ontology plays a prominent role in the authors' research, the main contribution of this research is to describe a reference model for structuring SS knowledge and a mapping tool compliant with that model. Therefore, an in-depth analysis of the characteristics of ontology is not a priority of this section. Here, the authors will limit themselves to a brief explanation of terms needed for structuring the SS ontology.

An ontology consists of concepts and relationships among those concepts that are needed to describe the target world. One of the main components of an ontology is a taxonomy of concepts representing things existing in the target world that are determined to be important. These are organized by identifying *is-a* relationships between them. In an *is-a* relationship – for example, <Destruction of regional environment *is-a* Problem> – the generalized concept (*Problem*) is called a super concept and the specialized concept (*Destruction of regional environment*) is called a sub concept. Thus, an *is-a* hierarchy describes the categorization of concepts. *Problem* is subdivided into sub concepts such as *Problems of the earth system*, *Problems of the social system* and *Problems of the human system*. The introduction of other relationships refines the definition

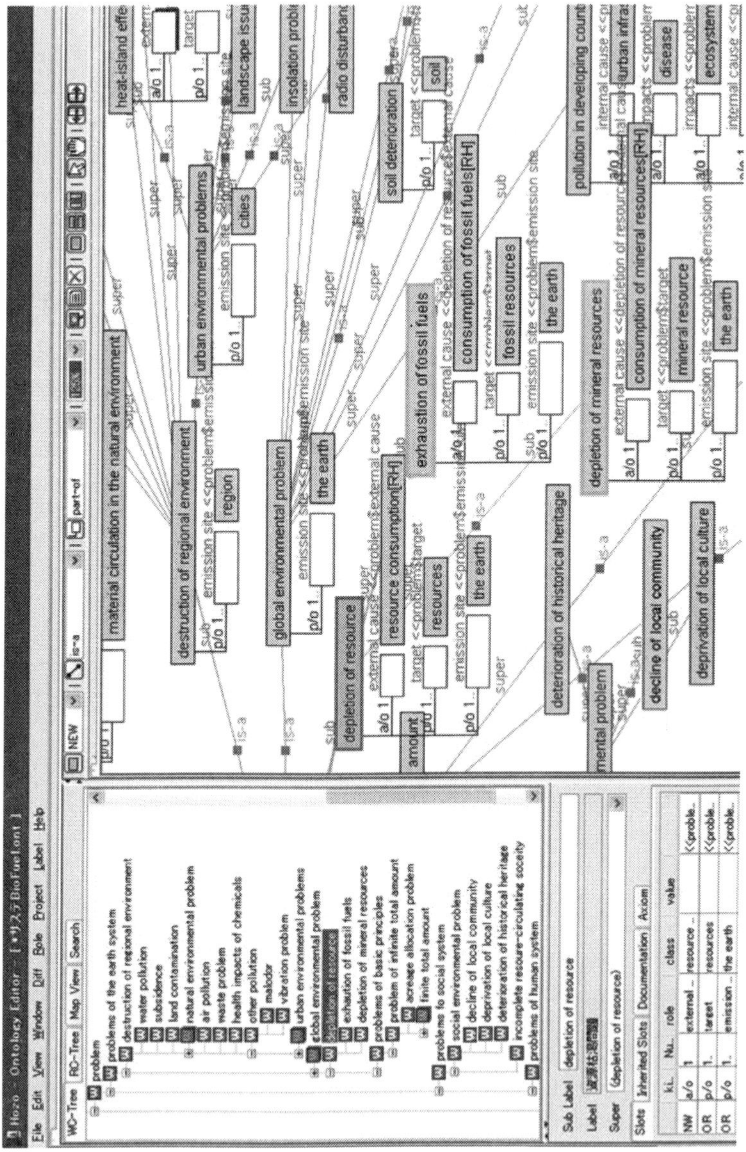

Figure 2.3.2 A snapshot of the ontology in Hozo.

of these concepts. For example, *part-of* relationships, which are also called *has-part* relationships, and *attribute-of* relationships are used to show the concept's parts and attributes, respectively. Examples of these relations are shown in Figure 2.3.2. In the figure, "p/o" denotes a *part-of* relation and "a/o" denotes an *attribute-of* relation. For example, *Depletion of resource* is defined with two parts and one attribute.

Top-level structure

Using Hozo, an ontology-building and application platform, the authors have developed a prototype of an SS ontology.[4] Here they briefly explain the top-level concepts and second-level concepts with the slots – representing concepts of parts and attributes – that are used to describe them. In the current implementation, the SS ontology has 562 concepts and 14 hierarchy levels.

Following the problem-solving approach of SS, *Problem* and *Countermeasure* are adopted as two of the SS ontology's top-level concepts. Additionally, when trying to solve a problem, a goal or goals for countermeasures must be set and both existing conditions and the impacts of the countermeasures must be assessed explicitly or implicitly. Post-assessment as well as pre-assessment may also result in finding new problems. For this reason, *Goal* and *Assessment* are included as top-level concepts of this ontology. Furthermore, *Domain concept* is set as another top-level concept to model concepts that are not covered by the above four categories. In the SS ontology, knowledge in the domain is not organized according to individual fields or disciplines, such as energy, climate, population, policy or law. Instead, it is organized according to more general concepts, such as objects, activities, situations and attributes, on the basis of ontological engineering theory (Mizoguchi, 2003, 2004a, 2004b).

In ontological engineering theory, an ontology is composed of domain-specific concepts under the upper-level concepts, which themselves are highly domain-neutral. In this way, the ontology is organized in a domain-neutral manner. In summary, the SS ontology consists of five top-level concepts: *Goal*, *Problem*, *Countermeasure*, *Assessment* and *Domain concept*. Although they are SS-specific, they are sufficiently generalized to be independent of the targeted domains. Furthermore, although concrete occurrences and activities can be the sub concepts of *Domain concept*, these concepts do not depend on the context of problem-solving. By describing the world using two types of super concept, domain-independent and domain-dependent, any types of countermeasures for sustainability that one would like to show can be represented. Domain-specific knowledge seen from a specific viewpoint can be represented by combining these concepts. Also, such a conceptual system can support the generation of ideas for new concrete countermeasures that were not conceived when the system was initially designed.

Problem

Problem is categorized into *Problems of the earth system*, *Problems of the social system* and *Problems of the human system.* As already mentioned, this structure follows the framework proposed by IR3S, the parent project of the authors' research. *Problems of the earth system* is further divided into *Destruction of regional environment*, *Global environmental problem*, *Depletion of resources* and *Problems of basic principles. Problems of the social system* is further divided into *Social environmental problem* and *Incomplete resource-circulating society.* The former consists of *Decline of local community*, *Deprivation of local culture* and *Deterioration of historical heritage. Problems of the human system* is substantially equivalent to *QOL problem.*

Goal

There are two possible approaches to defining the top-level concept of *Goal.* One is to describe a situation that people desire; the other is to describe an ideal social structure or system. The former approach often uses phrases such as *Global peace* or *Human happiness and well-being.* The latter approach includes goals that, for example, articulate the social structure for a *Resource-circulating society* (Ministry of the Environment, 2007) or specify the range of *Environmental carrying capacity.* The authors have named these two approaches *Situational goal* and *Structural goal*, respectively, and have incorporated both types of goals.

Assessment

Sub concepts of *Assessment* consist of *Assessment perspective*, *Value*, *Assessment indicator* and *Assessment method* (Rotmans, 2006; UNEP CBD, 2000). *Assessment indicator* was also subdivided into five types: *Qualitative indicator*, *Quantitative indicator*, *Warning indicator*, *State indicator* and *Indicators and time* (Munier, 2005). These conform to terms that are already established in the SS community.

Countermeasure

Countermeasure is divided into two major sub concepts: *Future-oriented countermeasure* and *Present/Ongoing countermeasure. Future-oriented countermeasure* includes *Scenario*, *Education* and *Plan. Education* is defined as a measure for training future generations who will be responsible for implementing necessary actions in the future. *Present/Ongoing countermeasure* focuses on the relationship between people and technology. Countermeasures in this sense consist of technologies, people and interconnections between all kinds of actions associated with technologies. Countermeasures concerning people, for example, include *Restrictions of their actions* and *Changes of their behaviour*. The sub concepts of *Present/Ongoing countermeasure* include *System-based countermeasure*, *Technology-based countermeasure*, *Action-oriented countermeasure* and *Conversion of styles*.

Domain concept

Domain concept is divided into three top-level concepts: *Substrate*, *Entity* and *Dependent entity*. *Substrate* consists of *Time*, *Space* and *Matter*, and *Dependent entity* consists of *Attribute* and *Quantity*. *Entity* is further divided into *Physical* and *Abstract* entities. *Physical* is divided into *Continuant* and *Occurrent*. By *continuant* is meant things that intrinsically require three-dimensional space to exist, and by *occurrent* is meant things that intrinsically exist in time–space. *Continuant* is divided into *Unitary* and *Non-unitary* objects. *Unitary* objects consist of *Agent*, *Artifact*, *Living organism* and *Natural structure*. *Agent* has two concepts called *Macro agent* and *Micro agent*. Concepts of systems, such as *Social system*, *Ecosystem* and *Industrial ecology* belong to *Macro agent*. *Occurrent* includes *Situation* and *Process*, with the latter including *Activity*, *Phenomenon* and *Circulation*. *Circulation* is divided into three concepts: *Material circulation in the natural environment*, *Material circulation based on economic activity* and *Circulation of life*.

2-3-5 Divergent exploration of SS knowledge (conceptual map generation)

The necessity of an ontology exploration tool from an ontological engineering point of view

An ontology is intrinsically generic and captures reality as objectively as possible; hence it is viewpoint-neutral. This very property enables an ontology to contribute to stable and long-lasting knowledge representation in the target domain. It is one of the critical attributes of an ontology. However, for practitioners and domain experts who are eager to use an ontology directly in solving real problems, such an ontology might be too generic and objective to allow them to utilize it to their satisfaction. This is because few experts are interested in generic information/knowledge that does not stimulate them to apply it. What interests them is information viewed from their chosen perspective or viewpoint. In short, although an ontology on its own may be too generic for domain experts to appreciate, it might prove useful as an information resource with the help of a tool that enables domain experts to explore only those portions they consider to be of interest. The authors therefore decided to develop an ontology exploration tool that allows domain experts to explore an ontology in accordance with their needs.

Ontology exploration

The use of ontology in the authors' research differs from that of other existing ontology-based applications, which employ ontology as a meta-schema to provide better information searches than conventional search methods. The authors use their ontology directly; that is, their aim is to build a tool that allows domain experts to explore the ontology itself, rather than explore cyberspace with the help of an ontology. When it is complete, the ontology will be composed of a number of concepts with relations among them. This should be a significant source of knowledge, particularly since, in keeping with the authors' design philosophy for the SS ontology, it includes goals, problems, countermeasures and assessments, together with a great many real-world objects, processes and phenomena. Furthermore, these concepts are tightly connected with one another. Thanks to the non-domain-specific organization of these concepts, the tool can guide explorers to many kinds of causal and other chains across domains. Here the authors describe their basic idea of ontology exploration at Layer 2 in the reference model.

They understand that ontology exploration is undertaken mainly to comprehend the target world represented implicitly in the ontology by viewing it from various viewpoints and aspects. The importance of exploring multiple viewpoints amid the "sea of concepts" is exemplified by the well-known controversy over biofuels. The utilization of biofuel contributes to the reduction of CO_2 from the global system point of view,[5] but contributes to increasing poverty through rising food prices from the economic point of view. These conflicting effects of biofuel use are elucidated in a view-dependent way.

Ontology exploration tool

Based on the above observations, the authors developed a tool at Layer 2 of the reference model for allowing domain experts to perform divergent exploration of the SS ontology. Figure 2.3.3 shows how the tool generates a map from any specified concept following specified links that might lead to unexpected but interesting concepts.

The divergent exploration of an ontology can be performed by choosing arbitrary concepts according to the explorer's intention to obtain

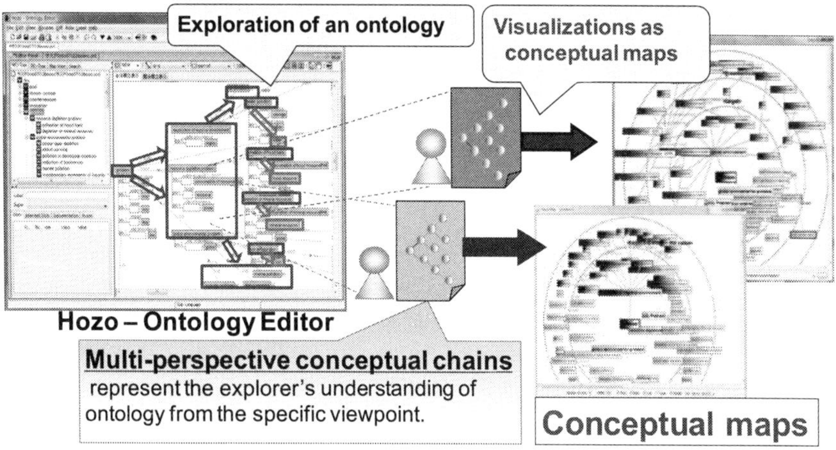

Figure 2.3.3 Ontology exploration tool.

what the authors call "multi-perspective conceptual chains". The viewpoint for exploring an ontology and obtaining multi-perspective conceptual chains is defined as the combination of a *focal point* and an *aspect*. The multi-perspective conceptual chains are visualized in a user-friendly format, i.e. a conceptual map. These terms are defined as follows:

focal point: a concept defined in an ontology from which the exploration starts;

aspect: focused links that the exploration follows from a concept to other concepts;

multi-perspective (multi-aspect) conceptual chain: a sequence of concepts traced from a focal point following focused links;

conceptual map: a drawing of all the multi-perspective conceptual chains from a focal point.

Aspects specify the manner in which the user explores the ontology. Because an ontology consists of concepts and the relationships among them, aspects can be represented by a set of methods for extracting concepts according to their relationships. The authors classify these relationships into four types and define two methods of extraction for each relationship according to the direction to follow (upward or downward). Although these are termed "relationships" in the Hozo ontology model (Mizoguchi et al., 2007), some of them correspond to properties in the OWL web ontology language (see Table 2.3.1). Users can constrain the links to follow by specifying kinds of link in the commands listed in (1)–(8). Similarly, users can constrain the types of concept to reach by specifying types of concept.

Table 2.3.1 Correspondence between concept types and aspects

	Related relationships		
	Hozo	OWL	Kinds of extraction
(A)	*is-a* relationship	rdfs:subClassOf[a]	(1) Extraction of sub concepts (2) Extraction of super concepts
(B)	*part-of* /*attribute-of* relationship	rdf:Properties which are referred in	(3) Extraction of concepts referring to other concepts via relationships (4) Extraction of concepts to be referred to by some relationship
(C)	*depending-on* relationship		(5) Extraction of contexts (6) Extraction of role concepts
(D)	*play (playing)* relationship		(7) Extraction of players (class constraints) (8) Extraction of role concepts

[a] In the web ontology language OWL, this represents an *is-a(sub-class-of)* relationship.

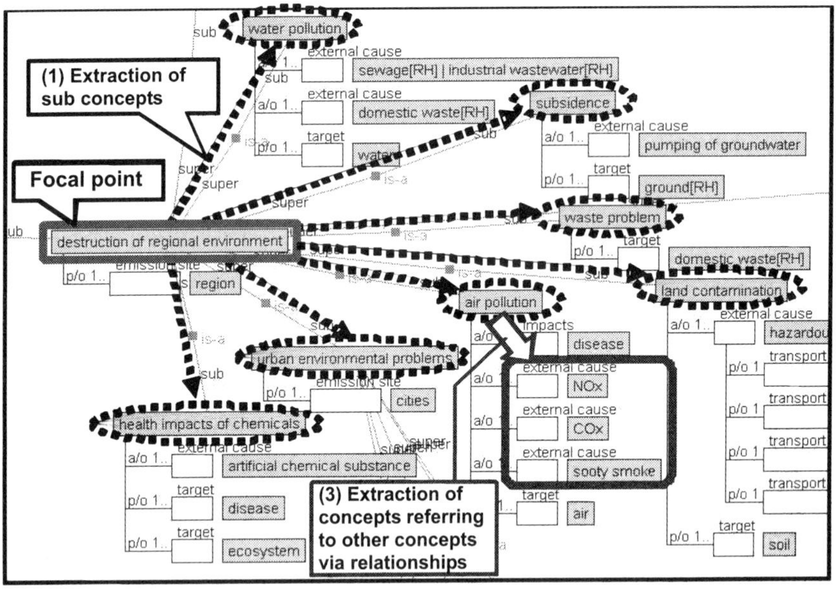

Figure 2.3.4 The mechanism of exploration.
Note: Please see page 471 for a colour version of this figure.

In the example shown in Figure 2.3.4, *Destruction of regional environment* is set as the focal point. Following an *is-a* link by choosing (1) – Extraction of sub concepts – as an aspect, seven concepts such as *Air*

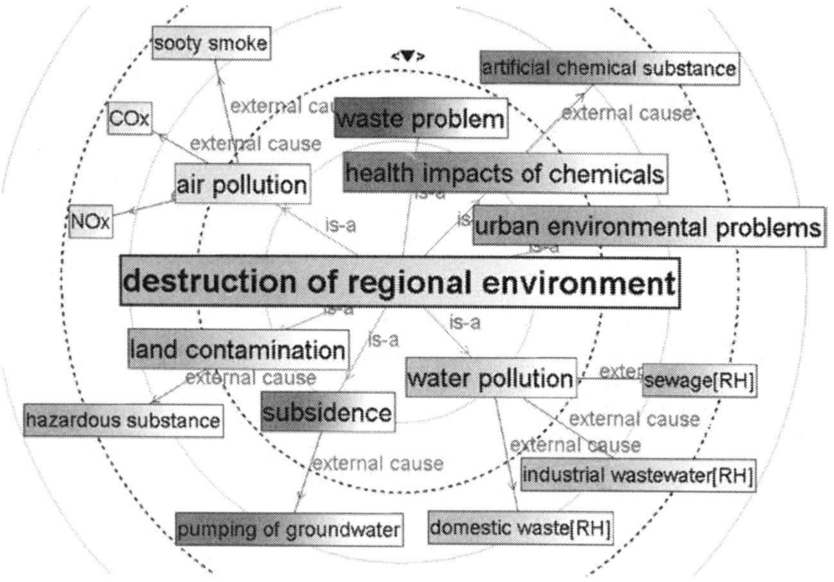

Figure 2.3.5 An example of a conceptual map.
Note: Please see page 472 for a colour version of this figure.

pollution, *Land contamination*, etc. are reached. If the user further follows the *External cause* link by choosing (3) – Extraction of concepts referring to other concepts via relationships – as an aspect, then he/she reaches *NOx*, *COx* and *Sooty smoke*. On the other hand, if the user imposes no constraint on the links to follow, then he/she will get *Disease* and *Air* instead of these three. As a result of this concept extraction, the system generates conceptual chains that match the user's interest and visualizes them as a conceptual map in which the focal point is located at the centre and the conceptual chains are represented as a divergent network (Figure 2.3.5).

Another example of map generation (ontology exploration) is presented in Figure 2.3.6 to show the importance of aspect-based control. Both of the maps in Figure 2.3.6 are generated when a user starts exploration of the ontology from the focal point *Biomass utilization*. The right-hand map has been generated by following links with the aspect of *Fossil fuel*. The map shows that CO_2 *emission reduction* is achieved through the use of biomass fuel. The left-hand map, on the other hand, has been generated by following links from the same focal point with the aspect of *Influence*. This map shows that hunger in developing countries will result from food price inflation caused by increased demand for biomass. Maps such as these can reveal conflicting impacts by highlighting the differences between aspects.

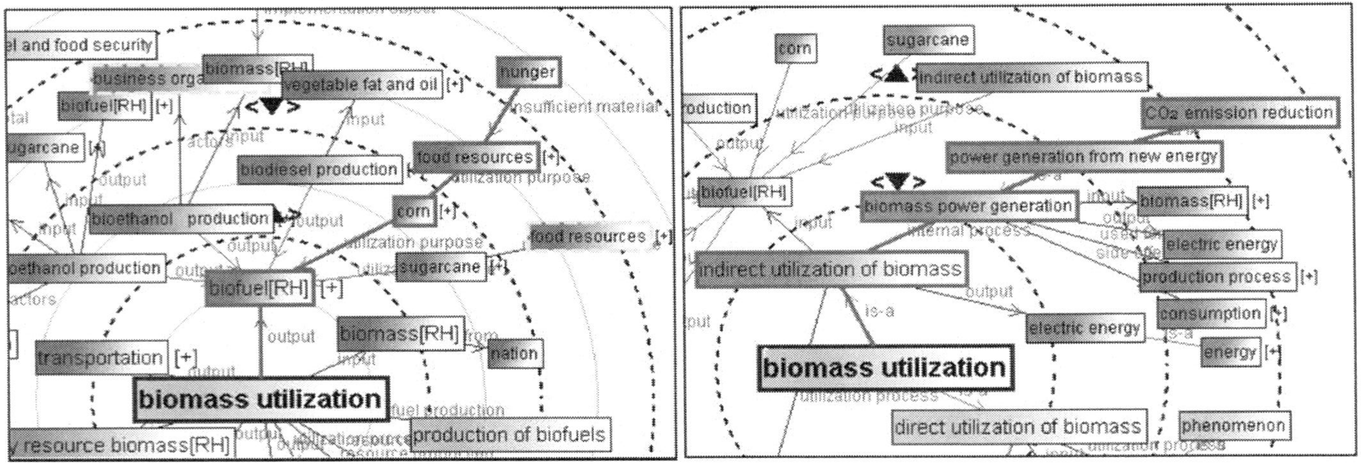

Figure 2.3.6 Comparison of two maps derived from *Biomass utilization*.
Note: Please see page 473 for a colour version of this figure.

Change of viewpoint

In addition to the focal point and aspect settings, another method of specification used in ontology exploration is *viewpoint control*, by which we mean control of the spatio-temporal parameters of exploration. For example, users might wish to explore the influence of global warming in terms of several temporal granularities, such as a 20-year, 50-year or 100-year period, to see when the influence on human life and regional environments becomes visible. In another case, they might wish to know how the impact would vary according to the spatial extent, for example on a regional scale or a global scale. In order to provide such functions, the authors have extended their ontology to provide spatial extent and time-scale attributes for slots.

Extension of functions for increasing the usability of the tool

The authors' tool also has the following supplementary functions to support users in exploring an ontology. Through these functions, multiple conceptual maps can be generated from the ontology based on various viewpoints to help users understand knowledge more systematically across domains.

(1) Functions for ease of interpreting generated maps
- Display of different kinds of link traced in different colours to clarify the correspondence between links and aspects.
- Display of the directions of the links traced.
- Variation of the colour and size of focal points and extracted concepts to highlight key concepts.
- Menu and command displays enabling interactive operations to facilitate user exploration.
- A function for finding all the paths between two specified concepts to enable more focused exploration.

(2) Functions for comparing multiple objects
- Highlighting shared concepts between two maps generated by the same focal point but with different aspects, to contrast common elements between the two.
- Highlighting multiple paths specified by users for comparison.
- Saving the generated maps for later reference.

(3) Other supplementary functions of the tool
- Highlighting of the focused conceptual chain.
- Control of the range of exploration.
- Linking conceptual maps with other ontology-based systems.
- Searching all conceptual chains that can be generated from a concept.

- Changing the viewpoint of a conceptual map and restructuring the map from another viewpoint.
- Comparing conceptual maps for understanding the difference between viewpoints.

2-3-6 Dynamic adaptation of the reference model

The underlying assumption of the reference model is that, for SS knowledge-structuring to be adaptive to various users' requirements, it needs the help of a software tool that can make the static data structure virtually dynamic. This is why the authors' reference model has two layers for dynamic adaptation: divergent exploration at Layer 2 and context-based convergence at Layer 3.

Divergent exploration at Layer 2

Divergent exploration allows users to acquire many paths from focal points they set to any concepts they reach by exploring the ontology. Such exploration may consist of ad hoc rather than pre-set procedures in order to increase opportunities for unexpected findings. Users are expected to attempt as many explorations as possible.

Note here that the exploration procedure does not entail formal reasoning. Even if the path-finding procedure for exploration looks similar to formal reasoning, it is not. Indeed, it is intentionally made informal to enable users to find unexpected paths rather than the more foreseeable ones. As discussed above, the authors designed the ontology without setting any boundaries between SS domains. This design rationale is critical for permitting exploration across domains. SS knowledge-structuring should not be domain-specific, but should allow the seamless integration of all related domain knowledge. However, it is apparent that such integration would be almost impossible through a static organization of knowledge across so many different domains. Static organization cannot cover all the possible interdependencies among knowledge items in all SS domains because it is impossible to enumerate these in advance. The authors have therefore introduced dynamic structuring to overcome this limitation. The performance of dynamic structuring depends on the quality of the ontology, which must contain well-balanced and fine-grained concepts in all the SS domains, as well as common concepts. These concepts are all interrelated and may reveal interesting and/or unexpected findings to explorers.

Context-based convergence (Layer 3)

Context-based convergent thinking should be undertaken after the user has spent sufficient time in exploration at Layer 2. Through exploration, the user will have acquired a sizeable number of interesting paths that are recorded in the system. Whereas divergent exploration is performed in a rather ad hoc way to explore as much as possible of the "sea of concepts" in the ontology, the convergent thinking undertaken at Layer 3 should generate tangible results that might be useful for further exploration at Layer 4. By convergent thinking is meant investigation in a more focused manner than the divergent exploration at Layer 2. The data obtained at Layer 2 may be highly varied and to some extent unfocused. To make optimum use of large quantities of such data, the procedures performed at Layer 3 follow the principle of the "trade-off". In real-world problem-solving, among the most typical problems (and the most difficult to solve) are those involving a trade-off. Although trade-offs are usually understood as domain- and context-specific, the authors believe they can be dealt with in a generalized way to enable the tool to detect numerous possible trade-off problems.

Needless to say, the authors' expectation is not to solve the trade-off problems thus identified. Rather, they are attempting to design a new functionality to support domain experts dealing with trade-off problems by providing them with useful information that would otherwise have been difficult for them to obtain. One of the important issues related to this effort is how to design generic formulas for trade-offs. Although this is critical to the authors' research, it remains unsolved to date. Functions that need to be developed at the next stage of this research are thus simply listed:

1. Compare data under specified constraints.
2. Detect hidden trade-off problems using generalized formulas for conflicts.
3. Analyse interesting and unusual paths.
4. Represent several circulations and compare them.
5. Coordinate with other information resources such as case reports and simulation models.

2-3-7 Conclusion

This section has discussed structuring knowledge in sustainability science to enable domain experts to explore the "sea of concepts". In order to make their research as generally applicable as possible, the authors proposed a reference model for SS knowledge-structuring and developed an ontology exploration tool within the framework of this model. It is hoped that the reference model will contribute to an appropriate understanding of the overall picture of SS, which consists of various domains. The SS ontology at Layer 1 provides a foundation for knowledge systematization in a domain-neutral way at the primitive level. At Layer 2, the ontology exploration tool enables researchers to explore the SS ontology from multiple perspectives and eventually to obtain an integrated understanding of SS knowledge from a diversity of viewpoints. So far this tool has been well received. In short, the authors have developed an infrastructure for structuring knowledge based on ontological engineering.

They are currently evaluating and improving the system through feedback from experts with the aim of providing a computational workbench designed to help SS experts solve "trade-off" problems more efficiently than before, using dynamic knowledge-structuring and ontology engineering technology. The key idea is a conceptualization of trade-off-related problems based on ontological engineering using conceptual chains collected through ontology explorations. The authors have tried to capture some examples of trade-off problems and represent them in conceptual maps. For their next step they are considering a general framework within which to treat these problems.

Acknowledgements

This work was supported by Grant-in-Aid for Young Scientists (A) 20680009 from the Global Environment Research Fund (Hc-082) of the Japanese Ministry of the Environment. The ontology of sustainability science on which this study was based was developed as a part of the IR3S flagship research project "Development of an Asian Resource-Circulating Society" undertaken by Osaka University and Hokkaido University, which was also supported by the Japanese Ministry of Education, Culture, Sports, Science and Technology (MEXT) through Special Coordination Funds for Promoting Science and Technology.

Notes

1. See the Environmental Information and Communication Network website at <http://www.eic.or.jp/>.
2. YATO stands for "Yet Another Top-level Ontology".
3. See Basic Formal Ontology at <http://www.ifomis.org/bfo>.
4. See Hozo: Ontology Editor – an Environment for Building/Using Ontologies, at <http://www.hozo.jp/>.

5. The authors understand this would not be true after careful quantitative examination. However, people initially believed it to be so, based on qualitative inference, which is what the authors' tool can do. In other words, maps generated by their tool have no guarantee of correctness. Instead, the tool generates possible conceptual maps to stimulate human experts' thinking.

REFERENCES

Brilhante, V., A. Ferreira, J. Marinho and J. S. Pereira (2006) "Information Integration through Ontology and Metadata for Sustainability Analysis", paper presented at the International Environmental Modelling and Software Society (iEMSs) 3rd Biennial Meeting, Summit on Environmental Modelling and Software, Burlington, VT, 9–12 July.

Friend, A. M. (1996) "Sustainable Development Indicators: Exploring the Objective Function", *Chemosphere* 33(9): 1865–1887.

Grenon, P., B. Smith and L. Goldberg (2004) "Biodynamic Ontology: Applying BFO in the Biomedical Domain", in D. M. Pisanelli (ed.), *Ontologies in Medicine*. Amsterdam: IOS Press, pp. 20–38.

Klein, J. T. (2004) "Interdisciplinarity and Complexity: An Evolving Relationship", *Emergence: Complexity & Organization* 6(1–2): 2–10.

Kraines, S., W. Guo, B. Kemper and Y. Nakamura (2006) "EKOSS: A Knowledge-user Centered Approach to Knowledge Sharing, Discovery, and Integration on the Semantic Web", in *Proceedings of the 5th International Semantic Web Conference (ISWC 2006)*, Athens, GA, 5–9 November, LNCS 4273.

Kumazawa, T., O. Saito, K. Kozaki, T. Matsui and R. Mizoguchi (2009) "Toward Knowledge Structuring of Sustainability Science Based on Ontology Engineering", *Sustainability Science* 4(1): 99–116.

Macris, A. M. and D. A. Georgakellos (2006) "A New Teaching Tool in Education for Sustainable Development: Ontology-based Knowledge Networks for Environmental Training", *Journal of Cleaner Production* 14: 855–867.

Ministry of the Environment (2007) *Annual Report on the Environment and the Sound Material-Cycle Society in Japan 2007*. Tokyo: Ministry of the Environment, Government of Japan.

Mizoguchi, R. (2003) "Tutorial on Ontological Engineering – Part 1: Introduction to Ontological Engineering", *New Generation Computing* 21(4): 365–384.

Mizoguchi, R. (2004a) "Tutorial on Ontological Engineering – Part 2: Ontology Development, Tools and Languages", *New Generation Computing* 22(1): 61–96.

Mizoguchi, R. (2004b) "Tutorial on Ontological Engineering – Part 3: Advanced Course of Ontological Engineering", *New Generation Computing* 22(2): 198–220.

Mizoguchi, R. (2009) "Yet Another Top-level Ontology: YATO", in *Proceedings of the Second Interdisciplinary Ontology Meeting*, Tokyo, Japan, 28 February–1 March, pp. 91–101.

Mizoguchi, R., E. Sunagawa, K. Kozaki and Y. Kitamura (2007) "A Model of Roles within an Ontology Development Tool: Hozo", *Applied Ontology* 2(2): 159–179.

Morioka, T., O. Saito and H. Yabar (2006) "The Pathway to a Sustainable Industrial Society – Initiative of the Research Institute for Sustainability Science (RISS) at Osaka University", *Sustainability Science* 1: 65–82.

Munier, N. (2005) *Introduction to Sustainability: Road to a Better Future*. Dordrecht: Springer.

Rotmans, J. (2006) "Tools for Integrated Sustainability Assessment: A Two-track Approach", *Integrated Assessment* 6(4): 35–57.

Suzuki, I., A. I. Sakamoto and H. Fukui (2005) "Development and Application of Ontology to Support Risk Communication in the Domain of High Level Radioactive Waste", *Environmental Information Science* 33(4): 9–17.

Tiako, P. F. (2004) "Conceptual Software Infrastructure for Sustainable Development", in *Proceedings of the IEEE International Engineering Management Conference*, Singapore, October.

UNEP CBD [United Nations Environment Programme Convention on Biological Diversity] (2000) "Ecosystem Approach", COP 5 Decision V/6, UNEP/CBD/COP/5/23. Decisions adopted by the Fifth Ordinary Meeting of the Conference of the Parties to the Convention on Biological Diversity (COP 5), Nairobi, Kenya, 15–26 May.

2-4

The application of ontology engineering to biofuel problems

Osamu Saito, Kouji Kozaki, Takeru Hirota and Riichiro Mizoguchi

2-4-1 Introduction

One of the most cited definitions of ontology is "an explicit and formal specification of a conceptualization" (Gruber, 1993: 199). Through conceptualization, relevant concepts are identified to explicitly describe a phenomenon in a formal machine-readable language. There are many applications of ontology engineering, such as building semantic web systems (Sabou et al., 2005), facilitating knowledge management (Brandt et al., 2008), supporting the integrated assessment of agricultural systems (Van Ittersum et al., 2008), structuring knowledge for sustainability science (Kumazawa et al., 2009) and developing a task-oriented mobile service navigation system (Sasajima et al., 2009).

The authors' research group has proposed a way to contribute to sustainability science by developing a sustainability science ontology and a tool that can generate comprehensive conceptual maps from multiple arbitrary perspectives of users (Kumazawa et al., 2009). Comprehensiveness is one of the critical research norms for sustainability science because the current piecemeal approach works poorly in the face of complex and evolving problems. Therefore, comprehensive solutions need to be applied to these problems (Komiyama and Takeuchi, 2006). Such comprehensiveness can be attained by the systematic reorganization of disparate existing fields and domains. The authors have therefore developed a sustainability science ontology and a tool to generate comprehensive views of relevant concepts.

Understanding sustainability science problems additionally requires consistent enquiry into a multitude of relevant domains and their networking concepts to flexibly adapt to dynamic changes both within and between domains (Kumazawa et al., 2009). The need for such consistent enquiry has been experienced through the recent controversy over biofuel sustainability. Biofuels (bioethanol and biodiesel) were initially considered beneficial for poverty alleviation and useful as a fossil fuel substitute. Governments and businesses made massive investments in fuel crop production and fuel refinery plants to achieve national targets or plans for expanding biofuel usage. However, the negative impact of these investments, including food insecurity, biodiversity loss and climate change, soon surfaced and were actively discussed worldwide (FAO, 2008; UN-Energy, 2007). A marked change in the perception of biofuels was observed because of some pertinent questions raised by recent analysis. For example, the 2008 report of the Food and Agriculture Organization of the United Nations (FAO) asserts that, "while biofuels will offset only a modest share of fossil energy use over the next decade, they will have much bigger impacts on agriculture and food security" (FAO, 2008: vii). Consequently, the economic and political climate for biofuels has since become markedly less favourable.

Komiyama and Takeuchi (2006: 3) define sustainability science as "a comprehensive, holistic approach to identification of problems and perspectives involving the sustainability of global, social, and human systems". Biofuel problems, in this sense, involve all three systems: climate change and food insecurity occur in the global system; agro-industrial development and structural change in agriculture and communities occur in the social system; and poverty and health problems occur in the human system. Developing the sustainability science ontology is a long-term and ongoing process. Thus, while working on the development of this ontology, the authors have chosen the issue of biofuels as a case study, and have expanded the functions of the mapping tool to meet the practical demands of researchers and policymakers in relevant fields. This section first describes the current state of the sustainability science ontology and mapping tool, and then describes the application of this tool to the biofuel issue to examine its effectiveness. Next, a methodology is discussed to improve the tool to facilitate more effective knowledge-structuring and exploration under certain conditions or particular interests.

2-4-2 A knowledge-structuring tool based on ontology engineering

System architecture

The authors proposed a reference model composed of five layers as a guideline for structuring knowledge in sustainability science and developed an ontology-based mapping tool for this purpose (see Figure 2.3.1 in Section 2-3). The reference model consists of layers corresponding to five kinds of information: raw data, the underlying static information structure, dynamic information reflecting individual perspectives, dynamic information organizing perspectives within context, and methodological information. The bottom layer, Layer 0, is the data layer and stores raw data corresponding to the real world. Layer 1, the ontology layer, stores the ontology for explaining and understanding the raw data in Layer 0. Layer 2 handles dynamic information that reflects individual perspectives. The main task supported by this layer is the divergent exploration of the conceptual world realized in Layer 1 (Figure 2.4.1).

At Layer 3, the investigator can set the specific context of a problem and can obtain "multiple convergent conceptual chains" (Klein, 2004) in accordance with the given context. At Layer 4, using all the information and knowledge obtained from the lower layers, the investigator pursues

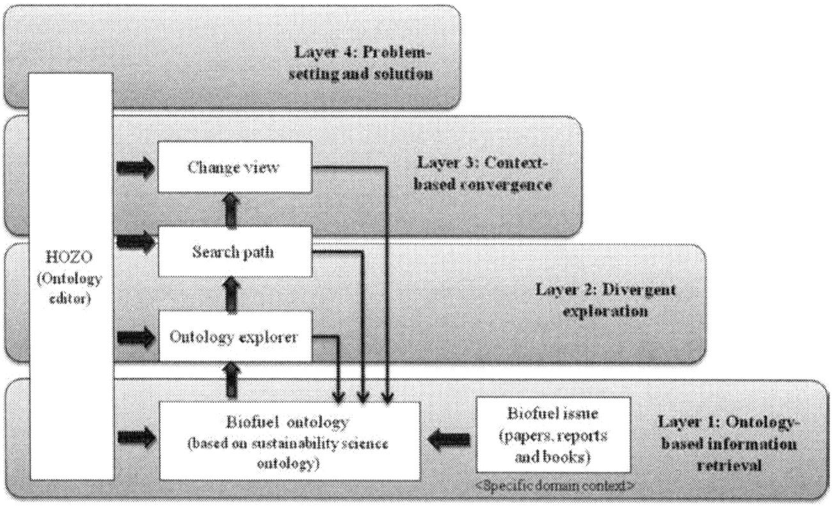

Figure 2.4.1 System architecture in relation to the reference model.

essential problem-solving tasks such as setting conditions for solving a problem, or searching for a new problem, as well as information integration and abduction.

Biofuel ontology

The authors built a biofuel ontology by extending an existing sustainability science ontology (Kumazawa et al., 2009) for structuring problems in biofuel usage. The sustainability science ontology covers all forms of biomass and bioenergy, including liquid biofuels, biogas and solid biomass for heat and power generation. The FAO (2006) defines biomass as material of organic origin in non-fossilized form, such as agricultural crops and forestry products, agricultural and forestry wastes and by-products, manure, and industrial and household organic waste. Biomass can provide different forms of energy, including heat, electricity and biofuels. The term "biofuel" in this study refers to liquid biofuels, i.e. bioethanol and biodiesel, derived from biomass feedstock such as maize, sugarcane, potato, paddy rice, wheat, sugar beet, sweet sorghum and oil crops including oil palm, rapeseed and jatropha. A greater range of lignocellulose materials, or so-called second-generation feedstock, is expected to be used for biofuel production in the future (Royal Society, 2007). These materials include woody plants, perennial grasses, algae and residues from agriculture and forestry industries.

Information and documents were collected and analysed through literature review (CBD, 2008; De Oliveira et al., 2005; FAO, 2006, 2008; Fargione et al., 2008; Farrell et al., 2006; Gerbens-Leenes et al., 2009; Hammerschlag, 2006; Hill et al., 2006; Martinelli and Filoso, 2008; Mattsson et al., 2000; Righelato and Spracklen, 2007; Royal Society, 2007; Searchinger et al., 2008; Sumathi et al., 2008; UN-Energy, 2007). One of the authors organized a bimonthly series of workshops on biomass and bioenergy, including biofuels, from August 2008 to March 2009, inviting leading experts from relevant fields in Japan. The authors also exchanged information with domain experts by participating in domestic and international symposiums and conferences.

In addition, one of the authors conducted a field survey in Malaysia, interviewing researchers and major stakeholders in biodiesel production and the development of feedstock plantations in that country and visiting oil palm and jatropha plantations in the states of Sabah and Sarawak in May 2009 (Plate 2.4.1).

Sustainability criteria and certification systems developed by the Roundtable on Sustainable Palm Oil (RSPO) were also reviewed. The survey investigated how sustainable practices actually exist and function, how multinational corporations influence biofuel development, and how

Plate 2.4.1 Jatropha (left) and oil palm (right) in Sarawak, Malaysia.
Note: Please see page 474 for colour versions of these images.

associated problems such as deforestation, biodiversity loss, water contamination, soil degradation and conflicts with indigenous communities can be observed on a community scale.

Ontology explorer and map generation tool

An "ontology-based domain overview" can be defined as understanding the comprehensive concept structure of a domain by looking at its ontology from multiple perspectives. The authors therefore developed a tool that allows domain experts to trace multiple chains of connected concepts in various ways specified by them from a focal concept of interest to reach the concepts of another category in a divergent way. This focal concept is called the *focal point* in the tool. The user can choose any concept in the ontology as a focal point.

This tool is designed to make an ontology more easily accessible to domain experts and to help them explore it. The current version allows users to create a map by interactively selecting commands and types of *aspect*. In this case, an aspect represents a relationship that the user is interested in or wishes to pay particular attention to. Types of aspect include *is-a* and *part-of/attribute-of* relationships, relationships between role concepts and contexts, and those between role concepts and players. From a selected focal point, a user can explore a selected aspect step by step, and, after each step of exploration, can decide which way to explore next by carefully analysing the displayed map. Repeating this process, paths connecting concepts with the focal point are visualized on the map. These paths are called "multi-perspective conceptual chains".

A map view is generated from a combination of focal point and aspect and, although users may select the same focal point, the generated view will differ depending on which aspect a user chooses during exploration.

Search path

"Search path" is a useful function that enables users to search for conceptual linkages between specific concepts. For example, if a user wishes to know the linkage and relevant concepts between *Production of biofuels* and *Agriculture*, a specified map can be generated by selecting *Production of biofuels* as the focal point and *Agriculture* as a search word (Figure 2.4.2).

Change view

The tool often generates a map that is too crowded with relevant concepts to understand it intuitively (Figure 2.4.3). A "change view" function

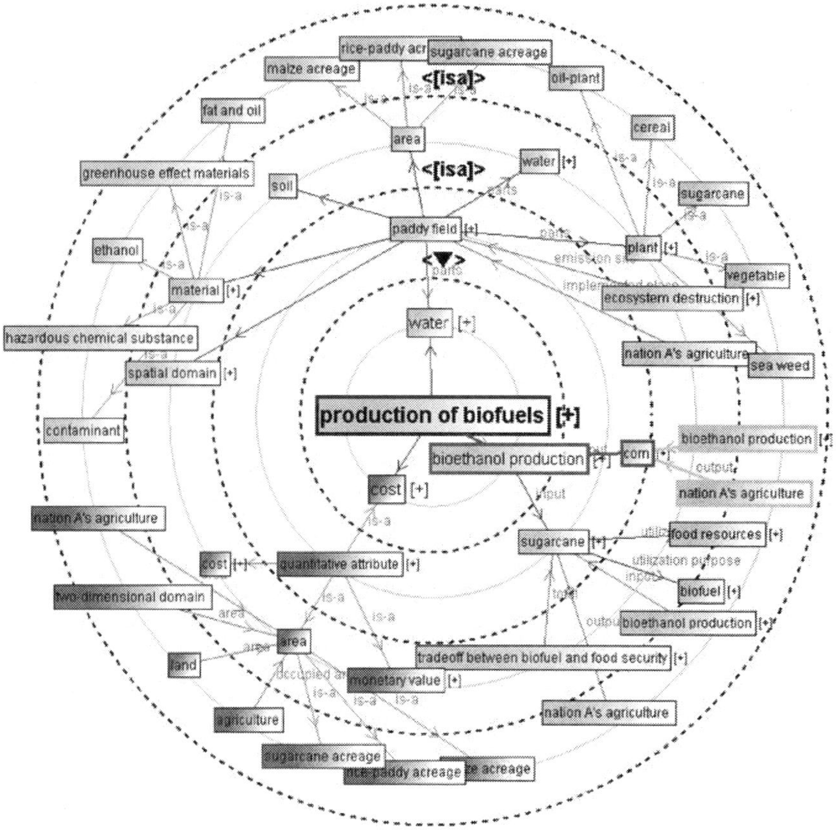

Figure 2.4.2 Map of conceptual chains generated by searching for *Agriculture*, with *Production of biofuels* as the focal point.

was therefore developed that enables the user to extract and highlight concepts of interest by specifying roles and classes of aspect.

2-4-3 The biofuel controversy

Although biofuels are considered an effective alternative to fossil fuel, many warnings and criticisms have been expressed over the expansion of biofuel production. UN-Energy (2007) highlighted the social, economic and ecological sustainability issues of the rapid development of bioenergy, including biofuels, in both small- and large-scale applications. Decision-makers considering adopting new policies or launching new investments in the biofuel sector need to consider the entire biofuel value chain, from production to use, and the associated social, economic and ecological impacts, from the local to the global scale.

Table 2.4.1 summarizes the key issues regarding potential positive and negative effects of biofuel addressed in recent years.

2-4-4 Mapping conceptual chains of biofuel issues

Biofuel and agriculture

Figure 2.4.2 shows a map view created by selecting *Production of biofuels* as the focal point and *Agriculture* as the search word for the "search path" function. From this map, a user can examine the land-use competition between different crops such as maize, rice, sugarcane, oil and vegetables. Key issues such as the trade-off between biofuel and food security, ecosystem destruction, hazardous chemicals, greenhouse effect materials, water and soil are also displayed on the map. By broadening the width of the map, the user can explore fringe concepts more deeply. The map generated by the tool stimulates the imagination of domain researchers based on the comprehensiveness and objectivity of the ontology. In addition, it is very important for decision-makers to seek compromises through transparent discussion of the trade-offs between different paths displayed on the map.

It takes more than an hour for a domain researcher to manually draw a conceptual map but, using the ontology-based map generation tool, it takes about five minutes to obtain a map like Figure 2.4.2. In addition, a user can compare two different maps from the same focal point because the map tool automatically highlights overlapping concepts.

Biofuel and the ecosystem

Figure 2.4.3 illustrates a map of conceptual chains generated by searching for *Ecosystem*, *Biodiversity*, *Nature conservation* and *Forest management*, with *Production of biofuels* as the focal point. There are too many relevant concepts shown in this map and it is difficult to understand the structure of the concepts. In such cases the user can employ the "change view" function. Figure 2.4.4 is a restricted view of the same map shown in Figure 2.4.3, where a user has restricted the map to the role concepts of *land* and *actor*. From this restricted map, the user can easily understand the specific conceptual chains from biofuel production to ecosystem, biodiversity, nature conservation and forest management with a focus on associated land-use types and actors.

Table 2.4.1 Positive and negative effects of biofuel

Energy services for the poor	(+/–) Competition of biomass energy systems with the present use of biomass resources (such as agricultural residues) in applications such as animal feed and bedding, fertilizer and construction materials[1] (–) In many developing countries, small-scale biomass energy projects face challenges obtaining finance from traditional financing institutions[1] (–) Liquid biofuels are likely to replace only a small share of global energy supplies and cannot alone eliminate our dependence on fossil fuels[2]
Agro-industrial development and job creation	(+) Biofuel is powering new small- and large-scale agro-industrial development and spawning new industries in industrialized and developing countries[1] (+/–) In the short to medium term, bioenergy use will depend heavily on feedstock costs and reliability of supply, the cost and availability of competing energy sources, and government policy decisions[1] (+) In the longer term, the economics of biofuel will probably improve as agricultural productivity and agro-industrial efficiency improve, more supportive agricultural and energy policies are adopted, carbon markets mature and expand, and new methodologies for carbon sequestration accounting are developed[1] (+) In the longer term, expanded demand and increased prices for agricultural commodities may represent opportunities for agricultural and rural development[2] (+) Biofuel industries create jobs, including highly skilled science, engineering and business-related employment; medium-level technical staff; low-skill industrial plant jobs; and unskilled agricultural labour[1] (+/–) Small-scale and labour-intensive production often leads to trade-offs between production efficiency and economic competitiveness[1]
Health and gender	(–) Market opportunities cannot overcome existing social and institutional barriers to equitable growth, with exclusion factors such as gender, ethnicity and political powerlessness, and may even worsen them[2] (–) Forest burning for the development of feedstock plantations and sugarcane burning to facilitate manual harvesting result in air pollution, higher surface water runoff, soil erosion and unintended forest fires[3,4] (–) Exploitation of cheap labour (plantation and migrant workers)[4] (–) Increased use of pesticides could create health hazards for labourers and communities living near areas of feedstock production[1,3]

Table 2.4.1 (cont.)

Agricultural structure	(–) The demand for land to grow biofuel crops could put pressure on competing land usage for food crops, resulting in an increase in food prices[1,2] (+/–) Significant economies of scale can be gained from processing and distributing biofuels on a large scale; however, the transition to liquid biofuels can be harmful to farmers who do not own their own land, and to the rural and urban poor who are net buyers of food[1] (–) Although global market forces could lead to new and stable income streams, they could also increase the marginalization of poor and indigenous people and affect traditional ways of living if they end up driving small farmers without clear titles from their land and destroying their livelihood[1]
Food security	(–) Demand for agricultural feedstock for liquid biofuels will be a significant factor in agricultural markets and world agriculture over the next decade and perhaps beyond[2] (–) Rapidly growing demand for biofuel feedstock has contributed to higher food prices, which pose an immediate threat to the food security of poor net food buyers in both urban and rural areas[2] (+/–) The effect of biofuels on food security is context specific, depending on the particular technology and country characteristics involved[1]
Government budget	(–) Because ethanol is used largely as a substitute for gasoline, providing a large tax reduction for blending ethanol and gasoline reduces government revenue from this tax, mainly targeting the non-poor[1] (–) Production of biofuels in many countries, except sugarcane-based ethanol production in Brazil, is not currently economically viable without subsidies, given existing agricultural production and biofuel-processing technologies and recent relative prices of commodity feedstock and crude oil[2] (–) Policy intervention, especially in the form of subsidies and mandated blending of biofuels with fossil fuels, is driving the rush to liquid biofuels, which leads to high economic, social and environmental costs in both developed and developing countries[2]
Trade, foreign exchange balance and energy security	(+) Diversifying global fuel supplies could have beneficial effects on the global oil market and many developing countries because fossil fuel dependence has become a major risk for many developing economies[1] (+/–) Rapidly rising demand for ethanol has had an impact on the price of sugar and maize in recent years, bringing substantial rewards to farmers not only in Brazil and the United States but around the world[1,2]

Table 2.4.1 (cont.)

	(–) The linking of agricultural prices to the vicissitudes of the world oil market clearly presents risks; however, it is an essential transition to the development of a biofuel industry that does not rely on major food commodity crops[1]
Biodiversity and natural resource management	(+/–) Depending on the types of crop grown, what they replaced and the methods of cultivation and harvesting, biofuels can have negative and positive effects on land use, soil and water quality, and biodiversity[1,3]
	(–) Problems with water availability and use may represent a limitation on agricultural biofuel production[1,3]
	(–) The introduction of criteria, standards and certification schemes for biofuels may generate indirect negative environmental and biodiversity effects passively in other countries[3]
	(–) If the production of biofuel feedstock requires increased fertilizer and pesticide use, there could be additional detrimental effects such as increased GHG emissions, eutrophicating nutrients and biodiversity loss[3]
	(–) Wild biodiversity is threatened by loss of habitat when the area under crop production is expanded, whereas agricultural biodiversity is vulnerable in the case of large-scale monocropping, which is based on a narrow pool of genetic material and can also lead to reduced use of traditional varieties[2,3]
	(+) If crops are grown on degraded or abandoned land, such as previously deforested areas or degraded crop- and grasslands, and if soil disturbances are minimized, feedstock production for biofuels can have a positive impact on biodiversity by restoring or conserving habitat and ecosystem function[3]
Climate change	(+/–) Full lifecycle GHG emissions of biofuel vary widely depending on land-use changes, choice of feedstock, agricultural practices, refining or conversion processes and end-use practices[1,2]
	(–) Land-use change associated with the production of biofuel feedstock can affect GHG emissions; draining wetlands and clearing land with fire are detrimental with regard to GHG emissions and air quality[2,3]
	(–) The greatest potential for reducing GHG emissions comes from the replacement of coal rather than of petroleum fuels[1]
	(+) Biofuels offer the only realistic near-term renewable option for displacing and supplementing liquid transport fuels[1]

Sources: [1]UN-Energy (2007), [2]FAO (2008), [3]CBD (2008), [4]Martinelli and Filoso (2008).
Notes: (+): positive effects; (–): negative effects; (+/–): both positive and negative effects.

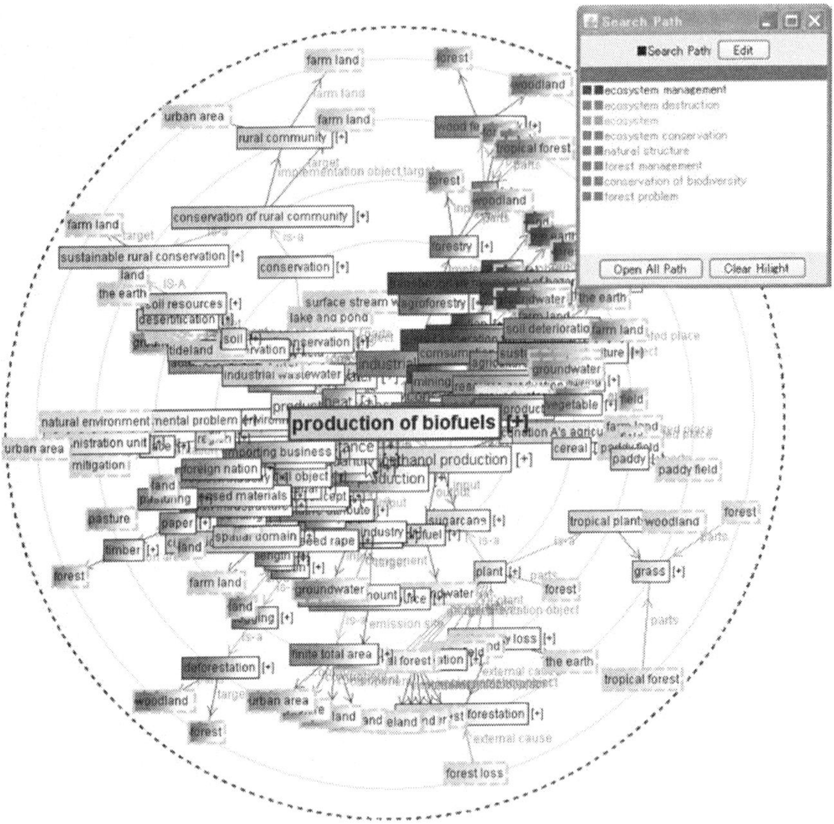

Figure 2.4.3 Map of conceptual chains generated by searching for *Ecosystem*, *Biodiversity*, *Nature conservation* and *Forest management*, with *Production of biofuels* as the focal point.

Evaluation of the mapping tool

The current version of the biofuel ontology contains 1,892 classes and 2,119 slots. To assess the effectiveness of the mapping tool, the authors analysed how many biofuel issues in Table 2.4.1 can be reasonably traced by the tool. For this analysis, 30 issues were randomly selected from the table. It was found that, when the search path function was used, the tool created 19 conceptual maps, each of which corresponded to a selected issue. An additional 3 maps (issues) were created by interactive manual exploration of the ontology. However, the tool could not describe appropriate conceptual maps for the remaining 8 issues because of a lack of relevant concepts in the ontology and defects in the exploration

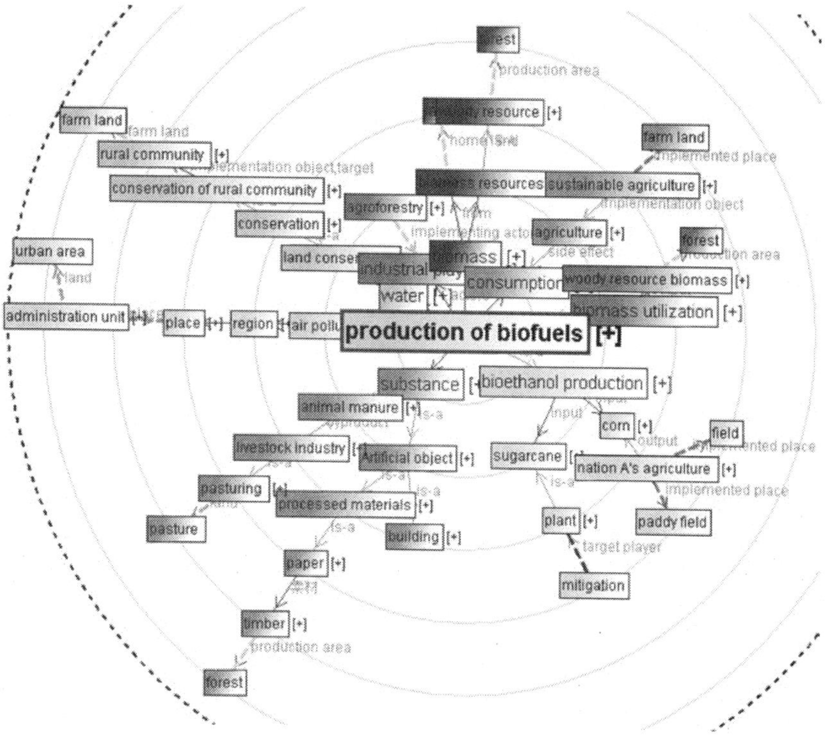

Figure 2.4.4 Restricted view of the same map as Figure 2.4.3, focusing on the roles of *land* and *actor*.

algorithm. In summary, the current mapping tool can describe 22 out of 30 selected issues (73 per cent) in Table 2.4.1.

The authors also asked four domain experts to use the tool and evaluate its practical performance. After basic instruction regarding its use, they created 13 conceptual maps (three or four maps per expert) within an hour in accordance with their specific interests. Then they chose 61 conceptual paths (linkages between concepts in a map) from the 13 maps; they explored and evaluated the paths with a four-level scale (4: very important or interesting; 3: important or interesting; 2: relevant, but neither important nor interesting; 1: wrong path). As a result, 30 paths (49 per cent) were graded as level 4, 22 paths (36 per cent) as level 3, 8 paths (13 per cent) as level 2 and 1 path (2 per cent) as level 1; thus 85 per cent of the selected paths were evaluated as level 3 or level 4. Although one should not exaggerate the tool's performance based on an experiment with such few samples, the experimental result suggests its practical applicability and effectiveness to some extent and provides useful feedback for its improvement.

2-4-5 Discussion

An ambitious attempt among the many applications of ontology engineering is the SEAMLESS project. This is an integrated assessment modelling project (Van Ittersum et al., 2008) that aims to provide a computerized framework to assess the sustainability of agricultural systems in the European Union at multiple scales. Based on a participatory and collaborative approach, the project creates a common ontology for models, indicators and raw data to harmonize and relate different concepts (Wien et al., 2007). Ontology is used to structure domain knowledge and semantic meta-information about components in order to facilitate the organization, retrieval and linkage of knowledge in the components (Van Ittersum et al., 2008).

Similarly, the authors' biofuel ontology has been developed as part of a research project entitled "Biofuel Use Strategies for Sustainable Development" (BforSD), which aims at conducting comprehensive analysis of biofuel use at the global, regional and national levels using the sustainability science approach (that is, repairing and balancing three fundamental systems: global, social and human). The biofuel ontology and the ontology-based map generation tool can act as an interface not only among scientists but also between scientists and stakeholders, particularly policymakers. Such interfacing effects are also considered in the SEAMLESS project (Janssen et al., 2007), but they have not yet been fully demonstrated. These projects show how ontology engineering can contribute to integrating assessment models in the sustainability domain, which is often characterized by interdisciplinary and cross-boundary characteristics, the interaction between nature and society, and the interplay between knowledge and action (Saito, 2008). One of the important advantages in using an ontology-based system is that the system can enhance understanding of a problem with respect to more diverse conceptual chains. For example, it can facilitate broader and more comprehensive understanding of the biofuel problem so as to arrive at more balanced problem definition and policy integration. Practical advantages as well as weaknesses need to be examined by applying the system to actual research collaboration or policy-making processes.

Conceptual mapping of sustainability was developed by the Global System for Sustainable Development (GSSD) at the Massachusetts Institute of Technology, which indicates "what to solve" in the domain of sustainable development. GSSD focuses on "the content-architecture – the levels, linkages, and complexities – that characterizes the domain of 'sustainability'" (Choucri, 2003: 1; Choucri et al., 2007). GSSD is also an ontology-based system for mapping sustainability. Its mapping system provides an internally consistent baseline for sustainable development and represents key aspects of the issue-area at hand, including attendant complexities and interconnections (Choucri et al., 2007). There are some similarities in the designing principles between the authors' system and GSSD, but a major difference lies in the fact that their system can support divergent knowledge exploration through interaction with the user and display various maps in accordance with the user's interests, whereas GSSD focuses more on developing a comprehensive knowledge platform for the provision of cyber content and material at various sub-national and national levels around the world.

2-4-6 Conclusion

The authors' research group has proposed another way to contribute to sustainability science by developing the sustainability science ontology and a tool that can generate comprehensive conceptual maps from users' multiple arbitrary perspectives (Kumazawa et al., 2009). This section has described the sustainability science ontology and the map generation tool, its application to the biofuel problem, and the advantages of the tool. Maps are obtained for the exploration of an "ocean of concepts" in the biofuel ontology. The authors found that the tool can generate a map that covers key concepts of the biofuel controversy. The map generated by the tool can stimulate the imagination of domain researchers because of the comprehensiveness and objectivity of the ontology. This tool also has the function of supporting divergent thinking, and can naturally be extended to development of a tool to aid convergent thinking by policymakers. The results obtained in this research represent the first successful application of ontological engineering to the biofuel problem and a first step towards developing a support tool for policymakers that stimulates their thinking process and encourages effective visualization to aid in harmonizing different policies into a unified policy.

Further improvement of the biofuel ontology is necessary because the quality of the map depends heavily on the completeness of the ontology. The general formulation of various trade-offs and functions for context-based convergence (Layer 3 of Figure 2.4.1) will be developed while putting this system into practical use to facilitate collaboration by researchers and the integration of a wide range of research results from the BforSD project.

Acknowledgements

This work was supported by the Global Environment Research Fund (Hc-082) of the Japanese Ministry of the Environment. This study was also supported by the programme "Promotion of Environmental Improvement for Independence of Young Researchers" under the Special Coordination Funds for Promoting Science and Technology provided by the Japanese Ministry of Education, Culture, Sports, Science and Technology (MEXT). The ontology of sustainability science on which this study was based was developed as a part of the IR3S flagship research project "Development of an Asian Resource-Circulating Society" undertaken by Osaka University and Hokkaido University, which was also supported by MEXT through Special Coordination Funds for Promoting Science and Technology. The field survey on oil palm and jatropha plantations in Malaysia was supported by Waseda University and Bridgestone Corporation through the W-Bridge Project.

REFERENCES

Brandt, S. C., J. Morbach, M. Miatidism, M. Theißen, M. Jarke and W. Marquardt (2008) "An Ontology-based Approach to Knowledge Management in Design Processes", *Computers and Chemical Engineering* 32: 320–342.

CBD [Convention on Biological Diversity] (2008) "The Potential Impacts of Biofuels on Biodiversity", Note by the Executive Secretary for the Conference of the Parties to the Convention on Biological Diversity, 19–30 May, Bonn, Germany, UNEP/CBD/COP/9/26, 24 April 2008.

Choucri, N. (2003) "Mapping Sustainability", Working Paper, Global System for Sustainable Development (GSSD). Available at <http://gssd.mit.edu/GSSD/gssden.nsf/> (accessed 26 May 2010).

Choucri, N., D. Mistree, F. Haghseta, T. Mezher, W. R. Baker and C. I. Ortiz, eds (2007) *Mapping Sustainability: Knowledge E-networking and the Value Chain.* New York: Springer.

De Oliveira, M. E. D., B. E. Vaughan and E. J. Rykiel Jr (2005) "Ethanol as Fuel: Energy, Carbon Dioxide Balances, and Ecological Footprint", *BioScience* 55: 593–602.

FAO [Food and Agriculture Organization of the United Nations] (2006) *Introducing the International Bioenergy Platform (IBEP).* Rome, Italy: FAO. Available at <ftp://ftp.fao.org/docrep/fao/009/A0469E/A0469E00.pdf> (accessed 26 May 2010).

FAO [Food and Agriculture Organization of the United Nations] (2008) *The State of Food and Agriculture 2008.* Rome: FAO. Available at <ftp://ftp.fao.org/docrep/fao/011/i0100e/i0100e.pdf> (accessed 26 May 2010).

Fargione, J., J. Hill, D. Tilman, S. Polasky and P. Hawthorne (2008) "Land Clearing and the Biofuel Carbon Debt", *Science* 319: 1235–1238.

Farrell, A. E., R. J. Plevin, B. T. Turner, A. D. Jones, M. O'Hare and D. M. Kammen (2006) "Ethanol Can Contribute to Energy and Environmental Goals", *Science* 311: 506–508.

Gerbens-Leenes, W., A. Y. Hoekstra and T. H. van der Meer (2009) "The Water Footprint of Bioenergy", *Proceedings of the National Academy of Sciences* 106(25): 10219–10223.

Gruber, T. R. (1993) "A Translation Approach to Portable Ontology Specifications", *Knowledge Acquisition* 5: 199–220.

Hammerschlag, R. (2006) "Ethanol's Energy Return on Investment: A Survey of the Literature 1990–Present", *Environmental Science and Technology* 50: 1744–1750.

Hill, J., E. Nelson, D. Tilman, S. Polasky and D. Tiffany (2006) "Environmental, Economic, and Energetic Costs and Benefits of Biodiesel and Ethanol Biofuels", *Proceedings of the National Academy of Sciences* 103: 11206–11210.

Janssen, S. J. C., J. J. F. Wien, Hongtao Li, I. N. Athanasiadis, F. Ewert, M. J. R. Knapen, D. Huber, O. Thérond, A. E. Rizzoli, H. Belhouchette, M. Svensson and M. K. Van Ittersum (2007) "Defining Projects and Scenarios for Integrated Assessment Modelling Using Ontology", MODSIM 2007 International Congress on Modelling and Simulation, Modelling and Simulation Society of Australia and New Zealand, December, pp. 2055–2061.

Klein, J. T. (2004) "Interdisciplinarity and Complexity: An Evolving Relationship", *Emergence: Complexity & Organization* 6(1–2): 2–10.

Komiyama, H. and K. Takeuchi (2006) "Sustainability Science: Building a New Discipline", *Sustainability Science* 1: 1–6.

Kumazawa, T., O. Saito, K. Kozaki, T. Matsui and R. Mizoguchi (2009) "Toward Knowledge Structuring of Sustainability Science Based on Ontology Engineering", *Sustainability Science* 4(1): 99–116.

Martinelli, L. A. and S. Filoso (2008) "Expansion of Sugarcane Ethanol Production in Brazil: Environmental and Social Challenges", *Ecological Applications* 18(4): 885–898.

Mattsson, B., C. Cederberg and L. Blix (2000) "Agricultural Land Use in Life Cycle Assessment (LCA): Case Studies of Three Vegetable Oil Crops", *Journal of Cleaner Production* 8: 283–292.

Righelato, R. and D. V. Spracklen (2007) "Carbon Mitigation by Biofuels or by Saving and Restoring Forests?", *Science* 317: 902.

Royal Society (2007) *Sustainable Biofuels: Prospects and Challenges*, RS Policy document 01/08. London: The Royal Society. Available at <http://royalsociety.org/> (accessed 26 May 2010).

Sabou, M., C. Wroeb, C. Goble and H. Stuckenschmidt (2005) "Learning Domain Ontologies for Semantic Web Service Descriptions", *Web Semantics: Science, Services and Agents on the World Wide Web* 3: 340–365.

Saito, O. (2008) "Promoting Collaboration of Sustainability Science and Ecology", Proceedings of International Conference on Sustainable Agriculture for Food, Energy and Industry (ICSA 2008), 2–6 July, Sapporo.

Sasajima, M., K. Furutani, Y. Kitamura, T. Naganuma, S. Kurakake and R. Mizoguchi (2009) "Prototyping of Task-Oriented Mobile Navigation System with Real Scale Mobile Services", Proceedings of Advances in Computer Science and Engineering (ACSE 2009), 16–18 March, pp. 205–210.

Searchinger, T., R. Heimlich, R. A. Houghton, F. Dong, A. Elobeid, J. Fabiosa, S. Tokgoz, D. Hayes and T.-H. Yu (2008) "Use of U.S. Croplands for Biofuels

Increases Greenhouse Gases through Emissions from Land-Use Change", *Science* 319: 1238–1240.

Sumathi, S., S. P. Chai and A. R. Mohamed (2008) "Utilization of Oil Palm as a Source of Renewable Energy in Malaysia", *Renewable and Sustainable Energy Reviews* 12: 2404–2421.

UN-Energy (2007) *Sustainable Bioenergy: A Framework for Decision Makers*. United Nations. Available at <http://esa.un.org/un-energy/pdf/susdev.Biofuels.FAO.pdf> (accessed 26 May 2010).

Van Ittersum, M. K., F. Ewert, T. Heckelei, J. Wery, J. Alkan Olsson, E. Andersen, I. Bezlepkina, F. Brouwer, M. Donatelli, G. Flichman, L. Olsson, A. Rizzoli, T. Van Der Wal, J. E. Wien and J. Wolf (2008) "Integrated Assessment of Agricultural Systems – A Component Based Framework for the European Union (SEAMLESS)", *Agricultural Systems* 96: 150–165.

Wien, J. J. F., M. J. R. Knapen, S. J. C. Janssen, P. J. F. M. Verweij, I. N. Athanasiadis, H. Li, A. E. Rizzoli and F. Villa (2007) "Using Ontology to Harmonize Knowledge Concepts in Data and Models", MODSIM 2007 International Congress on Modelling and Simulation, Modelling and Simulation Society of Australia and New Zealand, December, pp. 2055–2061.

2-5

Conclusion

Kazuhiko Takeuchi

As noted in Chapter 1 of this volume, sustainability science can overcome the conditions that contribute to the breakdown of global, social and human systems and the links among them only if it can mobilize all relevant fields of study in the effort to identify the phenomena and solve the problems that threaten the sustainability of these systems. To accomplish this crucial task, it is not enough simply to compile the fruits of research conducted under the old model of mutually isolated, compartmentalized academic disciplines seeking specific solutions to specific problems. A practical approach to integrating the sciences and the humanities must also be devised if such "integration" is not to remain an empty slogan.

Sustainability science is, in essence, a new field that engages in knowledge-structuring so as to provide a comprehensive overview of the vast accumulation of research results from disparate academic disciplines. Such an overview can be achieved only through a *transdisciplinary* approach that sublates traditional interdisciplinary approaches to the integration of knowledge on this scale. Sustainability science therefore requires the development of practical and effective methods for structuring knowledge. This chapter has provided presentations of two such methods, demonstrating that knowledge-structuring is indeed feasible. One approach, presented by Kajikawa in Section 2-1, employs citation network analysis of the vast body of papers published on sustainability-related topics to group the research domains that make up sustainability science. A second approach, described by Mizoguchi et al. in Section 2-3, maps the links among sustainability-related concepts according to a hierarchical structuring of these concepts produced through ontology engineering. The two approaches, which may be respectively described as inductive and deductive, are in fact complementary.

A look at the landscape of sustainability-related disciplines visualized by Kajikawa's method reveals that in such fields as agriculture, forestry, fishery and ecological economics much work on sustainability continues to be field-specific and monodisciplinary. The central position in this citation network of academic papers on the agriculture, forestry and fishery industries reflects the centrality in sustainability research of the topic of the sustainable use of renewable bio-resources. It also reflects the fact that the academic concept of sustainability stems from concepts (for example, maximum sustainable harvests) that originate in these fields. As more research is conducted on global-scale issues such as climate change and reduced biodiversity, we can anticipate an expansion of this focus to the physical sciences, engineering, medicine, law, political science, economics, sociology and philosophy.

If the overview of the academic landscape afforded by these tools is to be fully utilized, then scholars must commit themselves to creating a strategic alliance among these various disciplines as a basis for building a truly transdisciplinary field of sustainability science. This requires converting the results of inductive analysis into a framework of deductive knowledge. As Kajikawa argues, it is not enough to project trends from present conditions; the method known as backcasting must be used to scrutinize the current status of sustainability-related research in light of the future direction that must be taken by sustainability science. This normative approach is also the impetus for Kajikawa and Komiyama's proposal in Section 2-2 for "action-structuring", which hierarchically organizes and integrates the vast range of individual problem-solving actions into collective action for optimal effect.

One of the distinctive features of the ontology-based knowledge-structuring proposed by Mizoguchi et al. is its use of a computer to process and systematically visualize the myriad concepts associated with sustainability research and the complex web of relationships among them. The computer allows the meanings of these concepts to be graphically illustrated in terms of their interrelationships, which in turn facilitates more effective, coordinated application of the knowledge embodied in these linked concepts. Another important feature of this approach is that it enables the extraction of a view of these relationships from any position or viewpoint. Different viewpoints reveal different linkages among sustainability-related problems, making this an extremely useful policy-making tool if applied, for example, to problem-solving that involves the adjustment of costs and benefits among stakeholders.

Such an approach can contribute significantly to the participation of societal actors in efforts to achieve sustainability. One of the key characteristics of sustainability science is that it does not end with the unilateral process of dispensing research results to the general public. Rather, it depends on a bilateral process through which changes in social values and sustainability-oriented actions taken by the public in turn transform the character of sustainability science. In other words, sustainability science itself is an interrelationship-based discipline; it cannot evolve without this process of give-and-take with society at large. Ontology, which at first glance may appear to be a primarily technical methodology, is in fact a highly effective means of discerning the interrelationships among the societal actors relevant to sustainability.

Saito et al. have attempted to apply this ontological method in Section 2-4 to the problem of biofuel use. The results of their analysis reveal not only relationships among concepts (such as the linkage from biofuel production to a specific product, bioethanol, to raw materials such as maize and sugarcane), but also a link to the competition between biomass and food production, a link via water issues to women's rights in impoverished nations, a link via farmland to ecological destruction, and so on. Ontology thus affords a bird's-eye view of this complex web of problems as well as the means of considering multiple perspectives at the problem-solving stage.

This chapter has shown how computers can be used to structure the knowledge most relevant to sustainability science. However, it will take some time for the real impact of this process to manifest itself. As Kajikawa points out in his analysis, sustainability science has yet to be fully systematized; for the most part, sustainability issues continue to be treated as a subset of existing disciplines. When sustainability science acquires a genuinely transdisciplinary framework that integrates these existing disciplines, the value of methods for structuring the knowledge that provides objective support for that framework will become more apparent. To that end, what is needed is an ongoing "co-evolution" of the structuring of the knowledge and concepts of sustainability science.

3

Concepts of "sustainability" and "sustainability science"

3-1

The evolution of the concept of sustainability science

Motoharu Onuki and Takashi Mino

This section will examine the circumstances in which the study of sustainability science has been proposed and the basis for establishing the Integrated Research System for Sustainability Science (IR3S) by reviewing the historical evolution of the concept of sustainability.

3-1-1 Population issues

It is safe to say that the discussion of sustainability in society was initiated by Malthus (1798) in his work *An Essay on the Principle of Population*, in which he dealt with the relationship between population increase and increased food production. Mill (1848) also argued that the world's population and wealth cannot continue to increase indefinitely, and Hardin stressed in "The Tragedy of the Commons" (1968) that there is no technological solution to the issue of population. Just as Hardin's article presented an episode on the grazing capacity of pastures (commons) and a method for their use, Odum, in a paper published in 1971, addressed the concept of environmental capacity in the field of ecology and the fact that there is a limit to the number of living creatures that can sustainably inhabit a finite environment (Odum, 1971).

3-1-2 Limits to growth and steady-state economics

Amidst a proliferation of commentary on environmental pollution and population explosion, Meadows et al. (1972) argued in *The Limits to*

Growth that, if humanity continues along its present course, food shortages, natural resource depletion and environmental degradation will inevitably lead to a disastrous scenario involving radical attrition of the population. Two years after the book's publication the oil crisis occurred, triggering widespread public awareness in Japan of the limited resource and environmental capacity of the Earth. Though the oil crisis was caused not by a petroleum shortage itself but by a geopolitical event and it was an energy security issue, the general public in a country such as Japan that does not have sufficient natural resources in effect experienced the future running out of the oil.

Against the background of a sense of impending crisis concerning the future of humankind and the resulting spate of pessimistic declarations, Daly advocated in his book *Steady-State Economics* (1977a) the goal of intentionally avoiding quantitative economic growth. The notion of a steady-state economy had already been put forward by Mill in 1848. Mill predicted that the population would at some point settle into a "stationary state" owing to the inability of population and wealth to continue growing unabated forever. It was in response to this concept that Daly proposed the theory of a steady-state economy, based on his assertion that the throughput of people and physical goods must be capped at a reduced level in order to prevent depletion of the Earth's resources and destruction of the natural environment.

3-1-3 "Sustainable" as terminology

The term "sustainable" was initially used in the context of ecology, forestry and fisheries. However, as a consequence of the above reasoning, it also began to be used to describe the sustainability of humanity and society. Daly's essay "The Steady-State Economy: What, Why, and How?" (1977b) appeared in a book entitled *The Sustainable Society*.

The word "sustainability" gained overnight recognition when the World Commission on Environment and Development (informally known as the Brundtland Commission), convened by the United Nations, advocated the concept of sustainable development in its report *Our Common Future* in 1987. The report marked a watershed in that it defined sustainable development as "meeting the needs of the present without compromising the ability of future generations to meet their own needs" (WCED, 1987: 8) and put forward the following argument: there are limits to the environment's ability to meet these needs; we must preserve environmental capacity in order to satisfy the needs of future generations ("equity between generations"); the needs of the world's poor are also included among the needs of the present; and it is crucial to satisfy these needs (by rectifying the North–South divide).

The argument that "development", which describes a dynamic state, and "sustainability", which implies a static condition, are even compatible in the first place is still a point of contention. There has been agreement, however, on the point that sustainability and sustainable development carry the connotation of a world order that can eliminate poverty and the North–South divide while refraining from burdening future generations with adverse environmental effects and using up all of the Earth's resources in the present generation (Orr, 1992).

Still, the assertions of the Brundtland Commission have been criticized for adopting the position that economic growth is essential to mitigating the North–South divide and for not advocating curtailment of the environmental burden being imposed by "the North". In addition, debate still continues on the question of whether "development" should be perceived as economic development or human development.

The United Nations Conference on Environment and Development (also known as the Rio Summit) held in 1992 illustrated the importance of action based on the concept of sustainable development and resulted in the adoption of the Agenda 21 programme. The Johannesburg Summit held a decade later in 2002 emphasized the importance of education in facilitating action and proposed the Decade of Education for Sustainable Development, which is currently ongoing.

3-1-4 Sustainability science

In order to fulfil the basic needs of humanity without damaging the Earth's ability to sustain life, Kates et al. (2001), Clark and Dickson (2003) and Clark (2007) have all pointed out the importance of understanding the relationship between nature and society as a mutual one that is both dynamic and complex. They advocated the term "sustainability science" to describe this perspective.

The concept of sustainability was originally born out of concern with issues relating to natural resources and the environment. However, through publications such as *Our Common Future* and conferences such as the 1992 Rio Summit, sustainability came to be recognized as embracing the aim of realizing global social equity while also preserving the Earth's ecosystem.

The notion of interaction between human and natural systems in sustainability science can also be seen to have originated from apprehension about resources and the environment. There are also many similarities to the philosophy of adaptive co-management in terms of belief in a complex system in which any changes by humanity (i.e. society) to the mechanism or methods by which the system manages natural resources have an effect on the natural environment and resources and eventually incur a rebound effect on humanity/society, thus giving rise to the need for adaptation and cooperation.

Meanwhile, many of the objectives of the United Nations' Millennium Development Goals, which aim to realize social equity around the globe and across generations and expound the need for human development and the guarantee of human security, are dedicated to the realization of social equity. At Columbia University's Earth Institute, Sachs (2005) is among those who have called for the eradication of poverty by cancelling the debts of poor nations. It is also thought that the emphasis of sustainability science will gradually shift from issues of natural resources and the environment to how to move towards more fundamental human development.

3-1-5 Systematic improvements and reforms

Fukai (2005) classified and organized the various visions of a sustainable society according to the question of whether they attempt to improve capitalism and democracy from within their respective systems or whether they aim for the reform of these systems, and then pointed out the commonalities and differences in these approaches. According to Fukai, sustainable development as defined by the Brundtland Commission, the eco-economy described by Brown (2001) and eco-efficiency as proposed by the World Business Council for Sustainable Development are classified under systematic improvement theory, whereas Daly's steady-state economy (1977a, 1977b), eco-world government theory (Ophuls and Boyan, 1992), bioregionalism (Sale, 2000) and Bahro's theories on eco-anarchism (Dobson, 2000) are assigned to the systematic reform school of thought. The current inclination towards a greater capitalist system is seen as an issue for both the systematic improvement and systematic reform approaches. Whereas systematic improvement theory argues that it is possible to arrest and reverse this trend, or to promote economic growth from within the capitalist framework without increasing natural resource consumption or throughput, systematic reform theory argues that, without systematic reform, environmental destruction is inevitable regardless of how much we reduce resource waste and increase energy efficiency, because the inclination towards growth is an inherent trait of capitalism (Fukai, 2005). Systematic improvement theory also argues that economic growth without the accompanying exploitation and use of virgin resources is possible in the form of dematerialization, whereas systematic reform theory asserts that at least the developed nations need to make the transition to steady-state economics or degrowth. Fukai (2005) observed that discussions and communication between proponents of these two theories tend to be rigid and that there needs to be a bridge between them.

3-1-6 The conception of IR3S

Komiyama and Takeuchi (2006) proposed defining sustainability science as an academic field that points the way to understanding the diverse issues associated with sustainability in a holistic manner and to offering visions of the development of a sustainable society and methods for achieving it. To this end they launched the Integrated Research System for Sustainability Science (IR3S) and publish the journal *Sustainability Science*. The IR3S programme not only provides a forum for discussion and expands the concept of sustainability but also promotes sustainability science education and outreach.

REFERENCES

Brown, L. (2001) *Eco-Economy: Building an Economy for the Earth*. London: Earthscan Ltd.

Clark, W. C. (2007) "Sustainability Science: A Room of Its Own", *Proceedings of the National Academy of Sciences* 104(6): 1737–1738.

Clark, W. C. and N. M. Dickson (2003) "Sustainability Science: The Emerging Research Program", *Proceedings of the National Academy of Sciences* 100(14): 8059–8061.

Daly, H. (1977a) *Steady-State Economics*. San Francisco: W. H. Freeman.

Daly, H. (1977b) "The Steady-State Economy: What, Why, and How?", in Dennis C. Pirages (ed.), *The Sustainable Society: Implications for Limited Growth*. New York: Praeger.

Dobson, A. (2000) *Green Political Thought*, 3rd edn. London: Routledge.

Fukai, S. N. (2005) *Theories on a Sustainable World*. Kyoto: Nakanishiya Publishing (in Japanese).

Hardin, G. (1968) "The Tragedy of the Commons", *Science* 162 (3859): 1243–1248.

Kates, R. W., W. C. Clark, R. Corell, J. M. Hall, C. C. Jaeger, I. Lowe, J. J. McCarthy, H. J. Schellnhuber, B. Bolin, N. M. Dickson, S. Faucheux, G. C. Gallopin, A. Grubler, B. Huntley, J. Jager, N. S. Jodha, R. E. Kasperson, A. Mabogunje, P. Matson, H. Mooney, B. Moore, T. O'Riordan and U. Svedin (2001) "Environment and Development: Sustainability Science", *Science* 292 (5517): 641–642.

Komiyama, H. and K. Takeuchi (2006) "Sustainability Science: Building a New Discipline", *Sustainability Science* 1: 1–6.

Malthus, T. (1798) *An Essay on the Principle of Population*. London. Available at <http://www.esp.org/books/malthus/population/malthus.pdf> (accessed 27 May 2010).

Meadows, D. H., D. L. Meadows, J. Randers and W. W. Behrens III (1972) *The Limits to Growth: A Report for the Club of Rome's Project on the Predicament of Mankind*. New York: Universe Books.

Mill, J. S. (1848) *The Principles of Political Economy*. Amherst, NY: Prometheus Books, 2004.

Odum, E. P. (1971) *The Fundamentals of Ecology*. Philadelphia: W. B. Saunders.

Ophuls, W. and A. S. Boyan (1992) *Ecology and Politics of Scarcity Revisited: The Unraveling of the American Dream*. New York: W. H. Freeman.

Orr, D. W. (1992) *Ecological Literacy: Education and the Transition to a Postmodern World*. Albany, NY: State University of New York Press.

Sachs, J. (2005) *The End of Poverty: How We Can Make It Happen in Our Lifetime*. London: Penguin.

Sale, K. (2000) *Dwellers in the Land: The Bioregional Vision*. Athens, GA: University of Georgia Press.

WCED [World Commission on Environment and Development] (1987) *Our Common Future*. Oxford: Oxford University Press.

3-2

Exploring sustainability science: Knowledge, institutions and innovation

Masaru Yarime

3-2-1 Diversity of knowledge in sustainability science

Global sustainability concerns long-term constraints on resources, including, among others, food, water and energy. The challenge of sustainability is the reconciliation of society's development goals with the planet's environmental limits over the long term (Clark and Dickson, 2003). The new field of sustainability science now being developed aims at understanding the fundamental character of interactions among natural, human and social systems (Clark and Dickson, 2003; Kates et al., 2001; Komiyama and Takeuchi, 2006). Sustainability science concerns various domains, including nature (for example, climate, oceans, rivers, plants and other components of the natural environment), technology (for example, machinery, chemicals, biotechnology, materials and energy) and society (for example, economy, industry, finance, demography, culture, ethics and history). The academic landscape of sustainability science likewise consists of clusters of diverse disciplines (Kajikawa et al., 2007).

It is of critical importance that these diverse types of scientific knowledge be integrated effectively in sustainability science by establishing interfaces with a certain degree of affinity to disciplinary boundaries. There are, however, technical, economic, legal and institutional barriers and obstacles discouraging such knowledge integration (Maurer, 2006). Researchers are under increasing pressure to publish articles in scientific journals in their own specialism, without much incentive to collaborate with researchers working in different academic fields. At the same time, intellectual property rights have taken precedence since the 1980 passage in the United States of the Bayh–Dole Act, which allows universities to apply for patents based on the results of research activities funded by the federal government. Because one of the prime motivations for scientific collaboration is to assemble the appropriate expertise for tackling cross-cutting problems (Shrum et al., 2007), it is crucial to identify and elaborate the problems and challenges that researchers intend to tackle in order to promote collaboration in the emerging field of sustainability science. What is needed is careful analysis and a solid understanding of the institutional conditions for research collaboration.

To examine the patterns of research collaboration in sustainability science, a preliminary bibliometric analysis has been conducted by region as well as by academic field (Yarime et al., 2010). Patterns of co-authorship of scientific articles on sustainability were analysed on the assumption that knowledge-sharing and communication are reflected in the co-authorship patterns of academic publications. The results show that an increasing number of countries are now engaged in research on sustainability. Moreover, a growing percentage of scientific publication is shown to be based on research through international collaboration. The number of countries engaged in international collaboration is also increasing, as is the diversity of countries involved. Research collaboration tends to be conducted between countries that are geographically close, creating regional clusters. The international collaboration network is basically fragmented into three regional blocs: Europe and Africa, North and South America, and the Asia Pacific region. This suggests that the creation, transmission and sharing of knowledge on sustainability tend to be confined within these regional clusters. Research collaboration in the newly emerging field of sustainability science, which aims at utilizing diverse types of information and knowledge related to sustainability, thus requires organizational and institutional arrangements for more effective collaboration.

Field specialization in sustainability research conducted through international collaboration was also examined. For example, the field most frequently addressed in research collaboration between Japan and China is water resources. Whereas this area is strongly emphasized in China, it is not one that receives particular attention in Japan. This is also the case in other areas, including plant science, ecology, soil science and environmental science. The sustainability-related fields of research collaboration between Japan and China mainly reflect areas emphasized in China. Bilateral collaboration is for the most part influenced by a research agenda driven by China's urgent needs. Whereas the necessity of environmental protection has been increasingly recognized and addressed in China, Japan has accumulated a substantial body of research findings and expertise in diverse areas of environmental issues, partly as a result of the country's many experiences of coping with various types of pollution and accidents in the past. Thus research collaboration between Japan and China tends to be aimed at addressing research needs in China, with support in the form of knowledge and expertise coming from Japanese researchers.

The focus of research activities on sustainability also differs significantly among countries. Since each country has its own particular focus in the research fields related to sustainability, the existence of regional clusters could pose a serious obstacle to collecting, exchanging and integrating diverse types of knowledge, which is of critical importance in establishing the transdisciplinary field of sustainability science. Given these patterns of international collaboration, one way of promoting the creation, transmission and sharing of knowledge on sustainability science could be to encourage research collaboration within existing regional networks at the initial stage and to try to establish interregional linkages later.

Several initiatives have already been launched to set up global schemes for research collaboration on sustainability science. Among the new types of organizational and institutional arrangement are the Alliance for Global Sustainability (AGS) – an inter-university research collaboration between The University of Tokyo in Japan, Massachusetts Institute of Technology (MIT) in the United States, the Federal Institute of Technology (ETH) in Switzerland and Chalmers University of Technology in Sweden – and the Forum on Science and Innovation for Sustainable Development hosted by the American Association for the Advancement of Science (AAAS). The International Conference on Sustainability Science (ICSS) has also been established to encourage communication and knowledge exchange on diverse issues linked to sustainability science. These emerging organizational and institutional arrangements will have significant implications for global sustainability, which requires the production, communication and integration of diverse types of knowledge and expertise.

It is also critical to gain recognition among academic colleagues of sustainability science as an academically established field. Institutionalization through the establishment of academic programmes, societies/associations and journals is helpful for this purpose. Incentives for researchers need to be adjusted to promote cooperation and collaboration among those in different faculties, which will require changes in the criteria for performance evaluation. Promotion and tenure structures need to be adjusted in many universities and research institutes so as to promote mobility and long-term career paths. It is also important to create and maintain effective feedback loops of knowledge through collaboration with diverse stakeholders in society, including industry, government and citizens. Involving stakeholders in society, however, may pose difficulties in producing rigorous results in the traditional scientific sense. The process of collaborating with stakeholders needs to be applied appropriately in education and research in sustainability science.

3-2-2 The development of concepts and methodologies in academic fields

The emerging field of sustainability science faces a serious challenge in its establishment as an academic field through institutionalization, networking and collaboration with stakeholders in society. Because sustainability science is aimed at understanding the fundamental characteristics of complex and dynamic interactions among natural, human and social systems, it is crucial to make effective use of knowledge and information on diverse aspects of sustainability. That necessarily requires a broad range of academic disciplines, including the natural sciences, engineering, the social sciences and the humanities. Thus many concepts, methodologies and practical tools have been proposed in sustainability science, which poses a significant challenge to its establishment as an academic field.

In the development of academic disciplines in the past, one can observe at least two types of evolution: one is concept-oriented, as in chemical engineering, and the other is problem- or use-oriented, as in agricultural science and health science. In the case of establishing chemical engineering as an academic discipline, it was of critical importance that diverse chemical processes were conceptualized in 1915 into "unit operations" such as drying, distillation, separation, extraction, evaporation, absorption and adsorption (Rosenberg, 1998). Based on this intellectual foundation, the School of Chemical Engineering Practice was established at MIT, followed by the establishment of an independent academic department in 1920. Then a standard textbook, *Principles of Chemical Engineering*, was published in 1923. The conceptualization of unit operations in effect functioned as a "focusing device" in elaborating the purposes of research in chemical engineering. Concepts, tools and methodologies were applied to actual problems in industry, and the knowledge and experiences obtained were fed back to education and research at universities, leading to the development and institutionalization of chemical engineering.

In developing sustainability science as a new academic field, it is important to investigate whether such conceptualization is possible or desirable, and, if so, how and what kind of conceptualization. On the one hand, sustainability science could be considered a new field of science, that is, an emerging scientific branch analysing new phenomena with new approaches, as is the case with nanotechnology and bioinformatics. On the other hand, because different disciplines are involved in addressing many issues related to sustainability, the concept of interdisciplinarity or transdisciplinarity has been emphasized by many researchers in sustainability science. In that sense it could be considered a field of meta-science, the science of connecting, integrating or transcending sciences. It then needs to be investigated how that could be possible theoretically and to elaborate how transdisciplinarity can actually be implemented in research and education. A given issue can be tackled through different approaches, which, however, are not necessarily connected or integrated, let alone transcended. It is urgently necessary to develop concepts and methodologies to implement inter/transdisciplinarity in research.

The complexity of coupled human and natural systems is discussed in an article by Liu et al. (2007). The authors identify some of the most prominent characteristics of inter-systemic interactions, including reciprocal effects and feedback loops, nonlinearity and thresholds, surprises, legacy effects and time lags, resilience, and heterogeneity. Although these characteristics certainly represent the complexity of coupled human and natural systems, the mechanisms that explain why these characteristics emerge through complex interactions have not yet been clearly elucidated.

In existing work in the field of sustainability science, the influence of science and technology studies (STS) can be observed to a significant degree in its conceptualization and terminology, including, notably, boundary organization/spanning, co-production of knowledge and hybridization of scientific and local knowledge. Also, case studies so far mainly concern geographically limited areas in such regions as Africa, Asia and Latin America. On the other hand, a detailed analysis of sustainability in industrialized countries, where industries are highly advanced and sophisticated technologies play a crucial role, has been relatively lacking in past literature on sustainability science. It would be very useful to conduct case studies in countries such as Japan by taking a specific technology or industrial sector (for example, electric vehicles) as a boundary condition (Dijk and Yarime, 2010; Orsato et al., 2011).

3-2-3 Sustainability science as the analysis of the knowledge-circulation process in society

Depending on the target of scientific activity, three stages can be identified in the evolution of science (Yoshikawa, 2006). The first phase is "survival science". This type of science basically concerns knowledge for fighting against natural disasters such as floods, typhoons and infectious diseases, and it has contributed significantly to the survival of human beings in the past. The next phase is "development science". At this stage, scientific knowledge has been utilized extensively for the exploration of territory and the utilization of materials on Earth. That process, however, has resulted in the fragmentation of knowledge into different scientific disciplines or fields, leading to the deterioration of natural as well as artificial environments without a systemic understanding of the whole. To address this challenge, "sustainability science" has been proposed since the 1990s by leading scientific organizations such as the International Council for Science (ICSU). This emerging phase of science is aimed at generating knowledge of the processes of change in the interactions of nature, humans and society. Novel approaches to collecting and analysing various large data sets are required to understand the rules and mechanisms of these complex, long-term processes of interaction.

Sustainability science can be considered as an academic field that analyses the processes of production, diffusion and utilization of various types of knowledge with long-term consequences for society. Three components can be identified in a knowledge-circulation system in society: knowledge, actors and institutions. Knowledge itself has aspects of content, quantity, quality and rate of circulation. Important aspects of actors are their heterogeneity, linkages and networks, and interactions among them. Institutions cover a diverse set of entities, ranging from informal ones such as norms, routines and established practices to more formal ones including rules, laws and standards. Sustainability science thus deals with dynamic, complex interactions among diverse actors creating, transmitting and applying various types of knowledge under specific institutional conditions. There are many phases that can be identified in the production, diffusion and utilization of knowledge by different actors, without necessarily involving coordination with one another. Gaps and inconsistencies inevitably exist among different phases in terms of the quantity, quality and rate of knowledge processed. This constitutes a major challenge in pursuing sustainability on a global scale.

A knowledge-circulation system consists of different phases. First of all, a problem affecting sustainability emerges. In this phase, knowledge of the natural sciences is essential for investigating and understanding the causes and mechanisms of the problem. Next, the problem is recognized by many people in society. The way that the problem is reported in the media, including newspapers and TV, significantly influences how the problem is recognized and interpreted in society. Knowledge of methodologies such as discourse analysis is useful in understanding this process. As the problem becomes widely recognized in society, research activities are initiated by scientists at universities and research institutes. The behaviour of scientists will be heavily influenced by the norms and incentives in their communities, which could be significantly different from those in industry. Studies of the sociology of science (Merton, 1973) and the economics of science (Dasgupta and David, 1994) have accumulated valuable findings about scientists' behaviour. Scientific investigation of the problem is followed or accompanied by technological development. In this phase, private companies play a major role in inventing and diffusing technological solutions to the problem. The research and development (R&D) activities of private companies have been studied extensively in the field of the economics of technological change and innovation studies. The technologies developed by industry are introduced in society and subsequently used by different stakeholders. This will have a variety of impacts, some of them unexpected. Assessments of environmental protection and safety, energy/materials flow analysis and lifecycle assessment are useful in tracing and understanding the impacts on society. Following these impacts, there will be feedback from the stakeholders in society. The reactions of various actors to scientific and technological developments have been studied in the STS field. Thus there are many phases of the knowledge-circulation process, with feedback among different actors but not necessarily much coordination with each other.

To fully understand the process of knowledge circulation in society, two types of knowledge structure need to be analysed. The first is the structure of the content of knowledge per se. The content of knowledge relevant to sustainability belongs to diverse disciplines, ranging from natural sciences and engineering to social sciences and the humanities. This content needs to be classified, codified and systematized for in-depth understanding of the knowledge-circulation process. The other knowledge structure concerns the production, transmission and utilization of knowledge. The types of actors and the patterns of their networks and collaboration involving academia, industry and the public sector can be identified, and the ways in which knowledge is created, diffused and used by these various actors in society need to be examined in detail. In elucidating the structure of the content as well as the production, transmission and utilization of knowledge, an approach that could be called an "ecology of knowledge" may prove useful. Analogous to ecology in the biological sense, this new approach is expected to contribute to understanding the structure of knowledge through theoretical modelling and empirical research on such aspects as the dynamics of growth, mechanisms of diffusion and scaling of diversity, uncertainty and stability (Storch et al., 2007).

3-2-4 The role of knowledge in innovation

In the era of knowledge-based economies, rapid knowledge creation and easy access to knowledge sources are necessary components of innovation (Foray, 2004). As the commercialization of knowledge has assumed greater importance in economic growth, collaboration across organizational boundaries has become more commonplace, although in many ways these might be conflicting goals, since commercialization typically requires secrecy, or at least highly controlled and limited collaboration across organizations. In fields where scientific or technological progress is developing rapidly and the sources of knowledge are widely distributed, no single organization has all the necessary skills to stay on top of all areas of progress and produce significant innovation (Powell and Grodal, 2005). Many recent studies point to the crucial role of inter-organizational networks in influencing the change and direction of technological development.

Reviewing the past findings of empirical research on the role of external sources of scientific, technical and market information in innovation, Freeman pointed out the vital importance of external information networks and of collaboration with users during the development of new products and processes (Freeman, 1991). It is argued that dense ties between partners in technology collaboration networks foster information diffusion and knowledge exchange, enhancing the technological performance and collaborating opportunities of the partners (Ahuja, 2000; Baba et al., 2010; Stuart, 1998; Uzzi, 1997). Other innovation studies explain the benefits of inter-organizational relationships in terms of mutual and interactive learning through networks (Gulati, 1999; Powell et al., 1996).

University–industry collaboration has been effective in promoting innovation through the institutionalization of knowledge creation and dissemination. In the synthetic dye industry in the nineteenth century, the establishment of networks linking academia, industry and the public sector led to changes in educational institutions and patent laws, and is a key factor in explaining the technological leadership of Germany (Murmann, 2003). Since knowledge of synthetic organic chemistry was such a critical resource for firms in the dye industry, strong connections to the holders of this knowledge were a key variable in the long-term success of individual firms. At the same time, the networks of close ties that were created between academic scientists, industrial technologists and government officials in Germany allowed them to build a stronger system of research and training.

3-2-5 Cases of innovation in photovoltaics and water technologies

Sustainability science has been developed to investigate the nature and characteristics of the complex and dynamic interactions among natural, human and social systems. To ensure steady progress towards sustainability on a global scale, it is of critical importance to establish academic as well as institutional frameworks for making appropriate use of knowledge to encourage innovations (Yarime, 2007, 2009a, 2009b, 2009c). A case study of the development and adoption of lead-free solders in Japan, Europe and the United States analysed the membership of R&D projects and consortia as well as scientific papers. The study showed how collaboration networks involving universities, public research institutes and private firms influence innovation through the co-evolution of technology and institutions (Yarime, 2009a). To understand the mechanisms of innovation for sustainability, the social process of the production, diffusion and utilization of various types of knowledge needs to be analysed in detail. The cases of photovoltaics and water treatment technologies suggest that gaps and inconsistencies in the knowledge-circulation system could pose serious challenges to the pursuit of sustainability innovation.

The development of photovoltaics indicates a transition in the knowledge system in the evolution of the innovation process. Many R&D activities on photovoltaics in Japan have traditionally been conducted through research projects and consortia involving universities, private companies and public research institutes, with financial support from the New Energy and Industrial Technology Development Organization (NEDO). These extensive activities are considered to have made an important contribution to the steady and solid accumulation and sharing of technological knowledge. Recently, however, there has been an explosion of investment in production facilities for photovoltaics by start-up companies mainly funded through financial markets by venture capital and private funds in the United States, Europe and China. That has had a significant impact on the price of photovoltaics and the extent of their diffusion, changing the nature of the innovation process. One could argue that this is a transition in the knowledge system from one based on R&D projects supported by the public sector for basic scientific knowledge to another based on investments in production facilities by private funds for societal diffusion. The pattern of innovation through university–industry collaboration, which functioned relatively well in Japan in the past as a way of creating scientific and technical knowledge, may not be working as well in utilizing financial knowledge.

The importance of knowledge in promoting innovation for sustainability can also be seen in the case of membrane-based water treatment technologies. Relatively strong market positions have been maintained in membrane technologies by Japanese companies, including Toray, Nitto Denko and Asahi Chemical Industry. These companies have not been tremendously successful in establishing a robust business model, however, since they are mainly producing components such as membranes and exporting them to other countries such as China. Various types of knowledge are required for sustainable water management, including demand prediction, water treatment technologies, water management systems, infrastructure, and laws and regulations. The traditional innovation system based on close university–industry collaboration for basic technological development may not function effectively in introducing technologies in the context of different countries such as China. European companies such as Veolia, Suez and Thames Water utilize various types of knowledge, including technology, management and operations, as a package and are actively expanding in countries around the world. In Japan, the water sector has been managed and operated for a long time by the private sector, which has thus accumulated a significant amount of knowledge and experience in the field. As privatization has been encouraged as a general trend in recent years, it is argued that this knowledge should also be effectively utilized for water management systems overseas through strategic collaboration between the public and private sectors.

3-2-6 Encouraging sustainability innovations for the future

The innovation systems approach may be useful in providing a systemic view of innovation (Edquist, 2005; Lundvall, 1992; Nelson, 1993). According to this approach, a system of innovation, whether it is national, regional/local or sectoral, may be considered to have three basic aspects: knowledge (science and technology), actors and institutions. As discussed above, a variety of issues could be included in sustainability, ranging from environmental protection to poverty and public health. In the context of a knowledge base, these diverse issues could be classified according to their nature and characteristics. The spatial dimension, for example, is crucial in the sense that the nature of innovation will differ markedly depending upon whether the relevant issue is limited to a local area or has an impact on a global scale. The temporal dimension is also important. An issue could be viewed in terms of a very short timeframe, say a few years, or over a very long period of time, in some cases a century.

Knowledge also varies across issues in terms of domains. One knowledge domain concerns the specific scientific/technological field at the base of innovation, such as biotechnology or information technology. Another knowledge domain concerns applications and demands for social solutions. Knowledge may also have different degrees of accessibility, cumulativity, codifiability/stickiness, localization, fragmentation and distance to commercialization. These characteristics vary significantly between issues related to sustainability, with implications for the promotion of innovation. These differences will have a considerable influence on the function of learning, the extent of barriers to entry and the types of major player.

A social process involves a variety of relevant actors, including both organizations and individuals. Heterogeneous actors, such as universities, public research institutes, private companies, governments and non-governmental organizations, have their particular ways of behaving, with distinctive functions and incentives often reinforcing each other, in effect producing a state of equilibrium. In addition, these diverse types of actor are in some kind of relationship with each other. Traditionally, the field of industrial organization has analysed the actors involved in the processes of exchange, competition and command. One may also examine the degree of formality of relationships, whether they consist of formal co-operation or informal interaction, for example. These dimensions of actors and their networks are useful in identifying and analysing the obstacles and challenges to promoting innovation to address issues related to sustainability.

With knowledge and actors identified, the institutional environment that surrounds them can also be discussed. Social issues vary greatly in their specific institutions, which may include norms, habits, practices, rules, laws, regulations, and standards. These influence the actions, behaviour and expectations of the actors involved, as well as their interactions. For example, the laws and regulations concerning intellectual property rights and market competition are very important institutional conditions. In addition, systems for education and human resource development are important, as well as channels for financial investment. It is also necessary to understand what kinds of mechanism govern the interactions between relevant actors: whether they involve commercial profits, intellectual priorities or ethical considerations, for example. These rules of the game influence the processes and outcomes of innovation for dealing with sustainability issues.

Having analysed these innovation systems, their nature, the knowledge involved, the characteristics of the relevant actors and their relationships and interactions, and the institutions surrounding them, points of intervention for promoting innovation to address sustainability challenges can then be discussed. The discussion may focus on different aspects of innovation systems. For example, one might be interested in how to promote R&D activities for knowledge creation in the traditional sense, or how to nurture an innovative mindset in actors as entrepreneurs, or what kinds of institution, including formal public policies, can be implemented effectively and efficiently, and in what way.

Here, the approach of innovation systems has been examined with a particular focus on their structure in a relatively static framework. It is also possible, and indeed desirable, to focus on the dynamic process of innovation systems. It would therefore be useful to identify the functions of innovation systems in a dynamic framework. Previous literature identifies some of these, including knowledge creation, influence on search direction, market formation, development of positive feedback, legitimization and resource mobilization (Bergek et al., 2008).

Some concrete examples of innovation in different domains of knowledge need to be analysed. In this regard it would be useful to collect and analyse some of the best practices with successful outcomes. By focusing on actual cases in, for example, the fields of environmental protection, poverty reduction and public health, future discussions can be focused on some of the most critical aspects of sustainability-related issues. One way to proceed is to identify the knowledge, actors and institutions relevant to selected issues and to construct a conceptual model that could be useful in understanding the differences and similarities among innovations corresponding to sustainability challenges. Based on such exercises, implications could be drawn for policy-making and institutional design for the future.

REFERENCES

Ahuja, G. (2000) "Collaboration Networks, Structural Holes, and Innovation: A Longitudinal Study", *Administrative Science Quarterly* 45: 425–455.

Baba, Y., M. Yarime and N. Shichijo (2010) "Sources of Success in Advanced Materials Innovation: The Role of 'Core Researchers' in University-Industry Collaboration in Japan", *International Journal of Innovation Management* 14(2): 201–219.

Bergek, A., S. Jacobsson, B. Carlsson, S. Lindmark and A. Rickne (2008) "Analyzing the Functional Dynamics of Technological Innovation Systems: A Scheme of Analysis", *Research Policy* 37: 407–429.

Clark, W. C. and N. M. Dickson (2003) "Sustainability Science: The Emerging Research Program", *Proceedings of the National Academy of Sciences* 100(14): 8059–8061.

Dasgupta, P. and P. A. David (1994) "Toward a New Economics of Science", *Research Policy* 23: 487–521.

Dijk, M. and M. Yarime (2010) "The Emergence of Hybrid-Electric Cars: Innovation Path Creation through Co-Evolution of Supply and Demand", *Technological Forecasting and Social Change* 77(8): 1371–1390.

Edquist, C. (2005) "Systems of Innovation: Perspectives and Challenges", in J. Fagerberg, D. C. Mowery and R. R. Nelson (eds), *The Oxford Handbook on Innovation*. Oxford: Oxford University Press.

Foray, D. (2004) *The Economics of Knowledge*. Cambridge, MA: MIT Press.

Freeman, C. (1991) "Networks of Innovators: A Synthesis of Research Issues", *Research Policy* 20: 499–514.

Gulati, R. (1999) "Network Location and Learning: The Influence of Network Resources and Firm Capabilities on Alliance Formation", *Strategic Management Journal* 20: 397–420.

Kajikawa, Y., J. Ohno, Y. Takeda, K. Matsushima and H. Komiyama (2007) "Creating an Academic Landscape of Sustainability Science: An Analysis of the Citation Network", *Sustainability Science* 2: 221–231.

Kates, R. W., W. C. Clark, R. Corell, J. M. Hall, C. C. Jaeger, I. Lowe, J. J. McCarthy, H. J. Schellnhuber, B. Bolin, N. M. Dickson, S. Faucheux, G. C. Gallopin, A. Grubler, B. Huntley, J. Jager, N. S. Jodha, R. E. Kasperson, A. Mabogunje, P. Matson, H. Mooney, B. Moore III, T. O'Riordan and U. Svedin (2001) "Environment and Development: Sustainability Science", *Science* 292(5517): 641–642.

Komiyama, H. and K. Takeuchi (2006) "Sustainability Science: Building a New Discipline", *Sustainability Science* 1(1): 1–6.

Liu, J., T. Dietz, S. R. Carpenter, M. Alberti, C. Folke, E. Moran, A. N. Pell, P. Deadman, T. Kratz, J. Lubchenco, E. Ostrom, Z. Ouyang, W. Provencher, C. L. Redman, S. H. Schneider and W. W. Taylor (2007) "Complexity of Coupled Human and Natural Systems", *Science* 317: 1513–1516.

Lundvall, B.-A., ed. (1992) *National Systems of Innovation: Toward a Theory of Innovation and Interactive Learning*. London: Pinter.

Maurer, S. M. (2006) "Inside the Anticommons: Academic Scientists' Struggle to Build a Commercially Self-Supporting Human Mutations Database, 1999–2001", *Research Policy* 35: 839–853.

Merton, R. K. (1973) *The Sociology of Science: Theoretical and Empirical Investigations*. Chicago: University of Chicago Press.

Murmann, J. P. (2003) *Knowledge and Competitive Advantage: The Coevolution of Firms, Technology, and National Institutions*. Cambridge: Cambridge University Press.

Nelson, R., ed. (1993) *National Innovation Systems: A Comparative Analysis*. New York: Oxford University Press.

Orsato, R. J., M. Dijk, R. Kemp and M. Yarime (2011) "The Electrification of Automobility: The Bumpy Ride of Electric Vehicles towards Regime Transition", in F. Geels, R. Kemp, G. Dudley and G. Lyons (eds), *Automobility in Transition? A Socio-Technical Analysis of Sustainable Transport*. London: Routledge.

Powell, W. W. and S. Grodal (2005) "Networks of Innovators", in J. Fagerberg, D. C. Mowery and R. R. Nelson (eds), *The Oxford Handbook of Innovation*. Oxford: Oxford University Press.

Powell, W. W., K. W. Koput and L. Smith-Doerr (1996) "Interorganizational Collaboration and the Locus of Innovation: Networks of Learning in Biotechnology", *Administrative Science Quarterly* 41: 116–145.

Rosenberg, N. (1998) "Chemical Engineering as a General Purpose Technology", in E. Helpman (ed.), *General Purpose Technologies and Economic Growth*. Cambridge, MA: MIT Press.

Shrum, W., J. Genuth and I. Chompalov, eds (2007) *Structures of Scientific Collaboration*. Cambridge, MA: MIT Press.

Storch, D., P. A. Marquet and J. H. Brown, eds (2007) *Scaling Biodiversity*. Cambridge: Cambridge University Press.

Stuart, T. E. (1998) "Network Positions and Propensities to Collaborate: An Investigation of Strategic Alliance Formation in a High-Technology Industry", *Administrative Science Quarterly* 43: 668–698.

Uzzi, B. (1997) "Social Structure and Competition in Interfirm Networks: The Paradox of Embeddedness", *Administrative Science Quarterly* 42: 35–67.

Yarime, M. (2007) "Promoting Green Innovation or Prolonging the Existing Technology: Regulation and Technological Change in the Chlor-Alkali Industry in Japan and Europe", *Journal of Industrial Ecology* 11(4): 117–139.

Yarime, M. (2009a) "Eco-Innovation through University–Industry Collaboration: Co-Evolution of Technology and Institution for the Development of Lead-Free Solders", paper presented at the DRUID Society Summer Conference 2009, Copenhagen Business School, Copenhagen, Denmark, 17–19 June.

Yarime, M. (2009b) *From End-of-Pipe Technology to Clean Technology: Environmental Policy and Technological Change in the Chlor-Alkali Industry in Japan and Europe*. Saarbrücken: VDM Verlag.

Yarime, M. (2009c) "Public Coordination for Escaping from Technological Lock-in: Its Possibilities and Limits in Replacing Diesel Vehicles with Compressed Natural Gas Vehicles in Tokyo", *Journal of Cleaner Production* 17(14): 1281–1288.

Yarime, M., Y. Takeda and Y. Kajikawa (2010) "Towards Institutional Analysis of Sustainability Science: A Quantitative Examination of the Patterns of Research Collaboration", *Sustainability Science* 5(1): 115–125.

Yoshikawa, H. (2006) "Academic Reform and University Reform: Sustainability Science", *IDE-Contemporary Higher Education* 5: 24–32 (in Japanese).

3-3

Multifaceted aspects of sustainability science

Kensuke Fukushi and Kazuhiko Takeuchi

3-3-1 Introduction

Sustainability science deals with problems containing various aspects that are interlinked with each other. In many cases, such problems take a relatively long time to solve, which makes the nature of the mitigation complicated. Regarding the climate change issue, the projection of Earth's temperature rise is a problem for physicists; however, estimates of the effects of temperature rise on water resources, crop production, sea-level rise, infectious disease, fisheries and forestry are carried out by scientists and engineers in various fields. In order to implement effective climate change mitigation programmes in society, suitable political systems and attractive business opportunities should be developed for an extended period of time. In addition, scientific findings and new technologies have to be accepted by society. In order to mitigate global warming effectively, various areas of academia, politics, business and society have to cooperate. None of them should be left out. Many problems today are of similar complexity and require communication and cooperation among different kinds of people. This can be expected to have a synergistic effect in promoting innovations and dramatically accelerating the solution rate.

For people working in academic institutions, such as university faculty members, such collaboration has not been a standard activity in the development of an academic career. In-depth, basic research projects are usually valued highly, whereas general and practical findings are not favoured for career development. This tendency occurs because general and practical findings are difficult to publish in top-flight professional journals. Gibbons (1994) defined conventional academic development as "mode 1 science". In mode 1 science, knowledge development is carried out within individual academic disciplines; problem statements and solutions are demonstrated within a limited academic society; publication of results is done through academic journals; and quality control is conducted by peer scientists. In "mode 2 science", knowledge development is carried out across the boundaries of academic disciplines; problem statements and solutions are demonstrated with the participation of industry, governments, citizens and not-for-profit organizations; publication of results is done through various media; and quality control is conducted by society. The most important aspect of Gibbons' analysis is that the academic results in mode 2 science are evaluated by society. Sustainability science is an academic alliance that has characteristics of mode 2 science. The major purpose of sustainability science is to integrate existing knowledge to develop novel solutions for a sustainable society and for nature.

3-3-2 Sustainability science and climate change research

If one searches for keywords related to climate change in academic databases developed and managed by Thomson, Reuters or Google, one finds research reports in many academic fields, including physics, geology, biology, chemistry, geography, oceanography, civil engineering, environmental engineering and science, material research, mechanical engineering, chemical engineering, agriculture, fishery, medicine, pharmaceutics, economics, ethics and philosophy, business, and theology. This is quite natural since all people and geographical areas are affected by climate change and the phenomenon is the primary interest of many researchers in many fields. Hiramatsu et al. (2008) investigated fund distribution for climate change research in Japan. They found that a relatively large amount of funding is allocated to climate change research in Japan, which may attract researcher involvement from different fields. For example, in water-related engineering (civil-environmental engineering), many researchers are working on water reuse and recycling. It is very important to develop such technology in water-scarce areas; however, such research activities are equally popular in water-rich areas since a number of large-scale funding opportunities are available in this field. Water scarcity due to climate change is very difficult to manage when it occurs in developing countries, especially in the least developed countries. These countries are usually busy dealing with urgent problems such as poverty, high infant mortality and recovery from disasters, and cannot even begin to address problems that may occur, or are likely to occur, in 100 years' time.

The assessment reports of the Intergovernmental Panel on Climate Change (IPCC) were developed by representatives of individual countries, and each line of the summary was carefully reviewed so as not to include sentences that would have a negative impact on their own country's activities. There is a particularly wide gap between developed and developing countries in the targets for reducing greenhouse gas emissions. In order to promote climate change mitigation and adaptation in developing countries, accelerating the amount of investment from developed countries has to be considered. Especially in the process of developing a low-carbon society, cities in developing countries may have an advantage since they have to build urban infrastructure almost from scratch. In contrast, most cities in developed countries already have urban infrastructures that consume energy and resources at a high rate. In order to change this existing infrastructure, a huge investment cost is usually necessary. If a low-carbon and resource-circulating society can be developed in developing countries through investment from developed countries, it would be beneficial for both developed and developing countries.

As discussed in this section, in order to implement global warming mitigation programmes, various academic disciplines have to tackle new problems within their fields, and they then have to link outcomes from research activities in order to yield the unique and outstanding tools, methods, systems, processes, indices and programmes that are needed to save the Earth.

3-3-3 Sustainability science and environmental pollution

The environmental Kuznets curve (Kuznets, 1955; see Figure 3.3.1) illustrates the trend of environmental pollution at various stages of economic development. Generally, this curve applies well to concentrations of pollutants in the atmosphere and of dissolved oxygen in urban rivers. The reason for this phenomenon is complicated; however, it is obvious that societal demand for a clean environment becomes greater as people attain higher income levels. According to the Kuznets curve, society has to experience a heavily polluted environment first in order to achieve high economic development and then a clean environment later. The income level producing the maximum pollution level will vary according to the kind of pollutant. In the case of air pollution, the following pollutants are going to be problematic, depending on the income increase: SOx and

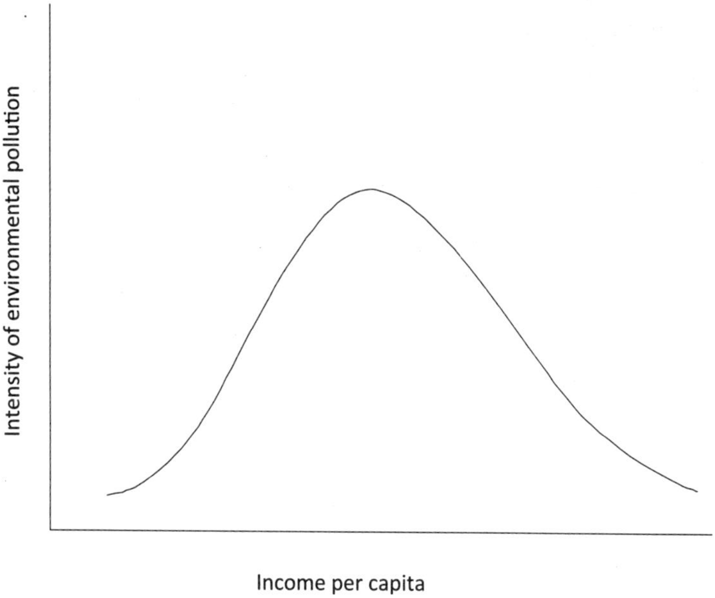

Figure 3.3.1 The environmental Kuznets curve.
Source: Kuznets (1955).

NOx, lead, particulates, polycyclic aromatic hydrocarbons and benzene. However, investigations have been carried out in order to avoid the process of development causing a polluted environment in various sectors.

The definition of sustainable development in developing regions is successive development with short-term goals. In order to maintain this "successive development", society has to find a target and a driving force to achieve that target. The most probable driving force is the economic one, and governments can offer economic incentives by providing an appropriate social system. For example, there are various subsidies for installing solar panels on the roofs of houses, and surplus electricity can be sold to a power company at a good rate (approximately twice the normal rate). Economic incentives are thus the major driving force for installing solar panels. Although this driving force may be economic, such economic incentives are induced by policy.

Climate change mitigation requires an approach of cooperation among technology, business and policy since it involves changes in society, including people's lifestyles. In order to minimize the peak level of pollutants in Kuznets' curve, a policy for prevention is vital. Academia, industries and government need to work together towards this objective.

3-3-4 Conclusion

Sustainability science is an academic field that aims to secure the sustainability of natural, social and personal systems and the peace and prosperity that human beings tend to seek. However, the Earth has undergone dynamic changes in the past. Ancient climate change, industrial revolutions, wars and information technology have drastically changed human life. As an integrated academic discipline, sustainability science is able to propose the correct direction and path for existing academic disciplines to solve complex problems and eventually lead society to a state of peace and prosperity. All academic fields may have the same ultimate goal as sustainability science, and that is why sustainability science, with its intrinsic nature as a discipline, requires collaboration among many academic fields.

REFERENCES

Gibbons, M. (1994) *The New Production of Knowledge*. London: Sage.
Hiramatsu, A., N. Mimura and A. Sumi (2008) "A Mapping of Global Warming Research Based on IPCC AR4", *Sustainability Science* 3: 201–213.
Kuznets, S. (1955) "Economic Growth and Income Inequality", *American Economic Review* 45(1): 1–28.

3-4 Conclusion

Kensuke Fukushi and Kazuhiko Takeuchi

This chapter has discussed the history, concept and characteristics of sustainability science. The development of sustainability science has occurred in many places in the world. The reason for this phenomenon is that the problems we face today are complex and a single academic discipline alone cannot solve them. As discussed in nearly every section of this chapter, such problems as climate change, biodiversity, urbanization and poverty require the participation of various stakeholders with differing areas of knowledge and specialization. The process of solving complex problems such as these usually takes a long time and the solution pathways need to be modified according to the state of the society or the natural environment.

By integrating the knowledge produced by existing academic disciplines, sustainability science can develop innovative solutions and propose pathways to them. However, existing academic disciplines are not well prepared to link with each other, and researchers who do interdisciplinary work of this sort often suffer from a lack of recognition. Sustainability science proposes that conventional disciplines open channels to link with each other. Such linkages are the only way to reach solutions to complex problems. The specific methods of linkage and integration have to be developed for individual cases, and professionals capable of doing such work need to be educated. Sustainability science will be evaluated by society for its necessity and its suitability for achieving common goals. The authors in this volume believe that the philosophy of sustainability science will come to serve as a universal concept for the social sciences, the natural sciences, engineering and other conventional academic disciplines in the near future, and that such a shift will contribute to the realization of a sustainable and peaceful society.

4

Tools and methods for sustainability science

4-0 Introduction

Hideaki Shiroyama

Sustainability science is a discipline that points the way towards a sustainable society (Chapter 1). Sustainability science is more than just the collection of specific solutions to specific sustainability issues in specific contexts. But it has to incorporate methods to identify problems and solutions relating to sustainability in specific contexts and to manage the process of transitions to the solutions. In addition, it has to incorporate tools to implement solutions through influencing the behaviour of actors in society. This chapter introduces tools and methods to identify problems relating to sustainability issues, to find solutions, to manage the process of transition to solutions, and to implement solutions through influencing the behaviour of actors.

Concretely, the following sections discuss problem-structuring methods (Section 4-1), technology governance (Section 4-2), policy instruments (Section 4-3), consensus-building (Section 4-4), public deliberation (Section 4-5), science and technology communication (Section 4-6) and global governance (Section 4-7). Problem-structuring methods are used to identify sustainability issues in societies, and consensus-building and public deliberation are methods of finding solutions to those sustainability issues and managing the transition process. Policy instruments are tools to implement solutions through influencing the behaviour of actors in societies, and global governance is also a tool to implement solutions through influencing the behaviour of actors at the global level. Furthermore, because science and technology are indispensable components of the problems and solutions relating to sustainability, technology governance and science and technology communication are introduced as methods for managing science-related and technology-related issues in the context of sustainability.

4-1

Problem-structuring methods based on a cognitive mapping approach

Hironori Kato

During the past decade, the definition of environmental problems has evolved to include problems associated with energy consumption, air quality, equity, safety, land-use impact, noise and the more efficient utilization of fiscal resources in urban and/or rural areas. However, not everyone shares the recognition of these problems as being "environmental". They may be recognized by different actors in different ways. Recent studies suggest that individuals' decisions often depend on the decision-making context, which is sometimes referred to as a framing effect (Tversky and Kahneman, 1981, 1986). The framing effect can also be observed in sustainability science, particularly in the problem identification process. In order to identify sustainability-related problems, public-policymakers need as accurate as possible an understanding of the many participants' problem identification perceptions with regard to the social/natural system. Additionally, they should analyse this problem structure from a multidisciplinary viewpoint. When more actors are involved in the system, their perceptions of problem identification become more difficult to comprehend. Inaccurate speculation and misunderstanding about a participant's problem perceptions may lead to a deadlock in building consensus. A well-designed and sophisticated method for understanding participant problem perceptions and providing feedback to stakeholders may contribute significantly to better planning and management of sustainable systems.

This section reviews past research on the process of problem-structuring and proposes a practical method of problem identification with cognitive mapping and structuring for sustainable planning. This method has been applied to the case of strategic regional transportation planning in the Kanto region of Japan (Kato et al., 2009).

4-1-1 Literature review

Several studies have considered problem identification and problem-structuring methods. They can be categorized into the following two types: soft operational research and transdisciplinary research studies. The former, which studies problem-structuring methods, includes studies by Ackoff (1979), Checkland (1983), Eden and Ackermann (2001), Friend and Hicking (1987), Howard (1993) and Mason and Mitroff (1981). The latter includes studies by Hansmann et al. (2003), Loukopoulos and Scholz (2004) and Scholz et al. (2006). A series of transdisciplinary research studies has been described as "embedded case studies" (Scholz and Tietje, 2002). The overall framework of the case study reported here is very similar to the embedded case-study approach. This similarity can be observed in the analytical process and collaborative method. With regard to problem-structuring methods, the method proposed in this study may be most similar to the approach of Eden and Ackermann (2001), who propose Strategic Options Development and Analysis (SODA). With SODA, they interview stakeholders to create cognitive maps, then integrate the maps into a comprehensive problem map to understand the overall problem structure. Although the cognitive map approach is useful for understanding each interviewee's perceptions, the completion of the maps generally requires much time and incurs enormous costs. In addition, the simple integration of different maps does not reflect interactions among the stakeholders. In this study, the proposed method improves the problem-structuring method of SODA by reducing the cognitive-mapping requirements and highlighting the interactions among stakeholders.

4-1-2 Proposed method

In general, decisions are made and policies are discussed in response to perceived differences between the desired state of affairs and the decision-maker's perception and/or interpretation of the actual situation. The proposed problem-structuring method considers the participant's perception of the problems. The principal goal of this method is to determine a potential policy agenda by understanding the problem perceptions of the stakeholders. An overview of the method is illustrated in Figure 4.1.1.

1. Selection of stakeholders

- The potential stakeholders are selected from a set of participants in relation to the problems.
- The potential policy targets presented by the authority or information obtained from local university professors are used for stakeholder selection.

2. Cognitive mapping and interviews with the stakeholders

- Hypothetical maps of the stakeholders are made via literature surveys or Web searches.
- The hypothetical map is sent to the interviewees in advance.
- The study team members sequentially interview the potential stakeholders.

3. Cognitive map modification

- The cognitive map is revised on the basis of interview results.
- The map indicates the stakeholder's perception with regard to a causal flow in relation to the problem, impact flow in relation to the actions, and interactions with the other stakeholders.

4. Problem-restructuring and policy agenda analysis

- The maps of the stakeholders are integrated via problem-restructuring.
- The main factors and drivers are identified from the interview results.
- The potential policy agenda is abstracted from a matrix of factors and drivers.

5. Comparison of stakeholders' recognition and their interaction analysis

- The problem recognition difference among the stakeholders and the interactions among them are analysed.

6. Feedback to the stakeholders

- The analysis results are presented to the stakeholders via a workshop.

Figure 4.1.1 Proposed problem-structuring method with cognitive mapping.

First, the stakeholders associated with the problem were selected. A "stakeholder" is defined as a participant who can influence or be influenced by the particular problem. The manner in which stakeholders are selected depends on the availability of data. In the case study shown later, problem system maps corresponding to policy targets were utilized in order to list stakeholder candidates. This is because it was possible to make the system maps from potential policy targets defined by the respective authorities.

Second, the potential stakeholders were sequentially interviewed in order to comprehend their problem perceptions, using the cognitive maps in the interviews. The concept of the cognitive map was originally introduced by Tolman (1948). It provides a communication medium for

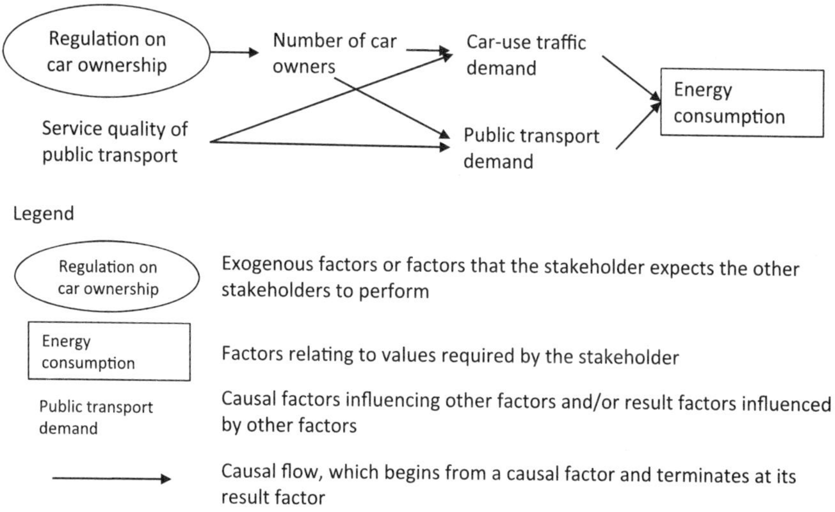

Figure 4.1.2 Example of a cognitive map.

people engaged in the analysis of a complex system (Aissaoui et al., 2003; Axelrod, 1976; Eden and Ackermann, 2004). The cognitive map is a graphical representation of an influence network among notions, with each notion described by a text. An influence is a causality relation from one notion to another. An example of a cognitive map is shown in Figure 4.1.2.

The map used in this study contains the following three types of items: non-highlighted items, which represent causal factors influencing other factors and/or result factors influenced by other factors; oval symbols, which denote exogenous factors or factors that the stakeholder expects the other stakeholders to perform; and rectangular boxes, which represent the factors relating to the values or goals required by the stakeholder. The arrow connecting items denotes a causal flow, which begins from a causal factor and terminates at its result factor.

For this case study, hypothetical cognitive maps were prepared by collating stakeholder profiles via literature surveys or Web searches. The stakeholders were then interviewed with the hypothetical maps. The interview consisted of structured questions and open questions. The structured questions included the following three key questions: what is (are) your or your organization's goal(s); what is (are) the major constraint(s) disrupting your activities towards the goal; what do you expect other stakeholder(s) to perform? On average, each interview took around two hours; after asking the structured questions, the interviewers asked the interviewees to respond freely about the hypothetical maps. In most cases, two or three people were interviewed; most of these people were chief executives or officials responsible for managing their organizations. The interviewers consisted of a study team from The University of Tokyo, including the author.

Third, the cognitive maps were revised on the basis of the interview results. Incorrect or less important factors or actions were deleted from the hypothetical map and additional factors or actions were inserted if necessary. Ultimately, the maps reveal the following three perceptions of the stakeholders: causal flows in relation to the problem, impact flows in relation to the stakeholder's current actions, and interactions with the other stakeholders.

Fourth, the stakeholders' cognitive maps were integrated into a unified problem structure, in this case for transportation problems. Potential problems pointed out during the interviews with the stakeholders were brought together. In order to determine the main factors, these problems were discussed in a meeting with experts. The main factors were determined from the stakeholders' behavioural goals (square boxes) or the factors near the goals in their cognitive maps that have social values. Additionally, the exogenous factors (referred to as "drivers"), which represent the background factors of the problems, were selected. The drivers were mainly selected from the oval-shaped factors in the stakeholders' cognitive maps. The potential policy agenda was then extracted from a matrix consisting of the main factors and drivers.

Fifth, the differences in problem recognition among the stakeholders were analysed. There are two approaches to performing this comparison: (1) a comparison from the viewpoint of the stakeholder's mission, time and spatial dimensions, and (2) a comparison from the viewpoint of the stakeholder's main concerns. These viewpoints are referred to as the stakeholder's "recognition". The stakeholders' interactions were further analysed by means of a reciprocal expectations matrix. The stakeholders' expectations of other stakeholders were analysed from the oval-shaped factors in the cognitive maps.

Finally, the analysis results were fed back to the stakeholders and the implications of the analysis were discussed. The case study included a feedback workshop.

4-1-3 Sustainability and transportation planning

Transportation is one of the most critical factors influencing global and local sustainability. Sustainable transportation systems contribute to the environmental, social and economic sustainability of the communities they serve. Transportation systems exist to provide social and economic connections, and people quickly take up the opportunities offered by increased mobility (Schafer, 1998). The advantages of increased mobility need to be weighed against the environmental, social and economic costs that transportation systems pose. Transportation systems have significant impacts on the environment, accounting for 20–25 per cent of world energy consumption and carbon dioxide emissions (World Energy Council, 2007).

Recently, the importance of interaction between transportation planning and other policy planning has been highlighted in the context of sustainable transportation planning. For example, transportation planning must be considered a part of the land-use planning and development process, which requires an integrated approach to analysis (Banister, 2000, 2003; Goodwin et al., 1991). On the basis of this understanding, several publications have investigated various approaches to transportation planning, including proposals for new planning processes and new technical tools (Kane and Del Mistro, 2003; Loukopoulos and Scholz, 2004; Walter and Scholz, 2007; Zegras et al., 2004). Most transportation researchers consider transportation planning to be a public matter. Transportation planning requires that multiple participants cooperate in and contribute to the planning process, and transportation planners must understand their interactions (Szyliowicz and Goetz, 1995; Wachs, 1995). Extensive interactions among the participants are beneficial to transportation planning. Most transportation researchers also agree that transportation systems are enormous and diverse, and include economic, social, environmental and technological subsystems. Transportation planning is inherently complex, and "problems" must be abstracted from these complex characteristics (Linestone, 1984). Furthermore, the complexity of issues involving transportation systems is thought to require a new planning methodology (Banister, 2003; Szyliowicz, 2003). The new methodology may include (1) the establishment of a vision, (2) understanding types of decisions, (3) assessing opportunities and limitations, (4) identifying near- and long-term consequences, (5) relating alternative decisions to goals, and (6) providing information to decision-makers and assisting them in establishing priorities (Meyer and Miller, 2001). The case study described below applies the proposed problem-structuring method, focusing particularly on problem identification and problem-structuring in transportation planning.

4-1-4 Case study

Case overview

The Kanto region is one of the nine main regions of Japan. It consists of seven prefectures: Tokyo, Kanagawa, Saitama, Chiba, Tochigi, Gunma and Ibaragi. The region covers about 10 per cent of the total area of Japan and comprises over 30 per cent of the total population. It contains several megacities, including Tokyo, Yokohama, Kawasaki, Saitama and Chiba, as well as rural areas on its fringes. The regional population – over 30 million in 2009 – has gradually increased as a result of immigration. Government agencies have predicted that the regional population will keep growing, although the total population of Japan has been decreasing since 2000. When population trends for the entire Tokyo Metropolitan Area are investigated in detail, it is observed that, although the urban population has increased, the rural population has decreased. One of the major reasons for this decrease in rural areas is that young people are moving out from the rural area whereas older people stay there.

With regard to the transportation planning system, no statutory planning existed for regional transportation systems in Japan as of 2009. Although some informal regional transportation plans exist, such as a regional railway master plan, they have no legal basis, particularly for policy implementation and budgeting, and no comprehensive approach. The strategic regional transportation plan considered in this case study may be regarded as an attempt to address the transportation system of the Kanto region from a comprehensive, holistic viewpoint. In 2003, the Kanto Region Transport Bureau under the Ministry of Land, Infrastructure, Transport and Tourism (MLIT) initiated discussions on a strategic transportation plan. The planning work was commissioned to the Kanto Regional Transport Council (KRTC), which consists of academic researchers and local business organizations. The author is a member of a working group organized by KRTC, which discussed the strategic regional transportation policy for two years. The policy, which was completed in 2005, included only a public transportation policy, not a multimodal transportation policy. It must be emphasized that, although information provided by KRTC has been utilized, the analysis in this section is completely independent of the Council's discussion.

Selection of stakeholders

First, problem system maps including hypotheses about regional transportation problems were prepared. It should be noted that these problem system maps were used only for selecting potential stakeholders and were not tested. The maps reveal causal chains, including causal factors and corresponding result factors. Causal flows were drawn from a specific problem in both the upstream and downstream directions. Causal relationships are represented on a system map using arcs and nodes. These maps can be used to identify stakeholders as well as their relationships within the system. The problem system maps in relation to each policy target were prepared according to the same concept as the stakeholder's cognitive map described below. The seven tentative policy targets indicated in KRTC's mid-term progress report were used to establish hypothetical regional transportation problems. These problems were: the poor quality of transportation services for elderly people; the low quality of local and inter-urban transportation networks; a shortage of transportation services in rural areas; an inefficient freight transportation system; serious environmental impacts from car emissions; less safe and secure transportation services; and insufficient promotion of tourism.

After the system maps were prepared, they were used to list stakeholder candidates; several potential stakeholders were identified in the Kanto regional transportation system. This may indicate that the transportation problems in the Kanto region are very complex. About 25 interviewees were selected from the following 12 stakeholders: three private rail operators, including one urban subway operator, a public highway corporation, three prefectural transportation authorities, a local bus operator, an automobile producer, a highways authority, a local tourism policy authority, and a local police agency. Although the users of transportation services are important stakeholders, they were not included in the list of interviewees. This is simply because it is difficult to identify who these users are. Although users were not interviewed, meetings were held with local residents to discuss regional transportation issues when the authors joined the taskforce meetings arranged by MLIT. Data collected by MLIT on citizens' views about regional transportation were also referred to.

Cognitive mapping and interviews with stakeholders

Hypothetical cognitive maps were prepared of stakeholders associated with regional transportation problems in the Kanto region, browsing the stakeholders' web pages to obtain data for the mapping process. The hypothetical maps were sent to interviewees before the interviews were conducted. The authors sequentially met with and interviewed representatives of these stakeholders; the interviews started in April 2004 and were completed in March 2005.

Cognitive map modification

Next, the stakeholders' activity targets were revised on the basis of the interview results. The less important items were also eliminated from the original maps when the interviewees clearly stated that these items were unimportant or not considered. Additionally, factors were included that the stakeholders considered as obstacles to achieving their targets. Factors were also included that these stakeholders expected the other

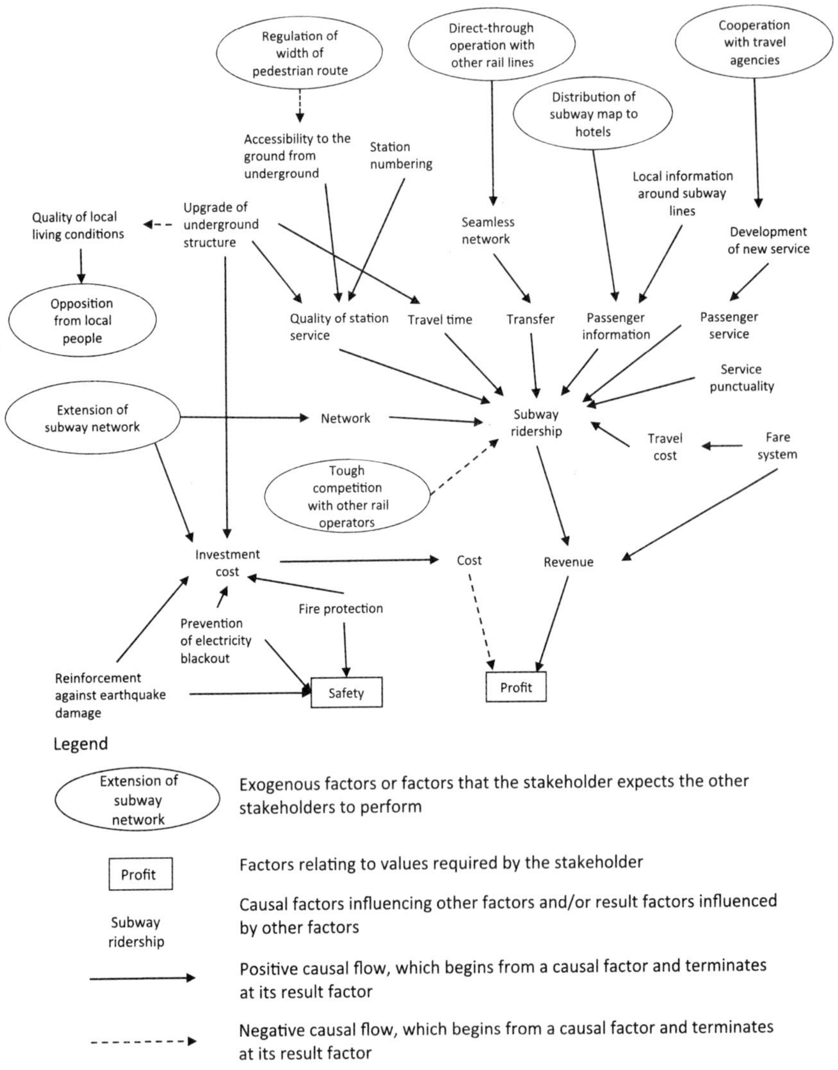

Figure 4.1.3 Cognitive map: Subway case.

stakeholders to perform. An example of a revised cognitive map is shown in Figure 4.1.3.

Problem-restructuring and policy agenda analysis

The potential policy agenda was analysed using problem-restructuring. The participation of experts and multidisciplinary discussions are critical for such analysis. The study team, comprising a transportation planner, a public policy analyst and the author, were involved in this analysis. They also utilized the comments of other experts, primarily transportation policy researchers.

Selection of the main factors

On the basis of the interview results, the following five main factors were selected for problem-restructuring:

- Quantity of transport: a lack of transportation supply, including poor transportation facilities in rural areas and insufficient investment in a ring-road network.
- Quality of transport: a low level of transportation service and infrastructure quality, including traffic congestion and low traffic speeds.
- Transportation marketing: insufficient efforts to increase demand for public transport or to promote demand for less-demanded services, including the promotion of foreign tourism and marketing of weekend transportation services.
- Environment: serious environmental impacts, including a lack of environmental technology/regulations and insufficient transportation demand management.
- Safety and security: dangers to traffic safety and security, including inadequate antiterrorism and disaster prevention measures.

Selection of the drivers

The following five potential drivers of change in the conditions influencing sustainability were selected on the basis of the interview results:

- decreasing younger population and growing elderly population;
- financial deterioration of the central and local governments;
- global competition associated with globalization;
- land-use changes (for example, recent movements of additional population into central business districts);
- greater social concerns with regard to the environment.

Most of these potential drivers were considered from short-term viewpoints. This is partly because the timescale of the stakeholders' problem perceptions is fairly short. Furthermore, it should be noted that many of the potential drivers exhibit low risks or a low probability of occurrence. This is probably because participants find it difficult to reach a consensus on taking action on low-probability, high-risk issues in their organization. These characteristics may bias the selection of drivers. Therefore, through discussions with experts, the following additional drivers were considered in order to balance any bias in the stakeholders' perceptions:

- Extensive changes in the industrial structure. A change in the industrial structure from conventional heavy industries to knowledge-based industries may influence freight transportation patterns. On the other hand, the low-income population has less opportunity to be employed in knowledge-intensive industries, whereas the high-income population may have greater opportunities. This may increase income inequality among the overall population. As a result, the social inclusion of the low-income population and improving accessibility to public services may be included among the more important policy agenda items.
- A greater number of women workers. A new working style or an increase in non-conventional workers may emerge from the labour shortage caused by the rapid ageing of the working population. An increase in job-sharing and in part-time employment means that there is a greater number of irregular commuters.
- More foreign workers. The globalization of the labour force is transforming the conventional work system, with increases in flexible working and English-language-based work. An increase in the number of foreign workers requires further investment in the internationalization of urban facilities, and urban security is also included as an important agenda item.
- Contribution to international society. A more internationally oriented perspective is demanded from international society. For example, because the Kanto region is one of the most successful public-transport-oriented megacities in the world, many developing countries request that Japan share knowledge about this system.
- Catastrophic natural disasters. Japan is prone to various kinds of natural disaster, including earthquakes, typhoons and floods. Since the Kanto region includes the national capital (Tokyo), robust measures to deal with natural disasters are critical.

Policy agenda analysis

The potential policy agenda devised by the study team is summarized in Table 4.1.1. The experts on the study team, including the transportation planner and the public policy researcher, extensively discussed this potential policy agenda. On the basis of these intensive discussions and analysis, certain new agenda items were identified that were not discussed in the stakeholder interviews. For example, the combination of "greater number of women workers" and "quality of transport" generates an agenda item in relation to "irregular transport service". The pairing of "contribution to international society" and "environment" generates "good practice as transit-oriented cities". The pairing of "decreasing younger population and growing elderly population" and "transport marketing" generates an agenda item associated with "transport for long-stay tourism" and "transport for in-home medical care". Although it may prove worthwhile to discuss these issues further with regard to transportation planning, they were not examined in detail in the present study.

Comparison of stakeholders' recognition and interaction analysis

The stakeholders' perceptions were analysed according to the following five viewpoints:

- the mission ("what is the stakeholder pursuing?");
- the networking range with other stakeholders ("who are the major stakeholders that a stakeholder pays attention to?");
- the spatial dimension range ("what is the spatial framework under which each stakeholder acts?");
- environmental conditions ("what environmental conditions, including technical, social and institutional conditions, are important for each stakeholder?");
- the time dimension range ("what is the time framework under which each stakeholder acts?").

Table 4.1.2 summarizes the comparisons of problem recognition by the stakeholders. First, stakeholders, even those in the same category, have different recognitions. For example, although all the railway operators recognize the importance of networking with other railway operators, their attitudes towards railway networking vary with their mission. Railway operator A shows a passive attitude towards collaboration with local governments and communities in terms of station-space use and tourism promotion. On the other hand, railway operator B recognizes the importance of increasing the value of the railway's neighbourhood and shows a proactive attitude towards collaboration with the local government and communities in terms of local development. The subway operator's concern regarding the local community is very limited; it is more concerned with the issue of connecting its underground facilities to the surface networks. The prefectural governments recognize their lack of capacity in relation to transportation policy. One reason for this is that they do not have their own regulatory authority in the local transportation market. Another reason is that there is limited cooperation among prefectural governments. There were also variations in the perceptions of stakeholders, even those belonging to the same category. Prefectural government A directly provides transportation services to its area of jurisdiction because it possesses municipal government functions. However, prefectural governments B and C do not provide a direct service because their prefectures include powerful municipal governments that are responsible for direct service provision. Prefectural government A shows a strong

Table 4.1.1 Potential policy agenda for regional transportation in the Kanto region

	Quantity of transport	Quality of transport	Transport marketing	Environment	Safety and security
Decreasing younger population and growing elderly population	Mobility improvement in rural areas; social inclusion of elderly people	Improvement of public transport service; irregular transport service	Transport for long-stay tourism, transport for in-home medical care		Traffic safety for elderly people
Changes in industrial structure	Investment in freight transport terminal	Information & communication technology aided transport service	Information & communication technology aided transport service	Promotion of environment-friendly technology	
Globalization	Investment in international port and airport	Efficient freight transport and accessibility to airports	International tourism promotion		Customs and security at airports and ports
Changes in land use	Transport system suitable for compact cities	Improvement of non-motorized transport service	Advertisement of non-motorized lifestyle		Security measures for traffic services
Greater number of women workers		Irregular transport service	Special service for women		Transport security for women
Greater social concerns about environmental issues	Transport demand management	Public transport promotion and efficient freight transport	Off-peak time transport marketing	Global and urban environmental problems	

Financial deterioration in governments	Infrastructure management in rural areas	Utilization of existing facilities	Mobility management		
Greater number of foreign workers	Transport service for foreign people	Transport service for foreign people	Transport service for foreign people		Security measures for traffic services
Contribution to international society		Good practice as transit-oriented cities		Good practice as transit-oriented cities	Terrorism prevention
Catastrophic natural disasters	Development of robust transport network				Disaster prevention policy

Note: Underlining indicates that the point was not mentioned by any of the interviewees.

Table 4.1.2 Comparison of problem recognition among stakeholders

	Missions	Interaction with other stakeholders	Spatial dimension of interest	Change in environmental conditions			Time dimension of interest				
				Technological	Social	Institutional	Critical factors	One year	Several years	Middle term	Long term
Subway operator	Safe and punctual operation	Competition and coordination with other railway operators	Own network and other railway networks connecting them	Common fare card system	Decentralization	The building standards act and the road act	Life-cycle of facilities and vehicles	++	++	+	0
Railway operator A	Sustainable and reliable operation; profit maximization	Competition and coordination with other railway operators and coordination with airline companies	Kanto region and neighbouring regions	Smart card technology	Ageing society and low birth rate	Deregulation and privatization	Life-cycle of facilities and vehicles	++	++	+	0
Railway operator B	Profit maximization and contribution to local community	Coordination with local governments	Local area along the railway lines	Smart card technology	Demand decrease	Fare regulation	Life-cycle of facilities and vehicles	++	++	+	0
Local bus operator	Sustainability of the company with local bus service	Coordination of timetable with railway operators	Local area along the bus routes	Smart card technology	Ageing society	Deregulation	Company's sustainability	++	++	++	++
Prefectural government A	Improvement in the local quality of life	Coordination with highway authority and transit operators	Own prefecture and airport	ITS technology	Decentralization	Decentralization	Other plans	++	++	++	0

Prefectural government B	Inter-district balance and better interaction with local council	Coordination with highway authority and transit operators	Own prefecture	ITS technology	Depopulation	Decentralization	Other plans	++	++	+	0
Prefectural government C	Inter-district balance and environmental protection	Coordination with highway authority and transit operators	Own prefecture	ITS technology	Depopulation of some districts	Lack of transport budget	Other plans	++	++	+	0
Highway authority	Reduction in traffic congestion, utilization of road spaces and revitalization of local economy	Coordination with policy agency	The whole of Japan	ITS technology	Privatization	Privatization	Long-term investment plan	++	++	+	0
Highway public corporation	Follow the highway authority control and local-user communication	Coordination with police agency and truck association	Own highway network	Anti-disaster technology	Laws on decision-making process in road construction and privatization	Laws on decision-making process in road construction and privatization	Highway authority's investment plan	++	++	0	0
Automobile producer	Sustainable car-oriented society and profit maximization	Coordination with vehicle regulator	The whole world	New fuel technology	Vehicle regulation	Vehicle regulation	Sustainability of car-oriented society	++	++	++	++

Notes: ITS = Intelligent Transportation System. ++ means that a stakeholder has strong concerns about the issues during a corresponding timeframe; + means that a stakeholder has some concerns about the issues during a corresponding timeframe; 0 means that a stakeholder has no concerns about the issues during a corresponding timeframe.

concern about the local transport service in its area, whereas prefectural governments B and C have other concerns, mainly in relation to the balance among the sub-regions in their area.

Second, different stakeholders have different recognitions. With regard to the spatial dimension, the automobile producer pursues a leadership role at the global level. On the other hand, the concerns of railway operator A, railway operator B and the bus operator are limited to their operating areas. With regard to the time dimension, private companies tend to have longer time recognition, whereas governmental units tend to have shorter time recognition. For example, the automobile producer has set 2050 as a target year for its business strategy. One of the major concerns of the local bus operator is a sustainable bus market for the next generation. On the other hand, the governments, whose officials usually change every two to three years and which have an annually fixed budget, cannot sustain a long-term perspective.

Interactions among stakeholders: Reciprocal expectations analysis

The stakeholders' networking range and their expectations of other stakeholders will now be examined. The relationships among stakeholders are listed in Table 4.1.3.

The table includes airline companies and citizens who have not been interviewed thus far. Their expectations are hypothetically described on the basis of results from interviews with experts. Each cell in Table 4.1.3 lists what the stakeholder in the horizontal row expects the stakeholder in the vertical column to do. Potential collaborations among the stakeholders are also shown in Table 4.1.3. First, potential collaborations are identified that are partially realized in practice. For example, there is a potential collaboration between railway operators and airline companies with regard to the use of a joint credit card system, tourism promotion and airport access. A potential collaboration also exists among governmental units. The highways authority can collaborate with the public transport authority with regard to the use of gasoline tax revenues. The reallocation of the tax revenues to the public transport system could be a potential compromise, at least between these two stakeholders. There is another potential collaboration between the local transport authority and the police agency with regard to transportation demand management (TDM), the strict enforcement of regulations on illegal parking, and the sharing of traffic data; however, their relationship is asymmetrical, that is, the police agency is expected by the other stakeholders to perform certain significant duties, whereas the police agency does not expect comparable commitments from the other stakeholders.

Second, potential collaborations are identified among the stakeholders that have not yet been realized in practice. For example, there may be a potential collaboration between the automobile producer, the highways authority and the police agency with regard to implementing countermeasures against global warming. The automobile producer owns the production technology, whereas the highways authority is responsible for infrastructure development and the police agency possesses the power to control traffic flow. The highways authority will benefit if regulations on illegal parking are tightened by the police agency and automobile technologies are improved by the automobile producers. This is because they can improve traffic capacity without any additional investment of their own. The police agency could also benefit by tightening traffic regulations, while it expects the highways authority to invest further in highways development. This is because the police agency has fewer resources than the highways authority. The automobile producer has incentives to support the highways authority as well as the police agency because it cannot earn profits from automobile users unless the road and traffic services are well managed.

Additionally, automobile producers could collaborate with public transport operators. This is because they realize that they cannot obtain societal support for a sustainable and global automobile society unless the automobile industry and the public transport system coexist effectively. Automobile producers could also collaborate with public transport operators in implementing countermeasures to deal with irregular transportation demands. The local bus operators and railway operators may develop new business as a means of utilizing their capacity by providing special services to young people who are not regularly employed, to elderly people and to women working part time. The automobile producers may find comparative advantages in an irregular transportation market because such demand requires greater flexibility.

Potential collaborations also exist among the many stakeholders in the long-stay tourism business. The public transport operators could collaborate with the automobile service providers (for example, car rental companies dealing mainly with tourists). There is also room for innovation by introducing a common season ticket or card system for long-stay tourists that can be shared among various transportation operators. The local governments could provide sufficient incentives to promote tourism because they can obtain taxes from tourist activities. The participation of local farmers and hotels may prove to be important for ecotourism and participatory tourism.

Feedback to and from stakeholders

After completing the analysis, a workshop was held with the stakeholders at the end of March 2005. Both the interviewees and other

Table 4.1.3 Interactions among stakeholders: Expectations of other stakeholders

	Stakeholders Y (who are expected to perform by stakeholders X)								
Stakeholders X	Public transport authority	Highways authority	Police agency	Local governments	Railway operators	Local bus operators	Airline companies	Automobile producer	Citizens
Public transport authority		Reallocation of gasoline tax revenue to public transport investment	Coordination for TDM	Joint subsidy to public transport	Service-level control	Service-level control		Compliance with vehicle emissions regulations	Support for transit-oriented policy
Highways authority	Cooperation in intermodal policies	Coordination among regional bureaus	Data-sharing coordination for reduction in traffic congestion and traffic accidents	Execution of highway investment plan	Station area development and rail crossing improvement			Improvement in vehicle safety	Support for highway investment
Police agency	Control of bus and truck operators	Highway investment							
Local governments	Joint subsidy to public transport	Coordination in highway construction	Coordination for TDM, the strict enforcement of regulations on illegal parking, and the sharing of traffic data	Coordination for transprefectural transport policies	Station area development, tourism promotion and investment in parking facilities	Coordination for bus-stop installation			Support for transit-oriented policy

Railway operators	Subsidy to railway investment	Subsidy to railway investment		Subsidy to railway investment	Coordination for smooth connections and introduction of common fare card		Joint credit card, tourism and airport access improvement		Support for station area development
Local bus operators	Subsidy to bus operation	Highway investment	Coordination for smooth bus operation	Coordination for bus-only lane	Smooth connections at railway stations	No over-competition			Greater use of local bus services
Airline companies	Support for airport access improvement				Joint credit card and airport access improvement	Airport access improvement	No over-competition		More use of airline service
Automobile producer	Strong leadership in transit-oriented transport policy	Highway investment	Control of illegal parking and coordination with other public sectors	Coordination with other public services				No over-competition	Support for sustainable car-oriented society
Citizens	Strong leadership in transit-oriented transport policy	Highway service improvement	Better road traffic management and road safety control	Local transport improvement	Railway service improvement	Bus service improvement	Airline service improvement	Cheaper and higher-performance vehicles	

Note: Each cell lists what the stakeholder in the horizontal line (X) expects the stakeholder in the vertical column (Y) to do. TDM = transportation demand management.

stakeholders were invited to the workshop, at which the results of the analysis were discussed. The total number of participants in the workshop was 10, and it lasted around two hours. The participants pointed out mistakes or misunderstandings in the revised cognitive maps they were shown. All the participants expressed strong interest, particularly in the cognitive maps of the other stakeholders. Furthermore, they commented that a reciprocal expectations matrix seemed to be very useful in their decision-making.

4-1-5 Conclusion

This section proposed a method of analysing the stakeholders' perceptions of problems and of structuring these problems by using cognitive maps. A case study was presented of strategic public transportation planning in the Tokyo Metropolitan Area, which suggests that the proposed method is fairly effective. The case study shows that the successful generation of a potential agenda is possible. The stakeholders' recognition was also compared by means of an analysis of the stakeholders' perceptions of problems. Additionally, interactions among the stakeholders were analysed by using a reciprocal expectations matrix.

Although the proposed method overcomes some of the difficulties of earlier problem-structuring methods, several research issues remain unresolved. First of all, the interviewees were initially selected from a list of potential stakeholders, but the ideal would be to interview all potential stakeholders. However, because this is a practical impossibility, how interviewees are selected should be examined. It may be preferable to select stakeholders through a public involvement process when applying this methodology in the real world. Second, the process as implemented in the case study was not completely open to the public. This is because the exercise is still at the trial stage. When the proposed method is applied to the real world, the process should be open as part of the public involvement process. Third, the interview results varied considerably with the interviewed individuals. Although an attempt was made to meet people who had sufficient knowledge and experience in their own organizations, the responses varied from one person to another even when two people belonged to the same organization. Methods of eliminating bias in interviewee selection should therefore be explored. Finally, the cognitive map varies with the individual who sketches the map, so the selection of a particular individual may also bias the results. The variations in maps among individuals should be investigated under conditions in which the same interview results are obtained.

REFERENCES

Ackoff, R. L. (1979) "Resurrecting the Future of Operational Research", *Journal of the Operational Research Society* 30: 189–199.

Aissaoui, G., S. Loiseau and D. Genest (2003) "Cognitive Map of Conceptual Graphs", in B. Ganter, A. de Moor and W. Lex (eds), *ICCS 2003. LNCS (LNAI)*, vol. 2746, Heidelberg: Springer, pp. 337–350.

Axelrod, R. (1976) *Structure of Decision: The Cognitive Maps of Political Elites*. Princeton, NJ: Princeton University Press.

Banister, D. (2000) "Sustainable Urban Development and Transport – A Eurovision for 2020", *Transport Review* 20: 113–130.

Banister, D. (2003) "Critical Pragmatism and Congestion Charging in London", *International Social Science Journal* 55: 249–264.

Checkland, P. B. (1983) "O.R. and the Systems Movement: Mappings and Conflicts", *Journal of the Operational Research Society* 34: 661–675.

Eden, C. and Ackermann, F. (2001) "SODA: The Principles", in J. Rosenhead and J. Mingers (eds), *Rational Analysis for a Problematic World Revised*. Chichester: Wiley.

Eden, C. and F. Ackermann (2004) "Cognitive Mapping Expert Views for Policy Analysis in the Public Sector", *European Journal of Operational Research* 152: 615–630.

Friend, J. K. and A. Hicking (1987) *Planning Under Pressure*. Chichester: Wiley.

Goodwin, P., S. Hallett, F. Kenny and G. Stokes (1991) *Transport: The New Realism*. Report to the Rees Jeffreys Road Fund. Transport Studies Unit, University of Oxford.

Hansmann, R., A. H. Mieg, R. W. Scholz and H. W. Crott (2003) "Shifting Students' to Experts' Complex Systems Knowledge: Effects of Bootstrapping, Group Discussions, and Case Study Participation", *International Journal of Sustainability in Higher Education* 4: 151–168.

Howard, N. (1993) "The Role of Emotions in Multi-organizational Decision-making", *Journal of the Operational Research Society* 44: 613–623.

Kane, L. and R. Del Mistro (2003) "Changes in Transport Planning Policy: Changes in Transport Planning Methodology?", *Transportation* 30: 113–131.

Kato, H., H. Shiroyama and Y. Nakagawa (2009) "Qualitative Survey in Transportation Planning: Cognitive Mapping Approach", paper presented at Transportation Research Board 88th Annual Meeting, Washington, DC, January.

Linestone, H. A. (1984) *Multiple Perspectives for Decision-making*. Amsterdam: Elsevier Publishing Company.

Loukopoulos, P. and R. W. Scholz (2004) "Sustainable Future Urban Mobility: Using Area Development Negotiations for Scenario Assessment and Participatory Strategic Planning", *Environment and Planning A* 36: 2203–2226.

Mason, R. O. and I. I. Mitroff (1981) *Challenging Strategic Planning Assumptions*. New York: Wiley.

Meyer, M. D. and E. J. Miller (2001) *Urban Transportation Planning: A Decision-oriented Approach*, 2nd edn. New York: McGraw-Hill.

Schafer, A. (1998) "The Global Demand for Motorized Mobility", *Transportation Research A* 32(6): 455–477.

Scholz, R. W. and O. Tietje (2002) *Embedded Case Study Methods: Integrating Quantitative and Qualitative Knowledge*. Thousand Oaks, CA: Sage Publications.

Scholz, R. W., D. Lang, A. I. Walter, A. Wiek and M. Stauffacher (2006) "Transdisciplinary Case Studies as a Means of Sustainability Learning: Historical Framework and Theory", *International Journal of Sustainable Higher Education* 7: 226–251.

Szyliowicz, J. S. (2003) "Decision-making, Intermodal Transportation, and Sustainable Mobility: Towards a New Paradigm", *International Social Science Journal* 55: 185–197.

Szyliowicz, J. S. and A. R. Goetz (1995) "Getting Realistic about Megaproject Planning: The Case of the New Denver International Airport", *Policy Sciences* 28: 347–367.

Tolman, E. C. (1948) "Cognitive Maps in Rats and Men", *Psychological Review* 55(4): 189–208.

Tversky, A. and D. Kahneman (1981) "The Framing of Decisions and the Psychology of Choice", *Science* 211: 453–458.

Tversky, A. and D. Kahneman (1986) "Rational Choice and the Framing of Decisions", *Journal of Business* 59: S251–S278.

Wachs, M. (1995) "Planning, Organizations and Decision-making: A Research Agenda", *Transportation Research A* 19: 521–531.

Walter, A. I. and R. W. Scholz (2007) "Critical Success Conditions of Collaborative Methods: Comparative Evaluation of Transport Planning Projects", *Transportation* 34(2): 195–212.

World Energy Council (2007) *Transport Technologies and Policy Scenarios to 2050*. London: World Energy Council.

Zegras, C., J. Sussmand and C. Cooklin (2004) "Scenario Planning for Strategic Regional Transportation Planning", *Journal of Urban Planning and Development* 130: 3–13.

4-2

Technology governance

Hideaki Shiroyama

4-2-1 Introduction

The development and diffusion of innovative technologies is indispensable for sustainable development. However, the development of technology is also accompanied by various risks and social problems, as well as benefits. And, as the scope of those issues has grown wider, the range of interested actors has increased accordingly (Shiroyama, 2007b).

For example, the development of nuclear physics and nuclear energy technology has had the benefit of securing sources of energy, but has also been accompanied by constant safety risks and the security risks of nuclear proliferation. As another example, the development of the life sciences and genetic engineering has raised issues of safety and ethics, and the application of this technology to food production (for example, genetically modified crops) increases those concerns. There is particularly strong awareness of these issues in regard to experimentation in gene therapy: the genetic manipulation of human beings, a measure for human sustainability. Moreover, certain pervasive technologies – such as nuclear energy and genetically modified crops – are being assessed by a variety of actors and from diverse points of view, such as the perspective of economic efficiency, in a manner that goes well beyond purely scientific and technological logic. Scientists and engineers may brand such talk as irrational and rumour-mongering, but this is the reality of society, and avoiding the use of specific technologies to avoid economic loss is highly rational as far as society is concerned (for instance, for electricity utilities and governments in the case of nuclear energy, and for agricultural producers and agricultural policymakers in the case of genetically modified crops).

As long as society decides to make use of technologies with diverse social implications for society that involve risks as well as benefits, there is a need for societal systems for managing the development and utilization of these technologies. In other words, technology governance is required. This section will outline how technology governance might be organized, describe the functions that are required of it, and analyse the nature of governance for sustainability in which technology governance tools are used.

4-2-2 What is technology governance?

Technology has many implications for society. For this reason, society has to assess the various problems and issues for deliberation that exist at the interface between society and technology. This function of societal assessment requires certain mechanisms and, to cope with the diverse issues, a specific style of institutional design is vital. These mechanisms and their institutional design are what constitute technology governance. A variety of actors, including experts in various fields, different levels of government (international organizations, national government and local government), groups (such as professional groups and employers' associations) and citizens, must then collaborate, share the effort of technology governance and – although they will sometimes come into conflict – manage the various problems at the interface between society and technology.

Governance and traditional government are often thought of as opposing forces. Government is taken to mean the official institutions for governing, whereas governance is understood to encompass a wide range of systems (including social customs and markets) that are outside the official institutions of government – in other words, "the whole range of institutions and relationships involved in the process of governing" (Pierre and Peters, 2000: 1) and "self-organizing, interorganizational networks" (Rhodes, 1997: 53). Whereas the organization of government is based on an internal vertical hierarchy, governance allows for structures that include horizontal relationships among entities such as various societal groups and companies, and among various levels of government.

A wide range of actors has come to be involved in technology as a reaction to the numerous social implications of technology in specific societal contexts. Scientists and engineers play a major role as individuals as well as forming various independent professional organizations. The role of companies in the introduction of technology to society is also significant. In recent years, companies have played a noteworthy role through corporate social responsibility. In government, on the other hand, although standardization has a significant role at the international level, there are many matters that national and local governments must deal with in accordance with local conditions. Thus it could be said that, within the domain of technology, there is more of an appearance of governance than of government.

4-2-3 Risk management

Clarifying risks and benefits

The development of technology can entail an increase in various risks as well as benefits. To cope with this state of affairs, risk assessment and risk management are being attempted by various segments of society (Shiroyama, 2007a). Risk assessment generally involves multiplying the probability of the occurrence of damage by the scale of the damage. Scientific knowledge based on immunological data and animal test data is essential for this assessment. As a matter of course, the scope of risk assessment can vary greatly, depending on whether it is based on the number of dead or on the number of victims (including the sick and injured), and whether a qualitative distinction is drawn between large-scale catastrophic disasters and smaller disasters. Risk management, on the other hand, refers to the activity of deciding where to draw the line and what level of risk to allow – based on risk assessment – before proceeding with an overall project.

When making risk management decisions, it is necessary to consider how the risks are balanced by the benefits of the technology concerned. Without taking such benefits into account, it would be impossible to understand why the car – which risk assessment regards as entailing a high level of risk in numerical terms – is accepted by society. When benefits are assessed, the question of distributive implications (that is, to whom the benefits accrue) is also important. Even if the overall benefits are considerable, society may reject a technology if those benefits are directed mainly towards a particular sector. It is recognized that society has not readily accepted nuclear power generation or genetically modified foods, despite the fact that their risks are assessed as low. One reason that can be cited for this is that it is corporations that are the direct beneficiaries of these technologies (in terms of perception, at least).

Often, some risks are ignored or exaggerated. When a company engages in technological development, it is possible that it will not publicly disclose the relevant risk information – even if it is aware of the risks that accompany the technology – out of concern for its return on investment from the development of the technology. When a company conducting technological development on-site fails to disclose information, it is extremely difficult – at least in the short term – for society to obtain this information separately and independently. On the other hand, the main thrust of opposition to a particular technology (which might even emanate from a competing company) may exaggerate some of the risks. The problem in this situation is how to conduct comprehensive and balanced risk-mapping. Even for experts, the perceived areas of risk vary among different fields of specialization.

Benefits, too, can be inadequately presented or exaggerated. In the cases of genetic modification technology and nanotechnology, it is a long way from these technologies to concrete benefits for society. Certainly the arguments can be made that the introduction of genetically modified crops will allow increased volumes of food production in developing nations, which will alleviate poverty, or that the introduction of medical diagnostic technology employing nanotechnology will enable preventive medical care based on simple continual monitoring, leading to reduced medical costs. However, a number of variables external to the introduction of a technology must come into play in order for its effectiveness to be realized. This leads to discontent on the part of technology developers because, when a technology is assessed, the risks alone are adequately addressed whereas the benefits are not. On the other hand, technology developers tout the effectiveness of a technology in their quest to obtain research funding and – because there are many variables in play that can influence its effectiveness – may be apt, so it is claimed, to exaggerate this effectiveness. Moreover, there is uncertainty about both risks and benefits – uncertainty over scientific understanding as well as uncertainty over the utilization of the technology.

As regards expectations of risk assessment from science, society often expects a definitive answer despite the fact that science inevitably involves a degree of uncertainty. It is of course possible that the uncertainty will recede as science progresses; however, it will be difficult to eliminate it completely, as the case of climate change shows. For society, the question thus arises of how to assess an acceptable level of uncertainty. The choice between the "precautionary principle" and the "no-regrets policy" expresses a difference in attitudes towards this uncertainty. The precautionary principle refers to taking preventive control measures (even while uncertainty remains as to whether anything will happen) because, if something does happen, the resulting damage will be enormous. The no-regrets policy, in contrast, refers to taking only meaningful measures (even if nothing is going to happen), instead of reacting during the period of uncertainty on the assumption that something will happen. Which of these two attitudes is selected is a policy decision for society.

There is also uncertainty over benefits. As mentioned earlier, one of the characteristics of technology is that it can be used for numerous purposes. There are many technologies that are used in ways that differ from those envisaged by their developers, as well as technologies used in a manner quite distinct from their original purpose. Technology developers sometimes advance the argument that, whereas it is easy to foresee certain risks in the initial experimental stage of any technology, the eventual benefits do not become clear until some time has passed (particularly in the case of revolutionary ground-breaking technologies), and that at the outset it is very difficult to explain the benefits even if asked to do so. However, it should probably be acknowledged that there are also risks that do not become evident for some time.

The multifaceted nature of risks and benefits

Both risks and benefits are multifaceted. For example, there are many cases in which a particular technology entails different risks and benefits after the international relations dimension has been factored in.

In domestic terms, nuclear technology is an energy technology that has the benefit of providing energy but also involves safety risks. With the addition of the international relations dimension, however, the picture changes. By reducing imports of oil (most of which comes from the Middle East) as a principal energy source, nuclear power generation has the benefit of increasing energy security (although maintaining this option requires imports of uranium). On the other hand, possessing the technology for nuclear power generation (particularly for the nuclear fuel cycle) increases the risk of nuclear proliferation at the international level.

The same applies to space technology. Normally, the benefits of maintaining the capability to launch satellites go no further than satellite communications and satellite broadcasting. However, factoring in the international dimension, space technology yields security benefits in the form of spy satellites. Additionally, in a domestic context, dual-use technology is normally technology for public civilian benefit, but, with the addition of the international dimension, there is a recognized risk that the diversion of space technology to military use might contribute to the spread of weapons of mass destruction.

The benefits of technology have also changed as a result of society's changing objectives. Sustainability requires the simultaneous attainment of various objectives. Up to now, for example, the provision of energy (that is, energy security) has been recognized as the sole benefit of nuclear power generation. However, as society has come to recognize global warming as a problem, the fact that this energy source does not emit carbon dioxide (CO_2) – a substance that causes global warming – has come to be recognized as an additional benefit.

Conversely, when coal-fired power generation technology is discussed in the societal context of global warming, emphasis is placed on the risk entailed in its high levels of CO_2 emissions. However, with factors such as rising oil prices adding to concerns over energy security, the use of coal is seen to have energy security benefits, since the coal-producing regions on which it relies are relatively spread out throughout the world.

Assessment of trade-offs

Diverse risks and benefits must thus be considered in the debate over the introduction of a new technology to society. Once this has been done, there is then the problem of what kind of societal assessment to make based on those risks and benefits. In the current context, there are various trade-offs that must be made when this societal assessment is performed (Graham and Weiner, 1995).

Risk trade-off refers to the fact that efforts made to reduce specific risks end up increasing other risks as a result. For example, if car bodies are made lighter in order to improve gas mileage, they become less collision-resistant and safety levels fall. In this case, the global warming and energy security risks are reduced but the safety risk increases. As another example, certain products used as substitutes for CFCs (which destroy the ozone layer) have led to reduced destruction of the ozone layer but have accelerated global warming. In this instance, the risk of destruction of the ozone layer and the risk of global warming are traded off against one another. Meanwhile, methyl bromide, which is used as a fumigant to lower food-related risks, increases the risk of destruction of the ozone layer. In this case the food safety risk and the ozone layer destruction risk are traded off against each other.

Risk trade-offs are also found in the introduction of renewable technologies. For example, wind power technology has benefits in terms of global warming by reducing CO_2 emissions and in terms of energy security because of its decentralized nature. On the other hand, it entails risks and negative effects with regard to supply stability, bird strikes and landscape values. In the case of biofuels, the benefits of energy security and possible CO_2 reduction are countered by the potential risk of food insecurity, especially in developing countries.

4-2-4 Assessing values and visions

When a societal assessment of a technology and associated policy measures is carried out, it is necessary to consider issues that relate to values as well as to assess risks and benefits. To be more precise, the two kinds of assessment should be carried out in concert with one another.

In making societal assessments of technology, a comprehensive assessment should be carried out after the risks and benefits have been broadly clarified. In the context of this comprehensive assessment, however, there is another important factor to be considered that will function as a kind of "trump card", whatever the other risks and benefits. This is the issue of values as they relate to individual rights and human dignity. Values frequently emerge as a key issue in the life sciences and in genetic engineering, which have advanced rapidly in recent years, but can also be prominent in fields related to sustainability.

For example, there is now a problem concerning population growth. This is an issue closely related to sustainability, but population tends to be treated separately from sustainability, partly because of the religious and other values-related issues involved. From the perspective of those who oppose the policies and soft technologies of family planning, these policies and technologies conflict with basic human rights and religious values.

Societal assessments of technology have also come to involve the issue of different images of society. With the growth of nanotechnology in recent years, interest has risen in fields that integrate areas such as nanotechnology, biotechnology and information technology – that is, in converging technologies. In response to this, research is progressing in the United States and Europe on the implications of those technologies for society. Research into these implications entails determining both the likely benefits for society and any concerns (for example, issues relating to the management of data collected using biosensors that employ nanotechnology, as well as privacy issues). It could be said that there are different kinds of technology assessment and that, in the course of this process, attempts have been made to differentiate between the respective aims of converging technologies in the United States and Europe. In the United States, the notion of converging technologies for "improving human performance" or human enhancement is asserted (Roco and Bainbridge, 2002: 1), whereas in Europe the concept of converging technologies for a "knowledge society" is stressed (Nordmann, 2004: 9). In other words, it could be said that in the United States this technology is being positioned as a means to improve such facets as human military capability and memory capacity, whereas in Europe the intention is to apply it for purposes that are more oriented towards societal cohesion.

Furthermore, the social vision in which each technology is framed (or the public policy purpose to which each technology is supposed to contribute) may have an impact on the range of stakeholders involved and the process of introducing the technology.

4-2-5 Promoting the generation of knowledge

The issues that have been considered so far – how society will make use of technologies, and what concerns society must take into account in its assessment of them – are premised on the pre-existence of technology and knowledge. However, the existence of knowledge and technology is not self-evident. For these to emerge, society must foster those groups of people identified as scientists and technologists and must stimulate their research activities. What kind of knowledge generation, then, deserves to be stimulated?

In this context it is necessary to revisit the role of the legal concepts of "academic freedom" and "freedom of research" (Yamamoto, 2007). These concepts have often been treated as justification for the concepts "science for the sake of science" and "research for the sake of research". However, they could instead be re-tasked as organizing principles for stimulating the generation of knowledge. In other words, simply carrying out research work under the direction of superiors in a hierarchical organization is insufficient for the generation of intellectual innovation. Certainly, implementation is a necessary component of research, and mechanisms to support this are essential. However, ideas – the essential components of research – are born of spontaneous investigative activity.

According to this way of thinking, academic freedom and freedom of research can, by enabling numerous trials and experiments in a bottom-up fashion, serve the function of stimulating intellectual innovation, which contributes to society. The construction of a voluntary network spanning several disciplines is vital to this process. In addition, the significance of ensuring diversity in scholarship and research is that this can lead to just such intellectual innovation. Free and autonomous forms of organization made up of the parties involved (such as researchers) that encourage spontaneous trials and communication are necessary to stimulate the generation of knowledge, and these are different from hierarchical organizations.

In the innovation and diffusion of new heat pumps for household use in Japan, for example, the networking of marginal actors in established organizations such as the Tokyo Electric Power Company (TEPCO) and the Central Research Institute of Electric Power Industry (CRIEPI) was very important. Institutional environments that facilitate horizontal communication and networking among marginal actors and relevant outsiders are essential for innovation (Juraku and Suzuki, 2008).

In fact, promoting the generation of knowledge is essential even for risk assessment. A system of laws on experimentation that will permit various types of experiments is necessary to stimulate production of the information needed for risk assessment. If such a system of laws is absent and experiments cannot be carried out, there is no alternative but to rely on the importation of the knowledge and information needed for performing risk assessments. It has been pointed out that, because safety regulations in Japan are often stringent, even the data needed to apply for approval and authorization under these safety regulations cannot be generated in Japan, and foreign experimental data are used instead. This situation does nothing to encourage accumulation of the information and knowledge on which risk assessments are based.

On the other hand, it cannot be said that academic freedom and freedom of research are tenets that command universal respect. For example, it is necessary to compare the risks involved in the areas of safety and security. This can involve deciding whether risks to safety ought to be given priority and academic or research freedom curtailed, or whether a shortsighted emphasis on safety and restrictions on research reduces the possibility of long-term innovation and increases society's vulnerability. Step-by-step clinical trials and the medical technology for medical and pharmaceutical product trials are perfect examples of this choice. If Japan is to be independent in areas such as nuclear power technology, a legal system that enables experimentation is a requisite in the quest for independent technological innovation.

A further issue is whether security risks should be emphasized and the publication of research (a key component of academic and research freedom) halted when there are fears that research results might be utilized by terrorists.

Institutional mechanisms for stimulating the generation of knowledge involve a number of other issues as well. Another bone of contention is whether intellectual property rights ought to be used to boost incentives for researchers. On the one hand, if intellectual property rights act as an economic incentive to spur researchers on to research success, the use of this mechanism will promote intellectual property rights. On the other hand, the use of intellectual property rights in this fashion will not work with people whose motivation to generate knowledge is not economic but rather the satisfaction of intellectual curiosity or the acclaim of their peers. In addition, there is the concern that it will be difficult to assemble knowledge composed of a variety of elements if intellectual property rights are established separately for each component element. The basis of the traditional research community used to be the active use of the academic commons. Within research communities there has always been an ethical emphasis on giving credit for an invention where it is due; however, the method that has come to be adopted involves sharing research results with the research community as soon as possible and allowing them to be used free, so as to stimulate the creation of further research results – and not to go to the lengths of obtaining intellectual property rights or keeping results secret. Whether to maintain the traditions of the academic commons or to make more use of intellectual property rights is a choice that will be pivotal in the generation of knowledge.

Other key issues include how to design structures for the provision of research funding and how to plan the evaluation of research results. To make effective use of academic freedom and freedom of research, it is not enough simply to preserve the autonomy of organizations; rather, it is essential to allocate human resources and financial resources that will enable such activities. If resource allocation is carried out by the government, it is inevitable that there will be a certain level of evaluation so as to maintain accountability. However, if only short-term evaluation of individual projects is carried out, the goals of preserving diversity and maintaining a foundation for wide-ranging intellectual innovation will not be achieved.

4-2-6 The art of *doushouimu* and the inevitability of value judgements in sustainability

So far, this section has outlined the substance and functions of technology governance. Now it will analyse how technology governance tools can be used in overall governance for sustainability and discuss the nature of such governance.

It is noteworthy that different actors within society hold different viewpoints. Hence it is important to understand the various frameworks within which perceptions of major issues are framed. There then has to be a platform on which these multiple viewpoints are shared and interests are coordinated. Problems of technology must not be left exclusively to experts in the science and technology fields in question, but must be opened up to other interested parties as well. In this process, dialogue between experts and citizens is important, but it is also crucial that there be dialogue among experts in different areas, and that a language be devised to enable them to understand one another. There is a need for stakeholder analysis as a means to this end and for leaders who will link together experts from various fields.

It is not necessary for all the actors involved in the decision-making aspect of governance to share a common vision. The notion of "sharing the same bed, dreaming different dreams" (*doushouimu* in Japanese) is an important one (Shiroyama, 2008). As has been emphasized, the actors in society have different viewpoints and concerns, and it is rare for the visions of these various actors to be in accord. For example, some actors may be interested in nuclear power or biomass energy technology as measures to combat global warming; others may be interested in these technologies as a means to achieve energy security. In such instances, however, even though the perspectives that inform the concerns of these actors differ, they will be able to form a united front in support of a particular technology choice. Conversely, clarifying the various benefits and risks for different actors through stakeholder analysis will not only provide the data required for decision-making but also enhance the potential for coalition formation among actors in keeping with this notion of "same bed, different dreams".

Understanding sustainable development as the co-evolution of different subsystems (Kemp et al., 2007: 78–79) shares the notion of "same bed, different dreams". Each subsystem has its own mission but those missions, even though some adaptations are needed, can coexist under the system-wide change for sustainable development.

This understanding follows the classic argument. As the 1987 report by the Brundtland Commission, *Our Common Future* (WCED, 1987), shows, sustainability requires the achievement of policy objectives in many dimensions, among them population and human resources, food security, species and ecosystems, energy, industry, and the urban challenge. Sometimes the achievement of one objective may have a negative impact on another objective, as when increasing food production for food security threatens to damage the ecosystem. Thus, finding a common framework for the simultaneous achievement of various policy objectives is necessary for sustainable development, and it is in such circumstances that the notion of "same bed, different dreams" has a place. Some "dreams", however, need to be adjusted because they may turn out to be a "nightmare" for others. In such cases a value judgement must be made to decide which value or dream should be discarded.

4-2-7 Conclusion

Technology governance requires an assessment of the social implications of technologies as a precondition for utilizing those technologies for sustainability. These social implications include multifaceted risks as well as benefits, some of which are inevitably uncertain. Because technology is an indispensable component of the problems and solutions relating to sustainability, technology governance is an important part of sustainability governance.

This section first outlined how technology governance might be organized and identified the functions that are required of it. These functions include risk management (which sometimes involves judgements on risk trade-offs), the assessment of values and the promotion of the production of knowledge. It then analysed the nature of governance for sustainability in which technology governance tools are employed. The notion of "sharing the same bed, dreaming different dreams" (*doushouimu*) is important here. Although the perspectives that inform the actors' concerns differ, the actors can form a united front in support of a particular technology choice precisely because the technology may have different social implications for different actors.

As to the nature of governance, technology governance for sustainability requires judgements about risk trade-offs and values, but it also facilitates coalition formation among actors under the conditions of *doushouimu*. In that sense, technology governance is political. This political aspect of technology governance needs to be transparent and should be recognized in the policy-making process involving technologies for sustainability.

REFERENCES

Graham, J. D. and J. B. Weiner (1995) *Risk vs. Risk: Tradeoffs in Protecting Health and the Environment*. Cambridge, MA: Harvard University Press.

Juraku, K. and T. Suzuki (2008) "R&D, Introduction and Diffusion Process of Energy Conservation Technologies and Implications for Public Policy: A Case Study of Residential High Efficiency Water Heaters", *Journal of Sociotechnology* 5 (in Japanese).

Kemp, R., D. Loorback and J. Rotmans (2007) "Transition Management as a Model for Managing Processes of Co-evolution towards Sustainable Development", *International Journal of Sustainable Development & World Ecology* 14: 78–91.

Nordmann, A., ed. (2004) *Converging Technologies – Shaping the Future of European Societies*. Brussels: European Commission.

Pierre, J. and B. G. Peters (2000) *Governance, Politics and the State*. New York: St Martin's Press.

Rhodes, R. A. W. (1997) *Understanding Governance: Policy Networks, Governance, Reflexivity and Accountability*. Buckingham: Open University Press.

Roco, M. C. and W. S. Bainbridge, eds (2002) *Converging Technologies for Improving Human Performance: Nanotechnology, Biotechnology, Information Technology and Cognitive Science*. NSF/DOC-sponsored report. Arlington, VA: National Science Foundation.

Shiroyama, H. (2007a) "Risk Assessment/Management and Legal Systems", in H. Shiroyama and Y. Nishikawa (eds), *Remaking the Law III: Law and the Development of Science and Technology*. Tokyo: University of Tokyo Press, pp. 89–114 (in Japanese).

Shiroyama, H. (2007b) *Governance of Science and Technology*. Tokyo: Toshindo (in Japanese).

Shiroyama, H. (2008) "Ensuring Sustainability and 'Doushouimu'", *TIGS News* 2: 2–3 (in Japanese).

WCED [World Commission on Environment and Development] (1987) *Our Common Future*. Oxford: Oxford University Press.

Yamamoto, T. (2007) "Scholarship and the Law", in H. Shiroyama and Y. Nishikawa (eds), *Remaking the Law III: Law and the Development of Science and Technology*. Tokyo: University of Tokyo Press (in Japanese).

4-3

Policy instruments

Mitsutsugu Hamamoto

4-3-1 Introduction

The concept of sustainability may lead to the need for constraints or targets (for example, safe minimum standards) to make the use of environmental resources more sustainable. However, setting such targets does not per se guarantee the attainment of sustainability. What is needed is change in the behaviour of economic actors so that targets consistent with the concept of sustainability are met. In order to alter the actors' behaviour, incentive mechanisms that lead them to modify their decisions on environmental resource use must be established within the framework of economic and social institutions. Such mechanisms may not be provided without policy intervention. Environmental policies – public policies for environmental quality improvement or sustainable use of environmental resources – must be employed to provide economic actors with incentives to make decisions with consideration for the environmental impacts their actions may have. Environmental economics has a major role to play in the design of such policies, and the choice of environmental policy instruments is crucial in designing environmental policies.

Comparative analysis of market-based and command-and-control instruments is one of the main issues addressed in the literature on environmental economics. Although most researchers have compared a single instrument with another one and assert the advantage of market-based instruments, there are many experiences of environmental policies combining several instruments. The Organisation for Economic Co-operation and Development (OECD, 2006) states that environmentally related taxes are often used in combination with one or more other instruments such as direct regulations, subsidies, labelling systems and voluntary agreements. In addition to the need for the combination of policy instruments, there is a growing recognition that environmental policies should be integrated with other relevant policies in order to make the nature of economic and other social activities consistent with sustainability. The environmental tax reform that has been carried out in several European countries is an actual example of such policy integration. There is also discussion of a rationale for integrating environmental policies with technology policies.

This section provides some key findings in the literature on the efficacy of environmental policy instruments and discusses issues concerning the effectiveness of instrument mixes and policy integration. Concern has been growing recently about the integration of environmental and other relevant policies, yet literature on this issue remains scanty. This section attempts to explore the rationales and effectiveness of combinations of environmental and other policies for enhancing economy-wide efficiency and for promoting environmental technology innovation.

4-3-2 The choice of environmental policy instruments

A considerable literature exists on comparisons of the costs of achieving specific environmental targets under different policy instruments. In order to minimize the cost of reducing pollution by a given targeted amount, it is necessary to meet the condition that marginal abatement costs for emissions reduction are equalized across all economic actors. In theory, this condition is satisfied when all agents face the same price for their contributions to emissions.

Environmental economists have investigated the cost-effectiveness of various environmental policy instruments. The instruments are classified into two broad categories: market-based instruments and command-and-control instruments. The former include emissions taxes, tradable emissions allowances and subsidies for emissions reduction; the latter include technology-based standards and performance standards. Market-based instruments are regulatory methods of establishing a common price on emissions. Compared with market-based instruments, command-and-control instruments are at a disadvantage in meeting the condition for cost-minimization. The cause of the disadvantage resides in limitations on regulators' ability to establish technology-based or performance standards that minimize pollution abatement costs: there are difficulties in collecting the information that is necessary to set such standards, and calculating the levels of standards to equalize marginal costs of abatement across all economic actors imposes enormous administrative costs. Thus, much of the literature supports the assertion that market-based instruments can achieve environmental targets more efficiently than command-and-control instruments.

4-3-3 Instrument mixes

The literature on instrument choice has tended to focus on the use of one instrument or comparisons of two or more instruments. In reality, however, there are many cases in which multiple instruments are adopted to control the discharge of a certain pollutant. One example is the combination of emissions taxes and subsidies for emissions reduction. The revenues from environmentally related taxes introduced in several OECD countries are earmarked for environmental purposes. The Dutch system of water management, which includes wastewater levies, uses revenues from the levies for pollution control (Andersen, 1999). In Sweden, a charge on nitrogen oxide (NOx) emissions from energy generation at combustion plants was introduced in 1992. Revenues raised from the charge are refunded to the plants in proportion to their production of useful energy. This charge system can provide plants with an incentive to reduce NOx emissions per unit of energy (OECD, 2007). In addition, there is a case of an emissions trading system working with command-and-control instruments. In the United States, the Sulfur Dioxide Allowance Trading Program was introduced to combat acid rain in a cost-efficient manner in the 1990s, and sulphur oxide (SOx) pollution has been controlled by direct regulations designed to achieve ambient standards for protecting local air quality since the 1970s (Tietenberg, 1995).

Several countries have adopted tax policies that are combined with voluntary agreements to reduce carbon dioxide (CO_2) emissions. The CO_2 tax policy in Denmark has a scheme in which firms entering into an agreement on energy efficiency improvement with the Danish Energy Agency receive rebates on their CO_2 tax payments. In 2001, the Climate Change Levy was introduced in the United Kingdom. Under this levy system, energy-intensive sectors can obtain an 80 per cent reduction in the tax rate if they enter into Climate Change Agreements on targets for carbon emissions reductions or energy efficiency improvements (OECD, 2006).

Introducing carbon taxes is expected to have negative impacts on energy-intensive industries. If governments seek to design climate policies using carbon taxes, they may have to consider the political feasibility or distributional effects, as well as the cost-effectiveness, of such policies. Carbon taxes combined with voluntary agreements, as in the above cases, make it possible to mitigate the distributional effects on firms and the adverse impacts on their international competitiveness caused by the introduction of taxes intended to reduce carbon emissions. However, the cost-efficient attainment of emissions targets is unlikely to be realized under these policy schemes because the marginal abatement costs of regulated entities cannot be equalized overall. Thus, when regulators use instrument mixes such as emissions taxes combined with voluntary agreements, they cannot avoid a trade-off between cost-effectiveness in achieving environmental targets (that is, efficiency) and the mitigation of adverse distributional impacts (that is, equity).

Environmental economists have discussed the efficiency of market-based instruments under uncertainty. Economic analysis of climate change suggests that the marginal cost curve for reducing greenhouse gas (GHG) emissions is very steep, whereas the marginal benefit curve for reducing emissions is fairly flat because the damage caused by climate change results from the overall stock of GHGs in the atmosphere, which is the accumulation of emissions over many years. In this situation, welfare losses are smaller under a tax policy than under a tradable permit system when the marginal costs are uncertain (Weitzman, 1974). Therefore, economic theory suggests that price approaches such as taxes would be more efficient than quantity approaches such as emissions trading systems for controlling GHG emissions. However, a tax policy makes firms bear not only abatement costs but also tax payments for residual emissions, which would bring about large transfers of income from firms to the government. This is the main reason that firms oppose emissions taxes. Thus the introduction of a tax for reducing carbon emissions is likely to face political difficulties (McKibbin and Wilcoxen, 2002).

Inefficiency under uncertainty and large transfers of income are the serious disadvantages of tradable permits and taxes, respectively, when these instruments are used alone to reduce GHG emissions. In order to overcome these drawbacks, a hybrid approach is proposed (McKibbin and Wilcoxen, 2002; Pizer, 2002): a tradable permit system (or a cap-and-trade system) combined with a tax that functions as a cap on permit prices. The aim of this approach is to prevent permit prices from soaring upward when abatement costs turn out to be much higher than expected. If the market prices of permits reach a specified level (known as a trigger price), firms can purchase additional permits at that price from the government. This system is called a "safety valve".

The Regional Greenhouse Gas Initiative (RGGI), an agreement on climate change policy measures among the north-eastern states in the United States, includes the establishment of a cap-and-trade system with a safety valve that is different from the above-mentioned one: if permit prices reach a trigger level for a sustained period, the compliance period will be extended. Furthermore, if this fails to be effective in reducing permit prices, regulated firms will be allowed to use offsets (that is, conducting GHG reduction projects in and outside the RGGI member states) for up to 20 per cent of their emissions.

Under a cap-and-trade system with a safety valve, firms can avoid unexpectedly high compliance costs, but a given emissions target is in effect abandoned when a trigger price comes into effect. Thus, regulators adopting instrument mixes such as tradable permits combined with emissions taxes may be faced with a trade-off between achieving environmental targets and mitigating the economic burden of compliance.

4-3-4 Policy integration for economy-wide efficiency

Emissions taxes and cap-and-trade systems with auctioned permits are policy instruments that raise revenues. These revenues can be used not only for environmental purposes but also for other policy objectives. Several European countries have introduced carbon taxes and used the revenues to reduce taxes on labour. During the 1990s, Finland, Sweden, Denmark and the Netherlands introduced carbon taxes accompanied by revenue recycling to reduce taxes on labour. Later, such environmental tax reforms were carried out in Germany and the United Kingdom. These cases can be viewed as an integration of environmental policies with tax policies.[1] The aim of such policy integration is to improve economy-wide efficiency as well as the quality of the environment by using environmental tax revenues to cut pre-existing distortionary taxes.

Recently, numerous studies have emerged on the relationship between environmental policy for reducing CO_2 emissions and the tax system. Some environmental economists suggest that carbon taxes can contribute not only to mitigating global warming but also to improving the efficiency of the tax system if carbon tax revenues are used to lower existing distortionary taxes such as capital and labour income taxes. This win–win property of carbon taxes is called the "strong double dividend". Although economists generally agree that revenue-recycling per se reduces the net cost of a carbon tax (a weak version of the double dividend), the strong double dividend hypothesis, which asserts that introducing a carbon tax accompanied by revenue-recycling could generate net economic gains for society in addition to the benefits from CO_2 emissions reductions, has elicited discussions on its validity.

Theoretical studies demonstrate that the strong version of the double dividend hypothesis fails to be supported. A carbon tax itself is distortionary because it increases the cost to firms of producing output, which will lead to slight reductions in the overall level of investment and employment. This is called the "tax interaction effect". If the carbon tax revenues are used to reduce other distortionary taxes, welfare gains will be generated: this is the revenue-recycling effect. Most studies using analytical and computable models show that the tax interaction effect outweighs the revenue-recycling effect (Bovenberg and de Mooij, 1994; Goulder et al., 1998; Parry, 1995). Thus, in theory, the net impact of a carbon tax is to reduce the level of investment and employment in the economy. This result does not absolutely deny the merit of the integration of environmental policy and tax system reform. Non-revenue-raising instruments such as emissions standards and freely allocated tradable permits cannot have the revenue-recycling effect. It could be concluded that, although the strong double dividend cannot be supported in general, the economic burden that carbon abatement policies impose on society can be mitigated if revenue-raising instruments such as carbon taxes or auctioned permits are adopted and the revenues are used to lower other distortionary taxes.[2]

4-3-5 Policy integration for environmental technology innovation

Environmental policy instruments and incentives to innovate

In the long run, it is critical that environmental policy instruments encourage technological innovation for protecting the environment. Environmental economists have investigated which policy instruments can provide stronger incentives to innovate. Their traditional approach to analysing the incentives for innovation created by policy instruments is to make comparisons of the cost savings from developing or adopting new technologies under various instruments. Downing and White (1986), using a simple model of a profit-maximizing polluter, show that market-based instruments are more attractive as incentives to innovate than are command-and-control instruments.

Milliman and Prince (1989) conduct a more comprehensive analysis of the relationship between environmental policy instruments and technological change. They use a model of identical firms in a competitive industry that has three stages of technological change: innovation, diffusion and optimal agency response. Their analysis shows that taxes and auctioned permits provide the greatest incentive to promote technological change and emissions standards provide the least. Montero (2002) compares the incentives for research and development (R&D) that are offered by market-based instruments (auctioned and freely allocated tradable permits) and command-and-control instruments under the conditions of oligopolistic output and permit markets. He demonstrates the possibility that command-and-control instruments provide greater R&D incentives than market-based instruments if R&D investment by a firm affects permit prices and its rival's costs through the permit and output markets.

Fischer et al. (2003) compare the incentives for innovation under emissions taxes, auctioned permits and freely allocated tradable permits using a model of identical firms in a competitive market that includes the possibility that a new technology an innovator has developed is imitated by other firms. The possibility of such imitation limits the ability of an innovator to fully appropriate the rents from its own innovation. Fischer et al. show that emissions taxes provide stronger (weaker) incentives to innovate than auctioned permits if the extent to which innovation can be imitated is less (greater), and that incentives for innovation under freely allocated tradable permits are weaker than incentives under both emissions taxes and auctioned permits, regardless of the degree of imitation. They also compare the welfare effects of the three instruments, finding that emissions taxes induce a higher amount of innovation and welfare compared with auctioned permits when the degree of imitation is not significant. In addition, a quantitative analysis conducted by Fischer et al. (2003) indicates that the differences between welfare levels under the three instruments and the first-best outcome grow larger as the imitation effect becomes more significant. This suggests that, in terms of knowledge spillovers, additional policy measures besides environmental policy instruments may be needed in order to induce socially optimal levels of innovation and welfare.

Market failures associated with technological innovation and diffusion

Knowledge production activities such as R&D are characterized by uncertainty and externalities. These characteristics are the generic sources of market failures. Uncertainty associated with R&D activity makes it difficult to finance R&D investment through capital market mechanisms. This leads to underinvestment in R&D. Externalities arise because knowledge has the basic character of a public good: the use of knowledge by an economic actor does not preclude other actors' use of the same knowledge. Therefore, new knowledge created by an innovator can bring spill-over benefits to other parties without compensating the innovator. This impairs the ability of innovators to realize a reasonable rate of return on their R&D activities. This is an appropriability problem that reduces innovators' willingness to invest in R&D.

Owing to uncertainty and externalities, market incentive mechanisms are thus likely to fail to produce the socially optimal rate and direction of innovative activity (Geroski, 1995). On the other hand, there are theoretical studies emphasizing the possibility that private rivalry for the rewards of a patent may lead to excessive investment in R&D. However, empirical studies show that the private rate of return on R&D is well below the social rate of return (Griliches, 1992), suggesting a tendency towards underinvestment in R&D. The discussion about market failures in knowledge production means that incentives to develop new pollution control technologies will be inefficiently low even if environmental policy instruments are appropriately introduced so as to internalize environmental externalities.

Sources of market failure can be found in the adoption and diffusion of new technologies. Imperfect information is one such source. Information may generally be underprovided through markets because it has the attributes of a public good. In addition, it may be costly for economic actors to learn of the existence of newly developed technologies and to learn how to use them or determine whether using them is profitable.

Another possible source of market failure is the potential for learning effects. When an actor decides to adopt a new technology and learns about it, the processes of adoption and learning generate information about the existence and benefits of the new technology. Other actors can use this information in considering whether to introduce the new technology. Thus, early adopters of a new technology can create a positive externality for later adopters. This effect is called "learning-by-using". Additionally, as firms produce more and more of a product with a new technology, their experiences in production will lead to improvements in several aspects: firms can produce more efficiently, or the quality of their products becomes higher. If these improvements benefit other firms without compensation, there can be an additional adoption externality through production experience, which is called "learning-by-doing".

Market failure associated with R&D activities can be addressed by technology policy instruments such as subsidizing private R&D activities, cooperative R&D ventures, government research projects and the design of patent rules. Policy instruments for addressing market failure related to technology diffusion include information dissemination, technology demonstration and subsidizing technology adoption through tax credits.

The rationale for combining environmental and technology policies

There are several reasons environmental technology – a specific area of technological change – should be supported through government intervention. Since the environment per se has public good attributes, environmental technology (like defence) is a suitable area for government efforts to promote innovation. Moreover, limits or prices on emissions are not enough to appropriately internalize environmental externalities, and these environmental regulations therefore provide weak incentives to develop new environmental technologies. Under such conditions, adopting a technology policy may be justified on the grounds that environmental technology innovation should be promoted (Jaffe et al., 2005).

The reasons emissions standards or emissions taxes tend to be set at levels that are far from socially optimal are primarily political. Especially when regulators attempt to introduce or tighten environmental regulations under conditions where pollution control technologies are immature, actors that must comply with the regulations will try to weaken (or postpone) them by exerting political influence. In such circumstances, policy intervention to promote innovation in environmental technology may have a role in softening opposition from these actors and make it possible for regulators to set emissions standards or emissions taxes at more desirable levels.

In the context of the climate change problem, the development of new technologies that can contribute to stabilizing atmospheric GHG concentrations in an economically feasible manner is essential to resolving the problem in the long run. When pollution control technologies are underdeveloped and the cost of abatement is substantial, establishing strict environmental regulations will be politically difficult. In this situation, governments have to set or strengthen environmental targets (for example, emissions caps) according to the availability of pollution control technologies that can reduce emissions at a reasonable cost. If new technologies that can lower abatement costs are developed and diffused, they will contribute not only to attaining certain levels of total emissions at lower cost, but also to making it possible to set environmental targets at more desirable levels. In other words, such new technologies may bring additional environmental benefits if proper policy responses are implemented. Thus, the innovation and diffusion of environmental technology will produce social returns in the form of abatement cost reductions and environmental benefits. Such social returns on environmental technology innovation will exceed private returns.[3] Therefore, the integration of environmental and technology policies is needed when pollution control technologies are immature.

There has been much debate concerning the regulatory choice between price and quantity approaches in designing climate policy. A recent movement towards the widespread acceptance and application of cap-and-trade programmes for controlling GHG emissions has been observed. Nordhaus (2007) criticizes quantitative approaches to slowing global warming, such as the Kyoto Protocol, for their disadvantages (for example, inefficiency under massive uncertainties, high volatility in the market price of carbon, and the generation of artificial scarcities to encourage rent-seeking behaviour) and recommends price approaches such as carbon taxes that are steadily raised over time so as to reflect the increasing prospective damage from climate change. Such price approaches may, however, encounter difficulties in increasing carbon taxes when abatement technologies that can reduce GHG emissions at reasonable cost are not available. Hence the integration of environmental policy using price approaches with technology policy would contribute to enhancing the political feasibility of raising carbon taxes.

4-3-6 Conclusion

Environmental policy instruments have a role in inducing economic actors to use environmental resources in a manner that is consistent with the concept of sustainability. Many studies have supported the advantage of market-based instruments over command-and-control regulations in terms of cost-effectiveness, and have recommended the use of a single market-based instrument such as an emissions tax or a tradable permits system. In the real world, however, there have been many cases of instrument mixes. The reasons that two (or more) instruments are mixed include equity considerations: the mitigation of distributional effects on firms affected by environmental policy or of adverse impacts on their international competitiveness. This suggests that, in addition to cost-effectiveness, policymakers have objectives such as equity or political feasibility in designing environmental policy, and that consequently they tend to be faced with trade-offs between efficiency (or achieving environmental targets) and equity.

Several market-based instruments can expand the range of possible designs for public policies to enhance the environment. When revenue-raising instruments such as carbon taxes or auctioned permits are adopted, the revenues can be used for lowering other distortionary taxes in order to mitigate the economic burden that carbon abatement policies impose on society. Another possible usage of the revenues is to finance subsidy schemes for the development and diffusion of environmental technologies.

Integrating environmental and technology policies can be theoretically supported by the existence of both environmental externalities and market failures associated with technological innovation and diffusion. In circumstances where environmental technologies are immature, such policy integration may be able to partly resolve political difficulties in introducing or toughening environmental policies.

Achieving a transition from carbon-intensive to low-carbon technologies is essential for a long-term solution to the problem of climate change. However, there are serious barriers to changing the development path of existing technologies because the current technological systems and social institutions are faced with the phenomenon of "carbon lock-in" (Unruh, 2000). In other words, technological and social systems based on fossil fuels that cause global warming and other environmental externalities have co-evolved so as to reinforce one another's advantages over other systems that consist of appropriate institutions to internalize environmental externalities and technologies that do not exploit environmental resources. Although existing institutions that have inadequately set limits or prices on the use of environmental resources should be corrected by introducing appropriately designed environmental policies, such policies also need to be integrally formulated with technology policies for fostering the development and deployment of low-carbon technologies in order to escape from carbon lock-in. However, changing the development paths of technological systems and social institutions may unavoidably impose a considerable burden on society, at least in the short run. Policy choices associated with the climate change problem determine the extent and distribution of that burden. What must be explored is the design of environmental and technology policy integration that makes it possible to escape from carbon lock-in while minimizing the burden in an equitable manner.

Notes

1. In the European Union (EU), concern has been growing about the integration of environmental and other relevant policies ("environmental policy integration"). The EU's Environment Action Programmes argue that environmental policy integration is needed in fields such as agriculture, energy, industry, transport and tourism. The aim of such policy integration is to change the nature of economic and other social activities in order that they may be consistent with the concept of sustainability. However, it is pointed out that the EU and its member states are faced with bottlenecks that obstruct progress towards achieving environmental policy integration (Lenschow, 2002).
2. A research project coordinated by the National Environmental Research Institute, Aarhus University, has analysed the effect of environmental tax reforms on economic growth and CO_2 emissions in European countries, finding that the tax reforms have contributed to both economic growth and reductions in CO_2 emissions (Andersen, 2007).

3. Japan's public policies for controlling SOx and NOx emissions during the 1960s and 1970s are characterized as the integration of command-and-control instruments and subsidies for promoting the development and diffusion of pollution control technologies. For details on this, see Hamamoto (2008).

REFERENCES

Andersen, M. S. (1999) "Governance by Green Taxes: Implementing Clean Water Policies in Europe 1970–1990", *Environmental Economics and Policy Studies* 2: 39–63.

Andersen, M. S. (2007) "Carbon-Energy Taxation Contributed to Economic Growth", National Environmental Research Institute, Aarhus University. Available from <http://www.dmu.dk/International/News/Archive/2007/CO2tax.htm> (accessed 31 May 2010).

Bovenberg, A. L. and R. A. de Mooij (1994) "Environmental Levies and Distortionary Taxation", *American Economic Review* 84(4): 1085–1089.

Downing, P. B. and L. J. White (1986) "Innovation in Pollution Control", *Journal of Environmental Economics and Management* 13: 18–29.

Fischer, C., I. W. H. Parry and W. A. Pizer (2003) "Instrument Choice for Environmental Protection When Technological Innovation Is Endogenous", *Journal of Environmental Economics and Management* 45: 523–545.

Geroski, P. (1995) "Markets for Technology: Knowledge, Innovation and Appropriability", in P. Stoneman (ed.), *Handbook of the Economics of Innovation and Technological Change*. Oxford: Blackwell, pp. 90–131.

Goulder, L. H., I. W. H. Parry, R. C. Williams III and D. Burtraw (1998) "The Cost-Effectiveness of Alternative Instruments for Environmental Protection in a Second-Best Setting", Resources for the Future Discussion Paper 98-22.

Griliches, Z. (1992) "The Search for R&D Spillovers", *Scandinavian Journal of Economics* 94 (Supplement): S29–S47.

Hamamoto, M. (2008) "Climate Change and Technology Policy: Double Market Failure and the Need for Policy Integration", in P. G. Caldwell and E. V. Taylor (eds), *New Research on Energy Economics*. New York: Nova Science Publishers, pp. 135–148.

Jaffe, A. B., R. G. Newell and R. N. Stavins (2005) "A Tale of Two Market Failures: Technology and Environmental Policy", *Ecological Economics* 54: 164–174.

Lenschow, A. (2002) "Greening the European Union: An Introduction", in A. Lenschow (ed.), *Environmental Policy Integration: Greening Sectoral Policies in Europe*. London: Earthscan, pp. 3–21.

McKibbin, W. J. and P. J. Wilcoxen (2002) *Climate Change Policy after Kyoto: Blueprint for a Realistic Approach*. Washington, DC: Brookings Institution Press.

Milliman, S. R. and R. Prince (1989) "Firm Incentives to Promote Technological Change in Pollution Control", *Journal of Environmental Economics and Management* 17: 247–265.

Montero, J.-P. (2002) "Permits, Standards, and Technology Innovation", *Journal of Environmental Economics and Management* 44: 23–44.

Nordhaus, W. D. (2007) "To Tax or Not to Tax: Alternative Approaches to Slowing Global Warming", *Review of Environmental Economics and Policy* 1(1): 26–44.
OECD [Organisation for Economic Co-operation and Development] (2006) *The Political Economy of Environmentally Related Taxes*. Paris: OECD.
OECD [Organisation for Economic Co-operation and Development] (2007) *Instrument Mixes for Environmental Policy*. Paris: OECD.
Parry, I. W. H. (1995) "Pollution Taxes and Revenue Recycling", *Journal of Environmental Economics and Management* 29: S64–S77.
Pizer, W. A. (2002) "Combining Price and Quantity Controls to Mitigate Global Climate Change", *Journal of Public Economics* 85: 409–434.
Tietenberg, T. (1995) "Tradeable Permits for Pollution Control When Emission Location Matters: What Have We Learned?", *Environmental and Resource Economics* 5: 95–113.
Unruh, G. C. (2000) "Understanding Carbon Lock-in", *Energy Policy* 28: 817–830.
Weitzman, M. L. (1974) "Prices vs. Quantities", *Review of Economic Studies* 41: 477–491.

4-4

Consensus-building processes

Masahiro Matsuura

4-4-1 Introduction to consensus-building processes

Definition of consensus-building

The term "consensus-building" has been used to describe a wide range of activities that seek agreement by multiple stakeholders and the general public. The ambiguity of its definition, however, has made this term popular among scholars and policymakers who are interested in decision-making processes in the public arena. In fact, even congressional lobbying is sometimes considered part of a consensus-building effort.

Negotiation theory, however, defines the term more precisely. According to this theory, each negotiating party makes a comparison between a proposed agreement and its BATNA (best alternative to negotiated agreement) (Fisher and Ury, 1991). If an offer is likely to provide better conditions than its BATNA, a rational decision-maker will accept the offer after adequate efforts to improve his or her gains from the negotiation (that is, haggling). Even if such an agreement does not achieve the original aspirations of all negotiating parties, each of them must be able to "live with" such an agreement (Susskind and Cruikshank, 2006).

Therefore, consensus-building can be defined as a range of activities that seek agreements that all stakeholding parties can "live with". Consensus-building is an indispensable component of sustainability science because it seeks self-enforcing mechanisms for a wide variety of stakeholders to coexist in the long term. This definition is particularly instructive in making a distinction between consensus-building efforts and social movements. The main purpose of a social movement is to advance a group's positions. Activists work to realize their ideals by mobilizing resources rather than by compromising with other parties. They pursue what they "live for" instead of what they can "live with".

Critics of consensus-building argue that such processes end in a number of compromises that are not satisfactory to all contending parties (Amy, 1990; Innes, 2004). This could in fact be true if the process is not managed properly. Advocates for consensus-building, however, argue for mutual gains through negotiation. According to Pareto's theory, additional value can be created through negotiation by trading issues that each party values differently. If such trade-offs are arranged in an optimal way, the condition of Pareto-optimality can be satisfied. In other words, advocates regard consensus-building as a kind of negotiation for finding mutual gain arrangements through exchange, rather than as a series of frustrating compromises.

In practice, consensus-building refers to a certain kind of decision-making process. It is often described by a five-step model, whose details are provided below (subsection 4-4-2).

Brief history

Although consensus-building has been practised in many parts of the world, the use of consensus-building processes (in the narrow sense) is primarily located in the United States, where the practice initially grew out of an American tradition of dispute resolution (Dukes, 1996). Techniques for dispute resolution were first applied to a long-standing dispute over the construction of a new dam on the Snoqualmie River in Washington State in the mid-1970s (Cormick, 1976; Mernitz, 1980). The dispute was successfully mediated in less than a year. Other applications of dispute resolution techniques to social issues emerged in various locations in the United States, such as in community dispute resolution centres (Susskind and McKearnan, 1999). These initial efforts slowly developed into the field of environmental dispute resolution. In 1986, Bingham published an analysis of 100 such cases.

The field grew substantially in the 1970s and 1980s. One landmark achievement was the enactment of the Negotiated Rulemaking Act in 1990. In order to forestall litigation against proposed regulations, particularly environmental ones, federal agencies adopted consensus-building processes in their rulemaking. Under the Act, agencies can invite stakeholding parties, such as industry and environmental representatives, to deliberations convened with the aim of preparing a draft regulation (Susskind and van Dam, 1996). The US Environmental Protection Agency has been particularly active in promoting its use. Another major institutional achievement was the creation of the US Institute for Environmental Conflict Resolution in 1997. The Institute, located in Tucson, AZ, is an independent organization that provides assistance in resolving environmental disputes that involve federal agencies. Many state governments also have offices of dispute resolution that provide similar services to their executive branch.

Overview of the practice

As the above history indicates, the theory and practice of consensus-building have been developed primarily in the United States in the last 30 years. Initially, its use was most prevalent on the eastern and western seaboards. However, the idea seems to be spreading around the world (MIT–Harvard Public Disputes Program, n.d.), although its expansion is limited compared with other ideas for policy-making processes, such as policy analysis and technology assessment.

Because consensus-building is often organized as an informal effort, no statistical data on its use in the United States are available. Regarding the number of practitioners, the Environment and Public Policy section of the Association for Conflict Resolution, a professional organization of dispute resolution practitioners in the United States, has 5,000 members. Aside from individual practitioners, a few organizations, such as the Consensus Building Institute and RESOLVE, provide assistance to government agencies and civil society organizations when they organize consensus-building efforts. The US Institute for Environmental Conflict Resolution offers technical and financial assistance when a dispute is concerned with issues of the federal government.

Although the idea of consensus-building emerged out of the practice of environmental dispute resolution, it has been applied to a variety of public policy issues, such as healthcare, energy and the pro-life/pro-choice debate on abortion.

Theoretical underpinnings: Negotiation versus deliberation

As mentioned at the outset of this section, negotiation theory is instrumental in defining the architecture of consensus-building processes. Consensus-building is often considered to be a kind of multi-stakeholder bargaining. This leads to a particular view of public decision-making processes. In the view of consensus-building, an effective decision consists of voluntary and unanimous consent by all stakeholders.

In contrast, theories of policy-making processes that draw on the Habermasian tradition construe public decisions as an outcome of public deliberation (Gutmann and Thompson, 1996, 2004; Mansbridge, 1980, 1999; Reich, 1988). In an ideal deliberation, members of the public discuss how well different public policy ideas serve the public interest without reflecting on their own personal interests. Deliberation is an opportunity for the public to identify what public interests are. Thus, it is not a forum for making trade-offs between interests. This frame of thinking about deliberation is completely different from the fundamental notion of consensus-building that seeks a Pareto-optimal solution through trade-offs between stakeholder interests.

Meanwhile, thinkers about consensus-building have recently begun to focus on this deliberative aspect of consensus-building processes. For instance, negotiations between stakeholders who have had no previous communication require the creation of inter-languages or boundary objects to mediate their communication (Fuller, 2009; Galison, 1997). This collaborative creation of new meaning through dialogue is a result of deliberation, not negotiation. Thus a consensus-building effort is not just a series of pure bargaining sessions; in practice it does include some aspects of deliberation. Its key distinction from other forms of deliberative democracy, however, is that consensus-building requires negotiation and problem-solving. Whereas the deliberative aspects are optional in consensus-building efforts, the negotiation aspects are a mandatory feature.

4-4-2 The five-step model of consensus-building

Introduction

When it is considered prescriptively, consensus-building is often perceived as a certain method of organizing and managing dialogue between different stakeholders. Although different models for consensus-building and environmental dispute resolution exist (Carpenter and Kennedy, 1988), this section will focus on the five-step method proposed by Lawrence Susskind and his colleagues (Susskind et al., 1999; Susskind and Cruikshank, 1987, 2006).

The five-step model encompasses the entire procedure of consensus-building from beginning to end. It starts with the convening step, in which issues and stakeholders are identified, and finishes with the implementation step, in which the stakeholder agreement is implemented and monitored. In this respect, consensus-building processes are more comprehensive than most other processes and techniques for problem-solving and deliberation, such as facilitation and design workshops. On the other hand, the focus of this model is on problem-solving not agenda-setting. Therefore, the issues to be discussed in actual stakeholder sessions are occasionally bound by the interests of a convenor who has the resources to convene stakeholders.

This method draws on a number of actual dispute resolution efforts, mostly in North America. However, this fact does not preclude its applicability to cases elsewhere. Although the step-based processes might seem very "American", it is imperative for users to adapt the processes for their own context even in North America. The difference between its applications in North America and elsewhere is the required level of adaptation.

Step 1: Convening

Consensus-building cannot be initiated if no one is interested in organizing such a process. A convenor who recognizes the need for resolving an issue through negotiation must initiate the process at the beginning of this convening step. In many cases, the convenor is one of the stakeholders involved in a (potential) dispute or a charitable foundation that has a strong interest in resolving a disputed issue.

The step begins with an informal assessment of the situation; typically the convenor conducts a brief assessment of the amenability of the issue at hand to consensus-building. If the convenor recognizes the need, the organization contacts a non-partisan neutral party to serve as an assessor, who then conducts an assessment of the situation as an independent party. In the United States, a number of organizations (both private and public) provide such assessment services. Examples are:

- Consensus Building Institute (Cambridge, MA)
- RESOLVE (Washington, DC)
- CDR Associates (Boulder, CO)
- Keystone Center (Keystone, CO)
- Institute for Environmental Conflict Resolution (Tucson, AZ)
- CONCUR (Berkeley, CA)

In addition, many state governments have offices of dispute resolution offering such neutral assessments for the public sector. Individual professional mediators provide assessment services as well.

The assessor prepares an assessment – often called a conflict assessment, stakeholder analysis or issue assessment – after a series of interviews with stakeholding parties. Interviewees are identified through the "snowball sampling" technique. An initial list of interviewees is usually provided by the convenor. At the end of each interview, the assessor asks, "Whom should I talk to?" Then the assessor goes to the people mentioned by the initial set of interviewees. The same question will be asked of this batch of interviewees as well. By repeating the procedure, a comprehensive set of interviewees can be recruited.

Interviews are usually conducted as confidential one-to-one sessions in order to uncover the hidden interests of disputing parties. In public disputes, stakeholders are often reluctant to reveal their real interests because of concerns about being seen as a weak party. Through confidential interviews, the assessor can gather the necessary information to assess the possibility of reaching a creative solution through stakeholder dialogues.

An assessment report describes key stakeholder categories, key issues to be negotiated and the likelihood of reaching an agreement. A "matrix" – showing categories of stakeholders and their interests – can be used to summarize these findings (Susskind and Cruikshank, 2006). By examining the matrix, the convenor and stakeholders can visually understand the nature of the conflict. A draft report is distributed to all interviewees for their feedback. In some instances, informal meetings of a few key stakeholders are organized in order to assess the likelihood of constructive dialogue and to obtain their feedback on a draft process design. This feedback is incorporated into a final report, which will be made public in most public cases.

In a few instances, the assessment report suggests that further public dialogue should not take place. For example, if a group of stakeholders is determined to reject the invitation to dialogue, and if the group is crucial to achieving a meaningful agreement, the assessor is obliged to report the truth and recommend discontinuation of the effort. One such example is an assessment on the future of Assembly Square in Somerville, MA (Consensus Building Institute, 2003). The assessment concluded that it was too premature to initiate a dialogue in the near future because of substantial gaps between developers, environmental groups and other stakeholders regarding the ways of meeting their short-term needs.

If an assessment concludes that a series of stakeholder dialogues will produce meaningful outcomes, the assessment report prescribes a suggested design for such dialogues, including key participants, draft agenda, timetable and ground rules.

After the assessment process is completed, the convenor must decide whether to continue with the dialogue. If they decide to do so, they should send invitations to key stakeholder representatives for a first meeting.

Step 2: Sharing responsibilities

At its first meeting, the stakeholder group must decide on its process for consensus-building. The most important decision is the choice of a facilitator/mediator. It is imperative that the group members reach agreement on the choice in order to generate a sense of ownership of the decision-making process. In reality, convenors must make the necessary arrangements before the first meeting. In non-confrontational cases, they can suggest a facilitator to the stakeholder group. In controversial cases, convenors might have to invite multiple candidates and let stakeholder representatives choose a facilitator. Such a choice is often made before the first full-group meeting by a steering group consisting of a handful of key stakeholders.

In this step, the participants also agree on the roles and responsibilities that each of them will take on. This procedure is particularly important in preventing confusion among them during the actual deliberation. Stakeholder meetings usually have only an advisory role. In many instances of public participation, however, participants often feel more empowered than they actually are. If participants are not fully informed of their role before deliberation begins, it can result in a disastrous outcome that will lead to serious dissatisfaction with such participatory processes.

Participants also discuss the draft agenda, ground rules and timetable in this step. Ground rules include acceptable/unacceptable behaviour in the meeting, the admission of observers and other rules for managing the meeting effectively. In some cases, participants are asked to sign a document to indicate their commitment to the ground rules (however, this is often considered culturally unacceptable in Japan).

The whole purpose of this step is to make stakeholders feel that they own the process. In traditional participatory processes, participants simply follow the rules and steps that were predetermined by their convenors. This arrangement makes them feel trivialized in the process design. If they "own" the process (that is, if they feel responsible for the design), they are more likely to commit to the agreement made through such a process.

Step 3: Negotiating

When everyone is at the negotiating table, deliberation begins. In this step, stakeholder representatives discuss their interests in the issue and explore possible trade-offs between issues.

Facilitators are responsible for managing the dialogue in the most constructive manner possible. They might use facilitation graphics and other techniques of meeting management. In a large meeting, a recorder – who keeps a record of the discussion on large sheets of paper – might assist the facilitator.

In order to minimize the risk of facing a deadlocked negotiation in which parties cannot explore trade-offs between issues, it is advisable for the group to leave every issue open to renegotiation until the decision phase. Parties might be tempted to discuss only one issue at a time and to resolve issues sequentially. This seemingly efficient approach precludes possible deals involving two or more issues. One party might be willing to offer further concessions on a particular issue in return for a compatible concession by other parties on another issue. The issue-by-issue approach precludes such creative trade-offs.

In this step, a "brainstorming" technique is often used to generate ideas for creative solutions. During a brainstorming session, participants must refrain from making critical or evaluative comments on what other participants suggest. By forestalling a critical or argumentative mode of debate, the group can seek Pareto-optimal options that bring about maximum possible benefits to the group as a whole.

Joint fact-finding

When an issue involves scientific and technical questions, joint fact-finding is often arranged as a part of the process. Its main purpose is to forestall advocacy science deployed for promoting certain public policy ideas. One classic example is a debate over the construction of a waste management facility at the Brooklyn Navy Yard in the 1980s (Klapp, 1989; Susskind and Cruikshank, 1987). Both the promoting agency and a local opposition group were aided by different groups of professional scientific advisers that produced contradictory assessments of the health impacts from the proposed facility. The dispute was deadlocked partly because the debate was centred on scientific assessment. Laypersons lacking in scientific knowledge could only take the side of one of these contradicting scientific conclusions without understanding the details behind these assessments. Such polarization of stakeholders under the auspices of divided scientific communities often ends up in an intractable dispute.

Joint fact-finding can be arranged by creating a group of independent scientists and technical experts who provide an integrated neutral assessment of the scientific and technical uncertainties that the stakeholder group has to deal with. The membership of this scientific subgroup has to be approved by all members of the stakeholder group.

This scientific subcommittee responds to all questions that the stakeholder group has in developing its agreement. In reality, the subcommittee meets with stakeholders in order to frame scientific questions in a proper way. Then the experts from different fields work together to develop an answer and respond to the group. The experts' reaction should not be overly conclusive. They have to explain the assumptions behind their analysis and be willing to provide sensitivity analysis. They should also have the ability to integrate the stakeholder group's local knowledge into their scientific analysis.

In other words, joint fact-finding is a kind of effort to create interaction between stakeholding laypeople and scientific communities. Whereas consensus conferences and similar efforts in the science, technology and social science communities seek to develop linkages between science and the public, joint fact-finding focuses particularly on stakeholding parties as representatives of the public.

Step 4: Deciding

Negotiation theory warns us of the difficulties in resolving the tension between "creating value" and "claiming value" (Lax and Sebenius, 1986). Negotiating parties can create joint gains (that is, achieve Pareto-superior results) through trading between issues that each party values differently. However, such joint gains have to be shared between negotiating parties. Although there have been many efforts to determine a "fair" division of such gains in mathematical terms, each party always strives for better terms of agreement in practical negotiations (Raiffa, 1982). This leads to a competitive mode of negotiation.

In order to forestall such a competitive mode, a skilled facilitator starts the discussion by exploring each stakeholder's interests behind his or her position. Once all participants recognize the chance of joint gains through exploring all possible trades (that is, "enlarging the pie" by joint actions), the discussion moves to the decision phase where the "pie" is divided (Bazerman et al., 2001).

It is a common tactic for a skilled negotiator to threaten his or her withdrawal from the final agreement at the last moment. In such instances, the facilitator has to assess the real chance of withdrawal and manage the discussion accordingly. It is a moment when the facilitator cannot be completely neutral in terms of the "division of the pie". It might be advisable for the facilitator to suggest an adjustment to the draft agreement in order to satisfy the objecting parties. In other instances, the facilitator might simply argue that the draft agreement is likely to provide better results than their BATNA.

When the parties cannot agree on issues associated with uncertainties and risks, the facilitator can suggest a contingent agreement (Susskind and Field, 1996). Such an agreement can be framed in a series of "what if" scenarios. For example, suppose a plant operator is confident about the effectiveness of its pollution prevention equipment, whereas local residents are worried about the health risks from possible pollution. In such an instance, the plant operator can promise to stop the plant and compensate all damage if certain substances in the plant effluents exceed certain levels. This kind of contingent agreement is useful especially when disputing parties have different assessments of uncertainties.

Step 5: Implementing

One particular feature of this five-step model of consensus-building is its attention to implementation. If any of the negotiating parties lacks the intention to actually implement the stakeholder agreement, the consensus-building effort becomes a substantial waste of time and other resources. The agreement must include mechanisms to make parties commit to its implementation. In other words, those who do not abide by the agreement should be penalized in some way.

In some instances, the agreement might have to be adjusted during its implementation because of unexpected changes in the surrounding circumstances. If that is the case, it is more advisable for the parties to renegotiate the terms than to penalize a few stakeholders by adhering to the rules.

Advocates of consensus-building argue for stability as one of four key ingredients of consensual agreements (Susskind and Cruikshank, 1987). One way of improving the stability of an agreement is to develop an institution for resolving emerging issues on a continuous basis. The concept of adaptive management is particularly useful in this context (Karl et al., 2007). When the uncertainties involved in the issues are high, it is more efficient for stakeholders to temporarily agree on short-term actions and seek further agreements as new facts are discovered than to debate over too-uncertain "facts".

4-4-3 Case studies

The Kita-josanjima intersection improvement in Japan

There are a number of examples of best-practice consensus-building processes in the United States, but first a Japanese project will be discussed – in which the author was involved as a participatory observer – as an example of a non-US case. Its details are documented elsewhere (Matsuura, 2008; Matsuura and Yamanaka, 2007). In 2004, a local field office of the Ministry of Land, Infrastructure, Transport and Tourism decided to design an improvement plan for a road intersection by adopting a consensus-building process as practised in the United States. It worked with local academics and a not-for-profit organization that provided nonpartisan support to the stakeholder group.

The stakeholder assessment was conducted in January and February 2005 by interviewing 54 individuals who seemed to have interests in the road intersection. On 22 July 2005, 21 stakeholder representatives, including the Ministry's representatives as well as one from the local police agency, had their first meeting. In the first meeting, a team of facilitators was suggested by the convenor and the group approved them. Technical consultants were also appointed. After four meetings, the group was able to reach an agreement that included eight improvement measures to the intersection (for example, improved lighting and realignment of the bike path). The Ministry implemented most of the agreement in 2007.

The key lesson from this case was the need for adaptation of the original ideas "imported" from the United States. Nine kinds of adaptation were identified in this first trial in Japan. For example, it is a common practice in the United States to list interviewees in the stakeholder assessment report. In the Japanese case, however, local residents did not want to have their names mentioned in the report, even though they were promised that they would not be identified with specific comments. One possible explanation for the difference is the traditional Japanese norm of obedience to authority. Those who appeared in the stakeholder assessment might be viewed by their neighbours as "strange" individuals who stood out by objecting to the government. Such cultural differences resulted in subtle adaptations of the consensus-building process.

Those who consider the use of consensus-building processes outside the United States should recognize the need for adaptation. However, there is no set recipe for such adaptations; they must be identified as the convenors and facilitators design and manage the process. On the other hand, too much adaptation can lead to a completely different participatory process that cannot be legitimately considered a consensus-building effort. For example, if certain categories of stakeholder are barred from participation because they do not politically align with the convenor, such a "consensus-building" effort can easily fail during the last implementation step. Opponents will try every available measure – for example, endless litigation or partnering with foreign news media and non-governmental organizations (NGOs) – to overthrow the agreement. Therefore, certain institutional changes might be necessary in countries and regions where democratic decision-making has yet to become a common practice (Matsuura, 2008).

Building coast-smart communities in Maryland

In the context of sustainability science, consensus-building processes have been applied in many different ways. In fact, negotiations related to the United Nations Framework Convention on Climate Change (UNFCCC) could be construed as a kind of consensus-building effort because all major emitters of greenhouse gases must reach consensus on a voluntary basis in order to achieve the goal of containing the temperature rise to a manageable level. Unfortunately, however, actual negotiations at the

Conference of the Parties meetings are substantially different from what the five-step model suggests.

This subsection will discuss a practical example of the use of a consensus-building approach to sustainability issues – the Building Coast-Smart Communities Interactive Summit held on 27 April 2009 in Maryland, USA. The state has 4,000 miles of shoreline, and sea-level rise is one of the most crucial impacts of climate change. The summit meeting, however, was not a forum for consensus-building. Instead, it was intended to stimulate interaction among professionals and stakeholders who have interests in the protection of the state's coastline.

The meeting was organized around a "negotiation simulation" exercise. Each participant was assigned to one of nine roles and negotiated with other participants who were assigned to different roles in order to experience the difficulties of consensus-building in a simulated setting. A professional facilitator, who is actually practising in Maryland, was also assigned to each group of stakeholders.

The goal of the simulated negotiation was to "reach agreement on a set of strategies for managing the climate change risks facing their coastal community" (Consensus Building Institute et al., 2009: 2). In order to complete the negotiation within a limited time (45 minutes), participants were provided with instructions that included a comprehensive list of 39 possible policy options; the imaginary agreement would be a package of policy options that every participant could agree with. Each option had a "score" and "cost" that each participant had to consider in proposing a package.

The nine roles represented in this simulated exercise were: county commissioner, county planner, local real estate development association, state government biologist, local Chamber of Commerce president, environmental advocate, local resident, farmer, and county emergency management director. Each role had a different preference as to the 39 policy options. These differences created room for negotiation and the need for consensus-building. More than 170 key stakeholders, including mayors, county commissioners, environmentalists, business leaders and state officials, participated in the exercise (Consensus Building Institute et al., 2009).

Meanwhile, the state government invited applications for a programme that provides funding and technical support for efforts to develop "coast-smart" communities. The simulation exercise served as a preview of the actual consensus-building efforts that were likely to be necessary in preparing such proposals.

Full-fledged consensus-building efforts are useful in preparing strategies for sustainable communities, but the consensus-building idea has to be acknowledged by key stakeholders. Negotiation simulations, involving actual stakeholders and a facilitator, can be helpful in paving the way for an actual effort by familiarizing them with the process as well as informing them of its usefulness. Although a new set of instructions could be prepared for each situation (like this Maryland case), conventional sets of materials are also available from the Clearinghouse for the Program on Negotiation (PON) at Harvard Law School.[1] Those who are interested in applying consensus-building processes to a particular issue of sustainability might consider organizing a similar simulated negotiation session in order to build momentum for convening actual stakeholders.[2]

Joint fact-finding on nuclear power

Energy is a sustainability-related issue to which the consensus-building approach has been applied for more than a decade in the United States. Partly because the US energy industry has involved a wide range of stakeholders since market deregulation, utility regulators have adopted consensus-building and similar negotiation-based procedures at the state and regional levels (Raab, 1994).

One recent example is the application of joint fact-finding to the debate over the use of nuclear power. This project was organized by the Keystone Center, a not-for-profit organization specializing in consensus-building and other processes to seek stakeholder agreements on public policy issues.

The project, which started in 2005, was initially proposed by the Center's Energy Board, which included members of Congress, utility representatives, NGO representatives, and other leading figures in the field. Following the board's discussion, the Center decided to launch the joint fact-finding effort with financial assistance from the Pew Foundation and several utilities.

The Center conducted an informal stakeholder assessment with approximately 30 interviewees. Drawing on this analysis, it organized a steering committee of nine key stakeholders and determined the issues to be discussed in the sessions to follow.

Participants in the full group comprised 27 representatives from a wide range of stakeholding organizations. According to the ground rules, stakeholder discussions were confidential in order to encourage participants to share sensitive information with others. In fact, according to the Center's staff, participants abided by the rules. Utility representatives provided data that seemed credible to all participants.

Four plenary sessions and more working group meetings were held during a 15-month period. According to the staff, both proponents and opponents seemed to be engaged in the working group discussions in a constructive manner. An experts' panel was set up in order to support the stakeholder group. Experts were asked to provide professional knowledge but to refrain from expressing their judgements on issues of controversy.

The results, published in July 2007, consisted of 35 consensual findings, including the following:

Economy: Before this joint fact-finding session, proponents and opponents had completely different assessments of the costs of producing electricity by nuclear power. Through this dialogue, they agreed that the cost would be 8–11 cents per kWh. This estimate was higher than proponents had suggested and lower than opponents had claimed.

Disposal of high-level waste: Participants agreed that geological disposal was the best option and proposed siting criteria. (Although the report does not mention it, the Yucca Mountain site does not match these criteria.)

Safety: Although the operational safety of nuclear power plants improved considerably after the Three Mile Island incident, the report raises concerns about proliferation.

4-4-4 Implications of consensus-building for sustainability

"Live with" as a norm for sustainability

The consensus-building approach assumes that every stakeholding party must be able to "live with" an agreement. This is one of its key distinctions from other participatory processes. The term "live with" implies that each party does not have to be able to realize all of his or her ideals and hopes through negotiation. Consensus-building processes are not a kind of debate in which one party takes all and the other loses everything. Instead, everyone has to receive more benefits from the agreement than unilateral actions would provide. This is a robust conception of associations between individuals. Consensus-building facilitates the formation of a voluntary association only when its members can benefit from participating in it. Otherwise, it is advisable for the individuals to take unilateral action and to reap the benefits from that.

Of course, such a liberal conception of association could encourage free-riding and opportunistic behaviour if the association is large in scale and lacking in effective rule-enforcement mechanisms. Consensus-building, however, discourages such opportunistic behaviour by trying to involve the whole range of stakeholders (including future generations) and to institutionalize a mechanism to enforce their agreement. The key difference between the liberal notion of individual freedom and consensus-building is that the latter is intended to encourage value creation through voluntary agreements, whereas the former is focused primarily on defending the individual's right to act unilaterally. The consensus-building approach offers practical solutions for reaching a sustainable agreement (but only if all stakeholders are involved) while allowing individuals the freedom to participate, or not participate, in the dialogue.

The proposal of this section is that the concept of "live with" – the core component of the consensus-building approach – be considered a normative component of sustainability, amidst the contemporary normative trend towards the protection of individual freedom. Arguments for limiting individual rights for the sake of environmental sustainability often lead to an intractable debate over what rights individuals have. Falling into the trap of this debate over rights reduces the chances of finding practical solutions to achieve the sustainability of humankind.

The consensus-building approach as an institutional capacity for sustainability

In practice, environmental and other kinds of disputes often escalate because of psychological dynamics and economic factors that compel stakeholders to adhere to their initial commitments and to try to "beat" the other side. Without efforts to resolve such disputes, societies can easily be torn apart and eventually suffer substantial costs from these intractable disputes, including lost opportunities for value creation through collaboration.

A society is more likely to be sustainable if resources for the consensus-building approach are readily available to its members so as promptly to forestall any potential psychological escalation. In fact, anthropologists have discovered that many communities around the world have institutional capacities for mediation and other techniques to resolve their disputes (Gulliver, 1979). In such communities, mediation is so embedded in their institutional systems that it is no longer recognized as a special skill but is viewed as a customary practice. On the other hand, mediation had to be recognized as a special technique in the United States – where the tradition of mediation did not exist among non-natives – and conscious efforts to institutionalize it took place only in the last few decades. For example, mediation techniques have been taught in high schools so that students can mediate disputes between their fellow students (a system known as peer mediation). Like peer mediation, the consensus-building approach can also function as an institutional capacity for different kinds of emerging communities of stakeholders, ranging from supranational organizations to grassroots non-governmental organizations. In other words, constructive mechanisms for deliberation are indispensable in making these organizations more sustainable.

Stakeholder involvement as a way to achieve sustainable development

The consensus-building approach seeks to involve the full range of stakeholding parties in the process in order to make the agreement fairer, wiser, more efficient and more stable (Susskind and Cruikshank, 1987). In this respect the approach is amenable to the concept of sustainable development. First, sustainable development must attend to the concerns of future generations, as prescribed by the World Commission on Environment and Development. Second, sustainable development often argues for the empowerment of under-represented citizens (for example, women and under-classes in developing countries) in the decision-making process.

Although some facilitators may be uninterested in who is at the negotiating table, advocates of consensus-building often argue for the full inclusion of all stakeholding parties. The interests of future generations and under-represented populations must be reflected in the process when they are likely to be affected by a possible agreement. Especially in the developing world, the consensus-building approach might require an expansion of the scope of participation beyond the conventional group of stakeholders dominated by local power holders. In this aspect, the consensus-building approach is not value free. It can be strategically deployed to encourage sustainable development in developing nations. On the other hand, its users must be open to adaptation of the approach in order to make the process acceptable to local stakeholders, as exemplified by the Japanese trial described above.

The importance of meta-governance

Introducing consensus-building processes can induce institutional changes at the local level. In particular, it can give certain groups of stakeholders – who were previously blocked from participating in public decision-making – access to the decision-making arena. Therefore, the use of the consensus-building approach itself is a matter of public decision. Although it might sound tautological, relevant stakeholders must agree to adopt the process precisely because it is an informal effort based on their voluntary agreement.

The decision to introduce a new consensual approach to making decisions on particular issues is a matter at the "meta" level of governance. At this level, institutions (legal, cultural, organizational) and resources (human, financial, knowledge) are key variables that determine the decision to use new processes.

In order to achieve sustainability through the introduction of consensus-building processes, it is not sufficient to have a few successful cases in the field. Advocates should mobilize at the meta level to institutionalize the consensus-building approach. This does not mean that consensus-building processes must be required for all kinds of decision-making under guidelines and public laws. In practice, the concept of "live with" has to be shared among the public as a kind of norm. An adequate capacity for managing consensus-building processes (for example, a group of skilled facilitators) must be developed before they can be applied to a wide range of issues. It may require adjustments in the local culture and institutions to allow the participation of certain categories of stakeholders. Any effort to achieve sustainability through the consensus-building approach must be attentive to such issues at the meta-governance level in order to prevent this approach from ending up in a few case studies without much "sustainable" impact in the field.

Notes

1. The PON Clearinghouse website is at <http://www.pon.org/>.
2. General instructions for the Building Coast-Smart Communities exercise are also available on its website (<http://maryland.coastsmart.org/>).

REFERENCES

Amy, D. (1990) "Environmental Dispute Resolution: The Promise and Pitfalls", in N. Vig and M. Kraft (eds), *Environmental Policy in the 1990s*. Washington, DC: CQ Press, pp. 211–234.

Bazerman, M., J. Baron and K. Shonk (2001) *You Can't Enlarge the Pie*. New York: Basic Books.

Bingham, G. (1986) *Resolving Environmental Disputes: A Decade of Experience*. Washington, DC: Conservation Foundation.

Carpenter, S. and W. J. D. Kennedy (1988) *Managing Public Disputes*. San Francisco, CA: Jossey-Bass.

Consensus Building Institute (2003) *Conflict Assessment on the Future of Assembly Square, Somerville, Massachusetts*. August.

Consensus Building Institute, MIT–USGS Science Impact Collaborative and Maryland Department of Natural Resources (2009) *Coast-Smart Communities: How Will Maryland Adapt to Climate Change?* 22 April. Available at <http://web.mit.edu/dusp/epp/music/pdf/CoastSmartGeneralInstructionsNoScorecard.pdf> (accessed 31 May 2010).

Cormick, G. (1976) "Mediating Environmental Controversies: Perspectives and First Experience", *Earth Law Journal* 2: 215–224.

Dukes, F. (1996) *Resolving Public Conflict: Transforming Community and Governance*. New York: Manchester University Press.

Fisher, R. and W. Ury (1991) *Getting to Yes*. New York: Penguin.

Fuller, B. (2009) "Surprising Cooperation Despite Apparently Irreconcilable Differences: Agricultural Water Use Efficiency and CALFED", *Environmental Science and Policy* 129(6): 663–673.

Galison, P. (1997) *Image and Logic: A Material Culture and Microphysics*. Chicago, IL: University of Chicago Press.

Gulliver, P. H. (1979) *Disputes and Negotiations: A Cross-Cultural Perspective*. New York: Academic Press.

Gutmann, A. and D. Thompson (1996) *Democracy and Disagreement*. Cambridge, MA: Belknap Press.

Gutmann, A. and D. Thompson (2004) *Why Deliberative Democracy?* Princeton, NJ: Princeton University Press.

Innes, J. (2004) "Consensus Building: Clarifications for the Critics", *Planning Theory* 3(1): 5–20.

Karl, H., L. Susskind and K. Wallace (2007) "A Dialogue, Not a Diatribe: Effective Integration of Science and Policy through Joint Fact Finding", *Environment* 49(1): 20–33.

Klapp, M. (1989) "Bargaining with Uncertainty: The Brooklyn Navy Yard Incinerator Dispute", *Journal of Planning Education and Research* 8(3): 157–166.

Lax, D. and J. Sebenius (1986) *Manager as Negotiator: Bargaining for Cooperative and Competitive Gain*. New York: Free Press.

Mansbridge, J. (1980) *Beyond Adversary Democracy*. New York: Basic Books.

Mansbridge, J. (1999) "Everyday Talk in the Deliberative System", in S. Macedo (ed.), *Deliberative Politics*. New York: Oxford University, pp. 211–239.

Matsuura, M. (2008) *Localizing Public Dispute Resolution in Japan: Lessons from Experiments with Deliberative Policy-making*. Saarbrücken, Germany: VDM Verlag.

Matsuura, M. and H. Yamanaka (2007) "Planning through Assisted Negotiation: Consensus Building for Traffic Safety", *Journal of the Eastern Asia Society for Transportation Studies* 7: 1546–1558.

Mernitz, S. (1980) *Mediation of Environmental Disputes: A Sourcebook*. New York: Praeger.

MIT–Harvard Public Disputes Program (n.d.) "Public Dispute Resolution Around the World", <http://web.mit.edu/publicdisputes/pdr/world.html> (accessed 31 May 2010).

Raab, J. (1994) *Using Consensus Building to Improve Utility Regulation*. Washington, DC: American Council for an Energy-Efficient Economy.

Raiffa, H. (1982) *The Art and Science of Negotiation*. Cambridge, MA: Harvard University Press.

Reich, R., ed. (1988) *The Power of Public Ideas*. Cambridge, MA: Harvard University Press.

Susskind, L. and J. Cruikshank (1987) *Breaking the Impasse*. New York: Basic Books.

Susskind, L. and J. Cruikshank (2006) *Breaking Robert's Rules*. New York: Oxford University Press.

Susskind, L. and P. Field (1996) *Dealing with an Angry Public*. New York: Free Press.

Susskind, L. and S. McKearnan (1999) "The Evolution of Public Policy Dispute Resolution", *Journal of Architectural and Planning Research* 16(2): 96–115.

Susskind, L. and L. van Dam (1996) "Squaring Off at the Table, Not in the Courts", *Technology Review* 89(5): 39.

Susskind, L., S. McKearnan and J. Thomas-Larmer, eds (1999) *The Consensus Building Handbook: A Comprehensive Guide to Reaching Agreement*. Thousand Oaks, CA: Sage.

4-5

Public deliberation for sustainability governance: GMO debates in Hokkaido

Nobuo Kurata

4-5-1 Introduction

Building a sustainable society requires public deliberation and participatory decision-making on environmental problems by citizens. Because top-down decision-making can lead to environmental discrimination, a democratic decision-making process is also indispensable in order to achieve environmental justice. Environmental discrimination is caused by the immaturity of democratic governance systems as well as by social discrimination. Debate on environmental problems inevitably involves conflicts over differing values, but solutions to such problems often require some kind of local knowledge. Citizens have the right to participate in decision-making concerning environmental policies and, as laypeople, may also be able to contribute to the solution of local environmental problems through their local knowledge.

Each actor considers environmental problems within his or her own frame. To increase understanding and trust among actors, informal meetings for the exchange of opinions on genetically modified organisms (GMOs) were held in Hokkaido, Japan, to supplement formal decision-making by committees of the Hokkaido prefectural government. Hokkaido Prefecture also held a GMO consensus conference in 2006–2007. This section will discuss these events to illustrate the use of public deliberation as well as the process of participatory assessment.

4-5-2 Environmental justice

The problem of inequality in the distribution of benefits as well as of disadvantages, risks and hazards is one of the most serious issues of environmental ethics. The issue is generally referred to as environmental justice or fairness and is, in fact, a form of social justice (Shrader-Frechette, 2002). Environmental problems involve relations not only between human beings and nature but also between people and people, people and society, and people and nations. There are also problems involving advanced countries and developing countries, the rich and the poor, companies and citizens, urban centres and outlying provinces, and producers and consumers. In all such cases, democratic decision-making systems are necessary for the protection of the socially vulnerable.

Democracy is also necessary to prevent environmental discrimination. Lack of access to information or withholding of the right to free speech can bring harm to a region's environment and its citizens' health. Protection of the regional environment requires governance of the bottom-up (or grassroots) type and local public deliberation for local decision-making.

Environmental discrimination is caused by social discrimination. Racial discrimination, for example, can cause health hazards. The level of damage caused by environmental pollution differs for people of different income levels. Average life spans, infant mortality rates and cancer mortality rates all vary depending on one's social class.

The biologically vulnerable (for instance, women, children, the elderly, foetuses, people with disabilities or diseases) are more likely to become victims of environmental problems. Likewise, the socially vulnerable, most often found among the poorest segments of the population, workers engaged in agriculture or fishery, or people in developing countries, are likely to be damaged more easily than others.

Take, for example, Minamata disease, one of the worst consequences of environmental pollution in Japan. Minamata disease, which is a typical example of pollution-related health damage in Japan, was first discovered in 1956, around Minamata Bay in Kumamoto Prefecture. It was caused by high consumption of fish and shellfish contaminated by methylmercury compound discharged from the Chisso Corporation's chemical plant. It is a disorder of the central nervous system that has various signs and symptoms, including sensory disturbance in the distal portions of four extremities, ataxia and concentric contraction of the visual field, and in some cases results in death. Foetal Minamata disease is also reported, which shows impairments similar to cerebral infantile paralysis as a result of the mother being exposed to methylmercury during pregnancy (Ministry of the Environment, 2002). The Japanese government and Chisso did little to prevent the pollution, which is said to have injured more than 10,000 people, of whom 2,000 died. Lawsuits and claims for compensation continue to this day. During the period that the damage occurred, the Japanese democratic system was immature, there was little public deliberation and the government did not respect the rights of local residents (Yoshida, 2007).

Another case is the Bhopal disaster that occurred at a Union Carbide pesticide plant in the city of Bhopal, India. On 3 December 1984, the plant released toxic methyl isocyanate gas, exposing more than 500,000 people to the chemical, of whom it is estimated that 25,000 died.

In developing countries, the system of governance is sometimes a form of bureaucratic authoritarianism. Bureaucracies wield tremendous power and industrial development is given top priority. Multinational companies exploit a double standard for industrial wastes, with strict standards imposed in their home country but not in developing countries. Dangerous factories that would not be permitted in advanced industrial nations are built in developing nations, and these factories often neglect the safety of local residents. Media in these countries do not report environmental problems and political arguments are restricted. Often the democratic governance system is immature. Priority is given to economic growth over human rights, political freedom or freedom of the press.

The governments of such countries attract the factories and garbage dumps of multinational companies at the expense of public health and the local environment. In these countries, labour is cheap and regulations on industrial plants are relatively lax. Because environmental problems go unreported and political discussions are banned, the victims cannot challenge the government and the demands of victims are not reflected in the administration, legislature or judiciary. In decision-making processes, power is concentrated in the hands of a small number of people. Thus environmental pollution is exacerbated by the immaturity of democratic political systems and the lack of public deliberation. Even when the government is not a dictatorship, there may be deficiencies in the procedures for information disclosure, with decision-making conducted in secret.

4-5-3 Deliberative democracy

In democratic decision-making systems, important elements include the decision-making autonomy of the people, disclosure of information with an adequate system for checking the administration, citizen participation in the decision-making process, and a system that ensures the citizens' right to be heard and to say "no".

Deliberative democracy is necessary not only to avoid environmental injustices, but also to make decision-making work fairly on local environmental problems. In a democratic society, deliberation must be the deciding factor in decision-making and the source of legitimacy must be the deliberative procedure (Dryzek, 2000; Fishkin and Laslett, 2003; Smith, 2003). Deliberative democracy is defined as "a form of government in which free and equal citizens (and their representatives) justify decisions in a process in which they give one another reasons that are mutually acceptable and generally accessible, with the aim of reaching conclusions that are binding in the present on all citizens but open to challenge in the future" (Gutmann and Thompson, 2004: 7).

4-5-4 Conflicts of values

In debates on environmental problems, there are often conflicts over values (Table 4.5.1). Because of the diversity of actors, these confrontations over values become very complex. Each stakeholder has a different sense of values and a different epistemic frame through which he or she views problems. To solve environmental problems one must therefore mediate value conflicts and compromise with one another. Top-down governance based on monistic values (or a simple utilitarian approach) is not adequate. To treat environmental problems, expert judgements are needed based on the scientific method of cost/benefit analysis.

Many local environmental problems are too complex to be judged only by experts, because these experts (natural scientists, social scientists, etc.) lack some of the local knowledge necessary to resolve these problems. Laypeople, on the other hand, may have such knowledge and relevant information. Natural scientists (for example ecology researchers) do not commit themselves to values. However, local environmental problems cannot be solved without committing to certain social values. The methods employed to conserve a local environment are deeply related to a certain set of values. Therefore, the solution of local environmental

Table 4.5.1 Structure of values concerning the environment

Values of local nature
Values of ecosystem / Values of individual animals and plants
Health and safety of residents
Quality of air and water
Industry, development, economic growth, consumption
Political liberty ("environmental fascism")

problems requires public deliberation and social decision-making of the participatory type, which involves laypeople and which ensures people have an equal right to express their opinions.

Decision-making on local environmental policies may not be adequate when conducted solely by experts and administrators because they lack important information (or local knowledge) necessary for this purpose. In such cases, decisions may be made on the basis of uncertain or inadequate information. Though a temporary "solution" may be advanced using a rational decision-making system, the solution will vary depending on the conditions and information provided (in other words, it is subject to the restriction and uncertainty of information). To reduce this uncertainty, the judgements of laypeople need to be taken into account in the decision-making process.

Decision-making that employs cost/benefit analysis or risk/benefit analysis by elite experts and administrators may not produce a definitive answer to a problem. Nonetheless, to increase the likelihood of finding effective solutions to environmental problems, some kinds of local knowledge must be factored into the decision-making process. It is therefore necessary to enlist citizens in local governance of the environment. These citizens (that is, laypeople) may be able to contribute to the solution of local environmental problems using their local knowledge. Citizens have not only the right but the responsibility to participate in decision-making concerning environmental policies. As the limitations of committees of "specialists" in government ministries and at the local government level become clearer, the necessity of public deliberation and participatory assessment by laypeople is increasingly recognized.

In democratic societies, the opportunity to express one's opinions and preferences must be afforded not only to scientists and the industrial sector, but also to citizens or laypeople. Citizen participation in decision-making processes is essential to a democratic society. Granted, it is difficult for citizens who are "laypeople" and lack "scientific" knowledge concerning environmental problems to judge the risks to their local environment. However, citizens with some knowledge and sufficient concern can render judgements about the direction of environmental policy from ethical and social viewpoints if they receive adequate information.

For example, conservation ecology concerns not only "ecology" as a natural science, but also the management of animals, plants and insects as a kind of social science. Various values, in addition to scientific knowledge, have to be taken into consideration in the management of wild animals, for example. When analysing the influence of alien species and planning measures to exterminate them, the effects of these measures on agriculture and fisheries in the region also have to be considered.

4-5-5 GMO regulation and public deliberation in Hokkaido

As an example of how public deliberation works, this section will now discuss the debate over genetically modified organisms (GMOs) in Hokkaido, Japan. The problems associated with genetically modified (GM) crops concern not only science and technology but also the politics and economics of agriculture. Agreement in GMO debates is difficult. Most Japanese consumers are opposed to GM food. These consumers have concerns about safety and the effects of GM food on human health. Yet GM soybeans are imported to Japan as materials for food, and GM maize is fed to domestic livestock. However, most consumers in Japan do not know this.

In debates over GMOs, the gap between actors is so wide, and their mutual antagonism so deep-seated and emotional, that it is almost impossible for concerned actors to discuss the issue. Each actor considers the problem through a different frame, and does not understand how the same problem looks through other frames.

In order to break through this impasse, it is necessary for people with opposing opinions to "sit at the same table". In Hokkaido, there were informal meetings for the exchange of opinions on GMOs as well as formal decision-making by committees of the Hokkaido prefectural government. Although it remains exceedingly difficult for the central government of Japan to build a countrywide consensus on the matter of GMO planting, informal meetings were useful in building such consensus on a local scale. These meetings, attended by both GMO proponents and opponents, contributed to finding points of compromise.

If actors sit at a table and begin to talk, mutual trust can arise and discussion becomes easier, even if agreement is still impossible. Meetings between scientists in favour of GMOs and opponents may facilitate compromise. If it is feasible to convene such a meeting, each actor may come to recognize that the same problem looks different according to one's frame, and begin to understand how it looks through other frames. The discussions at such meetings help the actors recognize the existence of differences in framing. Even if these frames are not shared by all participants, some transitions or bridging of frames will occur among them.

In Hokkaido Prefecture, an ordinance was enacted for the regulation of GM crop cultivation in 2005. Before and after enactment of the ordinance, some informal meetings for the exchange of opinions on GMOs were held, as well as meetings for formal decision-making by committees of the Hokkaido prefectural government. Informal meetings between proponent scientists and opponents facilitated mediation in a formal committee of the Hokkaido Prefectural Office, and this mediation

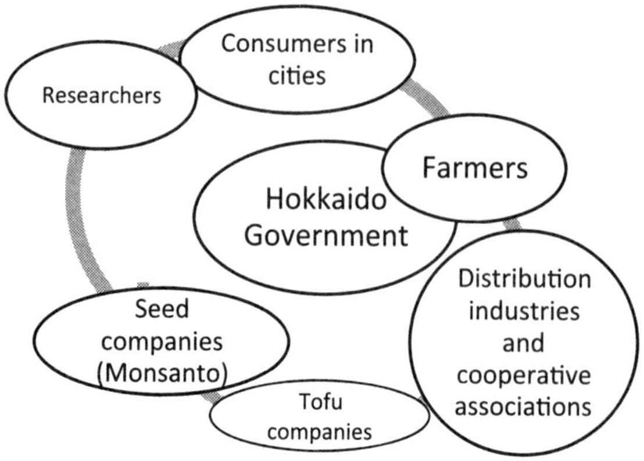

Figure 4.5.1 Actors in Hokkaido.

contributed to the regulation of GMOs in Hokkaido. In these meetings actors came to understand how other participants thought and some social learning took place, leading to the discovery of points of compromise between proponents and opponents. In addition to these meetings, Hokkaido Prefecture held a consensus conference on GMOs in 2006–2007.

The opinions of the respective actors on this issue are so divergent that it is inadequate to explain the situation with a simple chart of the "pros and cons" of GM planting. There are consumers who do not want to eat GM food, associations of consumers, farmers of organic agricultural products, farmers who would like to plant GM crops, and researchers planning to plant GM crops in open fields (see Figure 4.5.1). Standpoints varied among scientists as well. There were biotechnology researchers at the National Agricultural Research Center for Hokkaido Region who were planning to carry out GM rice trials in open farmland, as well as other biotechnology researchers who were proponents of GMO planting, but there were also ecology researchers who had concerns about biodiversity.

In the debate on GMOs the voices of farmers are seldom heard publicly, though they are regarded as the most important actors. However, the standpoints of farmers vary widely. In Hokkaido, for example, most farmers are engaged in traditional farming. Some are trying to grow GM plants (such as soybeans) or are considering the cultivation of GM crops, though the number of such farmers is small. On the other hand, others are converting to organic agriculture. One of the important features of the GMO debate in Hokkaido is the presence of farmers who would like to try, or have tried, to grow GM soybeans. In Japan overall, the number

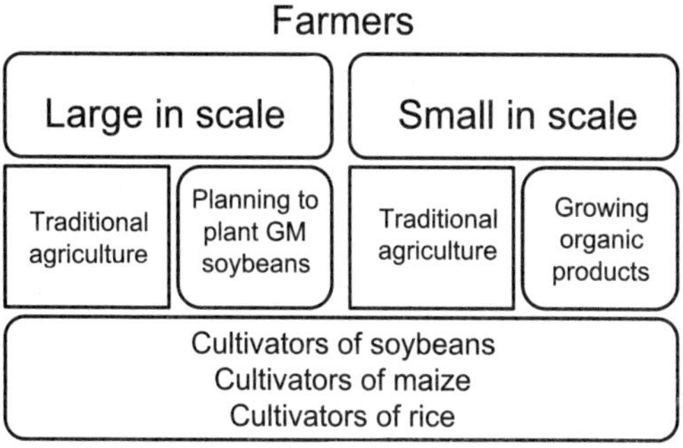

Figure 4.5.2 Farmers in Hokkaido.

of such farmers is very small, but in Hokkaido many farmers have adopted a large-scale style of agriculture and plant soybeans, so they are considering the cultivation of GM soybeans and maize (see Figure 4.5.2).

4-5-6 Outline of GMO regulation in Hokkaido (Table 4.5.2)

In 2003, the National Agricultural Research Center for Hokkaido Region planted GM rice for experimentation in an open field at the Center. Some groups of consumers and concerned citizens gathered at the Center and asked it to cancel the experiment, but the Center continued with it. However, scientists at the Center have not practised open trial cultivation since then.

In 2004, the Hokkaido prefectural government organized a committee to consider the conditions for planting GM crops. The Hokkaido Prefec-

Table 4.5.2 A brief history of GMO regulation in Hokkaido

Open experimental cultivation by National Agricultural Research Center for Hokkaido Region (2003)
GM guidelines by Hokkaido prefectural government (2004)
Committee for Planting Conditions for GM Plants (2004) (11 members: scientists, consumers, farmers, etc.)
Enactment of Hokkaido GMO regulations (2005) (Hokkaido Preventive Measure Ordinance against Crossing by GM Cultivation)
Hokkaido Food Safety and Reliability Committee (2006–) (15 members: academics, consumers and producers)
GMO Expert Subcommittee for the Scientific Study of Preventive Measures against Crossing or Commingling (6 scientists) (risk assessment)
Hokkaido GMO consensus conference held by Hokkaido prefectural government (Nov. 2006 – Feb. 2007) (proposals by 15 local citizens)

tural Office was trying to establish a Hokkaido food "brand image" of "safety and reliability", because agriculture is the prefecture's most important industry.

In March 2005, the Hokkaido Prefecture Legislative Assembly enacted an ordinance that regulates open cultivation of GM crops. Known as the Hokkaido Preventive Measure Ordinance against Crossing by GM Cultivation, the objective of the ordinance, which went into effect in January 2006, is to prevent the crossing of GM crops with non-GM crops and their commingling in production and distribution by coordinating the promotion of research activities using GM technology with agricultural production activities involving traditional crops, and also by coordinating bio-industrial activities pertaining to the development of bio-engineered plants and production activities pertaining to conventional agricultural products.

The ordinance calls for the adoption of a licence system for commercial cultivation of GM plants and a notification procedure for researchers who wish to engage in trial cultivation in open fields. Farmers planning commercial cultivation of GM crops must obtain a licence from the Hokkaido Governor's Office. In order to grow GM crops in outdoor fields for trial purposes, researchers have to submit notification to the Governor. Researchers must then have their plans reviewed by the Hokkaido Food Safety and Reliability Committee and receive approval from the Governor. Violation of the ordinance could result in imprisonment for up to one year or a fine of up to ¥500,000.

However, the Hokkaido ordinance is notorious among bio-scientists, many of whom regard it as unscientific and unsound as well as unnecessarily stricter than national regulations (under the Cartagena Protocol on Biosafety) for GMO planting. The Cartagena Protocol seeks to protect biodiversity from the potential risks of living modified organisms. They also claim that the regulation is a result of a failure in risk communication. To this criticism the Hokkaido prefectural government responds that the aim of the ordinance is to "set rules" for the trial and commercial cultivation of GM plants, not to "forbid" them. Is the ordinance in fact the consequence of a failure in risk communication and in public understanding of science by laypeople?

4-5-7 Public deliberation in Hokkaido

Though antagonism over the GM plant issue in Hokkaido was quite serious, it nevertheless seemed reasonable that, if a meeting of stakeholders were held, the actors might come to understand one another's views. Even if their opinions and interests differed, it would not be impossible for participants to discuss them around the same table. And even if an actual agreement proved unobtainable, it was still not impossible to build relationships based on mutual trust. In any event, a situation where proponents and opponents never meet or talk, and merely attack each other, could be avoided.

In the period before and after enactment of the Hokkaido ordinance, there were a number of informal meetings of this sort in Hokkaido. Researchers in biotechnology, researchers planning to plant GM rice, dealers in organic products, farmers growing organic food, consumers, and scientists studying ecology were able to exchange opinions at these meetings. Some proponents came to understand opponents' ideas and concerns. Conversely, some opponents began to trust the researchers at the National Agricultural Research Center, even if they could not reach agreement. A kind of social learning took place among participants. What made it possible to "set the table" for such discussions was a common agenda shared by proponents and opponents alike: the future of Hokkaido agriculture. When discussing the future of Hokkaido agriculture, some people say, "We don't want agriculture in Hokkaido to be controlled by foreign seed companies". This is linked to the question of whether farmers in Hokkaido choose the type of large-scale agriculture that involves extensive pesticide use, or small-scale agriculture that does not. In other words, the choice of growing GM crops or not entails a choice by Hokkaido farmers of what type of agriculture to pursue. The future of agriculture in Hokkaido was therefore regarded as an important theme of debate, in addition to the potential risks of GM food to human health and the environment. Hence it was easy to focus on this topic and share mutual concerns about it.

Some of these informal meetings were set up by the staff of the Hokkaido prefectural government. Although these meetings lacked the attributes of legitimacy, openness and transparency, they helped to remove some of the mental obstructions in each actor's mind that impeded the bridging of frames. These casual meetings had some influence on the public decision-making process (for example, the Committee for Planting Conditions for GM Plants). Some participants worked on the Hokkaido GMO consensus conference. One of the factors that facilitated compromise was the fact that the main target of the decision-making process was limited to agricultural policy for a local area (Hokkaido), and was not a GMO policy for the whole country. Countrywide consensus-building is very difficult on the matter of GMO planting, but such informal meetings were useful in building consensus on a local scale.

The talks in the informal meetings familiarized each actor with the opinions and interests of others and helped them understand how others with different frames thought of the same problems. Even if these frames could not be shared among all participants, some frame transition did occur. This frame transition by some actors and the bridging of frames through these casual meetings facilitated public deliberation and compromise in the committees. Compromise in the more formal meetings in turn contributed to the formulation of GMO regulations in Hokkaido.

4-5-8 The Hokkaido GMO consensus conference

It was in these circumstances that one of the members of the Hokkaido Food Safety and Reliability Committee proposed holding a consensus conference on GMOs in Hokkaido. A GMO consensus conference had already been held in Tokyo in 2000 with funding from the Ministry of Agriculture, Forestry and Fisheries. A consensus conference is a type of participatory technology assessment. Besides Japan, such conferences have been held in Denmark, the United Kingdom and the Republic of Korea. It is a method of reflecting laypeople's views in the process of policy-making related to science and technology (S&T). After listening to experts' lectures, concerned citizens (laypeople) discuss the usage of new technologies and prepare a statement together. The aim is not to deepen the understanding of citizens about science. Rather, it is a procedure that makes it possible for concerned people to participate in S&T policy-making.

Participatory technology assessments are those in which citizens who wish to participate in setting the direction of scientific enquiry and the use of new technology can express their opinions. Several methods and systems have been devised for this purpose, including consensus conferences, scenario workshops and dialogue forums (used in Japan for the nuclear power plants in Rokkasho and Onagawa). Such methods are effective in addressing problems with which various stakeholders are associated.

The Hokkaido GMO consensus conference was held four times between November 2006 and February 2007 (see Table 4.5.3).[1] It was the first consensus conference sponsored by a local government in Japan; other consensus conferences had been held by researchers or foundations. Out of 97 applicants, 15 citizen committee members were chosen from across Hokkaido, with consideration given to a balance of age, sex, occupation and region. They included farmers, engineers, housewives and high school students. The chief facilitator was an associate professor in the Communicators in Science and Technology Education Program

Table 4.5.3 The Hokkaido GMO consensus conference, 2006–2007

	Theme	Date	Contents
1st meeting	Learning	25 Nov. 2006 11:00–18:30	Lectures by eight "specialists" to 15 civilian participants. Specialists were: local TV director who produced documentary programme on GMOs in Hokkaido; scientist researching GM crops; scientist against GMOs; farmer against GMOs; leader of consumers against GMOs; farmer for GMOs; farmer who tried to grow GM soybeans in 2002; Hokkaido Prefecture official (explaining Hokkaido's GMO ordinance)
2nd meeting	Confirmation of knowledge	2 Dec. 2006 11:00–17:10	Brainstorming; discussion by all 15 members; group discussions (divided into 3 groups) with 1 facilitator and 2 sub-facilitators
3rd meeting	Formulating key questions	16 Dec. 2006 11:00–17:10	Discussion and group discussions Members formulated key questions with facilitators and submitted them to other "specialists".
1st day of 4th meeting	Answers to key questions	3 Feb. 2007 11:00–21:30	Eight other specialists answered key questions by members. Specialists were: journalist opposed to GMOs; scientist who tried GM rice experiments in Hokkaido; two researchers in biotechnology; official of Japanese government's Food Safety Commission; official of Hokkaido prefectural government; a professor of agricultural economics; the former Vice-Governor of Hokkaido Prefecture.
2nd day of 4th meeting		4 Feb. 2007 9:00–18:00	Group discussion and writing of statement. The 15 participants submitted "A View of Hokkaido Citizens on Genetically Modified (GM) Farming in Hokkaido: A Statement by

Table 4.5.3 (cont.)

	Theme	Date	Contents
			Civilian Members of the Consensus Conference". This statement was reported in the Hokkaido Food Safety and Reliability Committee (19 March 2007).

(CoSTEP) of Hokkaido University. The venue of the conference was the "Red Brick Building" of the Hokkaido prefectural government, a famous tourist attraction in Sapporo.

On the final day of the conference, agreement was not easily reached. Members were divided between two opinions. Conservative opinion (which was the majority opinion) expressed reservations about agriculture with GM crops in Hokkaido. There was also a minority affirmative opinion on GM farming. Because the time for the conference was limited, both of these opinions about GMOs were included in the final statement. The researchers planning to plant GM rice for experimentation in open fields agreed to the final statement, but farmers considering planting GM soybeans expressed some discontent with it.

Even if agreement was impossible, the consensus conference was effective as one means of clarifying the points of argument. In this sense, the conference can be regarded as a success. A key to the success of the conference was the limit on the range of discussion. At this conference the region under discussion was limited to Hokkaido and the main topic was agriculture in Hokkaido.

4-5-9 Conclusion

Building a sustainable society requires public deliberation and participatory decision-making on environmental problems by citizens. Harm may be done to the environment of a region and the health of its citizens if the public does not receive relevant information and people are not granted the right and the opportunity to express their opinions. Governance of the bottom-up type is therefore necessary for the protection of regional environments.

In this type of governance, people from various backgrounds need to be able to exchange opinions. If a place for debate can be designed, even when frames cannot be shared among all participants, some frame transition can be expected to occur. Frame transition of this sort led to compromise in more formal meetings, which in turn contributed to the formulation of GMO regulations in Hokkaido. This accomplishment suggests that decision-making of the participatory type is effective in achieving sustainability governance.

Note

1. The author served as an adviser to the conference.

REFERENCES

Dryzek, J. S. (2000) *Deliberative Democracy and Beyond*. New York: Oxford University Press.

Fishkin, J. S. and P. Laslett, eds (2003) *Debating Deliberating Democracy*. Oxford: Blackwell.

Gutmann, A. and D. Thompson (2004) *Why Deliberative Democracy?* Princeton, NJ: Princeton University Press.

Ministry of the Environment, Government of Japan (2002) "Minamata Disease: The History and Measures", <http://www.env.go.jp/en/chemi/hs/minamata2002/summary.html> (accessed 12 July 2010).

Shrader-Frechette, K. (2002) *Environmental Justice: Creating Equality, Reclaiming Democracy*. New York: Oxford University Press.

Smith, G. (2003) *Deliberative Democracy and the Environment*. London: Routledge.

Yoshida, F. (2007) "Environmental Restoration of Minamata: New Thinking Brings New Advances", *Sustainability Science* 2(1): 85–93.

4-6

Science and technology communication

Hideyuki Hirakawa

4-6-1 Introduction

The promotion of science and technology (S&T) communication is one of the top priorities on the current agendas of science, technology and innovation policy in industrial countries. On the one hand, S&T communication is expected to enhance public understanding of the science and technology that sustain our societies, drive our economies and affect our lives. The products and processes of science and technology are so prevalent in our societies that we cannot live without an appropriate understanding of their concepts, logics, mechanisms and effects. On the other hand, S&T communication is also expected to promote experts' and policymakers' understanding of public needs and concerns regarding the development and uses of science and technology. In contemporary democratic societies, policymakers cannot obtain legitimacy for their decisions on scientific and technological development without public consent.

The same applies in the context of sustainability. Science and technology could be a source of difficulty, but at the same time could provide powerful and irreplaceable means to achieve the sustainability of society. It is vital to direct their development properly, maximizing their benefits while minimizing their adverse effects on society and the natural world. To accomplish this it is necessary to enhance S&T communication to enable the exchange and sharing of knowledge, techniques, skills, experiences, insights, opinions and values, and the building of cooperative relationships among actors in various fields, such as policymakers, scientists and engineers, industries, civil society organizations and individual citizens (Kasemir et al., 2003).

This section will illustrate various methods of S&T communication and their implications for sustainability. It first introduces two models of S&T communication, their objectives and historical backgrounds. Then it presents two examples of Public Engagement with Science and Technology (PEST) methods currently employed worldwide. In conclusion, some challenges for S&T communication in the pursuit of sustainability are discussed.

4-6-2 Models of science and technology communication and historical background

Two models of S&T communication and their objectives

There are various styles of S&T communication, but they can be classified into two types according to their objectives. One is Public Understanding of Science (PUS) or Public Understanding of Science and Technology (PUST), whose principal objective is to inform the public, fostering their interest in science and technology and their knowledge of scientific and technological facts, laws, methodologies, concepts and relevant social issues (that is, scientific literacy). The communication styles of this type are characterized by one-way, unidirectional communication that transfers knowledge from techno-scientific experts to laypeople by means of school science education, open lectures put on by universities and academic societies, exhibitions and events at science centres, publication of scientific magazines and books, scientific TV programmes and so on.

The origins of PUST go back to the popularization of science movement in the nineteenth century and the rise of modern science education in the early twentieth century. More recently, the publication of a report by the Royal Society of the United Kingdom, *The Public Understanding of Science* (Bodmer, 1985) – often referred to as the Bodmer Report – triggered a PUST movement in the United States and Japan as well as in other European countries. In the United Kingdom, for example, practical educational and science popularization initiatives such as Science, Technology and Engineering Week were launched, many of them funded through the Committee for Public Understanding of Science (COPUS). Behind this effort was a serious concern on the part of the scientific establishment, policymakers and industries over the decrease in students enrolling in science courses at universities, the decline of adult interest in and knowledge of science and technology, and the prevalence of public scepticism about techno-scientific development, which had been reinforced in past decades by various disasters including environmental problems.

The approach of PUST is sometimes characterized by the widely held belief expressed by the term "deficit model" (Irwin and Wynne, 1996). This belief has two aspects. The first is the idea that the source of public scepticism towards science and technology is primarily people's lack of adequate knowledge about science. The second is the idea that, if the public uptakes sufficient knowledge about science and technology, the scepticism will eventually be dispelled.

Another type of S&T communication is Public Engagement with Science and Technology (PEST), which first emerged in the late 1990s. In contrast to PUST, which is based on unidirectional communication from experts to laypeople, PEST is characterized by its emphasis on bi- or multi-directional communication, such as "dialogue" among experts, ordinary citizens, policymakers and various stakeholders, and "public participation" or "public engagement" in decision-making concerning the development and use of science and technology. Its primary objective is to facilitate the exchange of knowledge, insights, opinions and values among various actors including lay citizens. By so doing, it aims to enhance their *mutual* learning from different perspectives on the social, political, economic, cultural, ethical and legal implications as well as technical aspects of science and technology.

There are various expected effects of this type of communication. First, ordinary citizens can not only acquire a technical knowledge of the science or technology in question but also learn about the different views and opinions of others on the same issues. Second, PEST facilitates the "visualization" of the problem structure. Through discussions among participants with different perspectives, various aspects of problems and ideas come to light. This process of problem visualization is important because some aspects and insights might be unavailable if the discussion is carried out only among experts and policymakers. Finally, PEST can also bring about political effects such as building mutual trust and cooperative relationships among stakeholders and providing a democratically legitimate basis for decision-making.

Historical background of changes in S&T communication

Theoretically speaking, the shift from PUST to PEST is characterized by a move of thematic focus from differences in knowledge content per se to differences in the problem framing through which one contemplates and argues an issue, as well as an emphasis on the social, cultural and political dimensions of the discourse on science and technology. How did such a shift occur?

The most decisive event was the UK government's announcement about the risk of Bovine Spongiform Encephalopathy (BSE) and its political consequences, namely the "BSE shock". Since the early years of the BSE saga in UK society, the government had repeatedly claimed that BSE posed no danger through the human consumption of beef and beef products, particularly based on the conclusions of a scientific risk assessment made by the Southwood Working Party (MAFF/DoH, 1989). On 20 May 1996, however, the government publicly acknowledged that BSE could infect humans and cause a similar disease called variant Creutzfeldt-Jakob disease (vCJD). As a consequence, profound public distrust of the government prevailed and people acquired a shared recognition of the fundamental scientific uncertainty that there could be unknown risks that even the best available science cannot foresee. In this regard, the PUST approach to dispelling public scepticism towards science and technology by providing the public with well-established knowledge and information lost its persuasive power, because the most important lesson from the BSE shock was that there could always be "unknown unknowns" (Grove-White, 2001; Wynne, 2002) outside the scope of current scientific knowledge: science does not even know what it does not know. As a result of the prevalence of this notion as well as distrust of government, PUST efforts completely failed to ease public anxiety and antagonism towards genetically modified (GM) crops and foods imported from the United States, which first appeared in European markets in 1996. The cause of public anxiety was not the lack of scientific knowledge, or a "knowledge deficit", on the part of the public, but their recognition of scientific uncertainty as well as the loss of public confidence in governmental decision-making, that is, a "democracy deficit".

The shift to PEST was a direct response by governments in the United Kingdom and other European countries as well as by scientific communities to this crisis of confidence in both science and politics triggered by the BSE shock and other problems. In fact, the UK Parliament published two reports calling for enhancing dialogue, participation and engagement in the fields of science and technology: *Science and Society: Third Report* (House of Lords Select Committee on Science and Technology, 2000) and *Open Channels: Public Dialogue in Science and Technology* (Parliamentary Office of Science and Technology, 2001). In the European Union, the European Commission published a report, *Democratising Expertise and Establishing Scientific Reference Systems: White Paper on Governance* (European Commission, 2001), based on recognition of the need for dialogue as follows:

> In short, we witness the paradox of expertise being a resource that is increasingly sought for policy making and for social choice, but one that is also increasingly contested. Efforts to restore the credibility of expertise, and trust in it, are vitally important. But they cannot be confined to "educating the public": the very process of developing and using expertise needs to be made more transparent and accountable, and sustained dialogue between experts, public and policy makers needs to be pursued. (European Commission, 2001: 2)

Similar changes also took place in Japan. One historic turning point was the year 1995, when various disastrous accidents and incidents related to science and technology occurred. In the early morning of 17 January 1995 the Hanshin-Awaji earthquake struck the Kansai area, including Kobe city, a major port and tourist destination, causing more than 6,000 casualties. On 20 March, members of the religious cult Aum Shinrikyo released lethal sarin gas in several downtown Tokyo subway trains during the morning rush hour. At the end of the same year, on 8 December, the Monju fast-breeder reactor in Tsuruga, Fukui Prefecture, suffered a severe sodium-leakage fire accident. As a result of these and similar events, Japanese policies related to science and technology, especially nuclear safety policy, have gradually changed to incorporate democratic values such as dialogue and consensus. As in European countries, an anti-GM movement also gained ground in the late 1990s, resulting in Japan's first government-funded participatory technology assessment (described later) of GM foods and crops in 2000. More generally, the Science and Technology Basic Plan in its 2nd (2001–2005) and 3rd (2006–2010) periods subsequently called for strengthening ties between science, technology and society by enhancing the accountability and outreach activities of research and promoting active public participation in science, technology and relevant policy-making. Since the autumn of 2005, several educational programmes for S&T communication in graduate schools have been subsidized by the government, such as the Communicators in Science and Technology Education Program at Hokkaido University, the Science Interpreter Training Program at The University of Tokyo, and the Master of Arts Program for Journalist Education in Science and Technology at Waseda University.

4-6-3 Variations in new styles of science and technology communication

In this subsection, examples are presented of two PEST methods currently employed worldwide. One is the science café and the other is the participatory technology assessment (pTA).

Science cafés

The science café is currently the most popular and simple style of PEST-type communication. It is a dialogue event on scientific topics held at a café or pub where ordinary people can get together and talk to one another in a relaxed atmosphere. The movement was initiated under the name *Café Scientifique* by journalist Duncan Dallas in Leeds, UK, in 1998 and has rapidly spread globally since the early 2000s. Dallas himself was inspired by the *Café Philosophique* cultural movement that had been popular in France since 1992. In Japan, the science café movement started around 2004, triggered in part by the publication of the *White Paper on Science and Technology 2004* by the Ministry of Education, Culture, Sports, Science and Technology (MEXT, 2004), which introduced the concept in a column. During Science and Technology Week in April 2006, the Science Council of Japan and the Japan Science and Technology Agency co-sponsored science café events in 21 cities around Japan. Nowadays more than 10 such events are held every week somewhere in Japan by various hosts such as universities, research institutions, academic societies, science centres, non-profit organizations and citizen groups.

The format of the science café varies. The most popular one is the British style, in which one expert gives a talk about a topic related to his or her own expertise for 20–30 minutes, followed by a break and a discussion with participants for 60–90 minutes. In France, on the other hand, to ensure a lively discussion, several speakers are invited from different disciplines and positions on the designated topic and talk for only a few minutes. In Denmark, as in France, several speakers are invited from different disciplines, such as science and the arts. They give a speech for 30 minutes, then take a break and have a discussion with participants.

Participatory technology assessment

Participatory technology assessment (pTA) is a branch of technology assessment (TA) practices. TA was initiated officially by the Office of Technology Assessment of the US Congress in 1972 and later spread to European countries. The traditional practices of TA were carried out by experts in academic fields related to the technology in question. In pTA, by contrast, it is a panel of laypeople that evaluates the technology. This format was first employed by the Danish Board of Technology (DBT) in Denmark in 1987 and has prevailed in other European countries since the mid-1990s.

What are the benefits of conducting pTA? One is that it enables light to be shed on various aspects of the problems regarding the technology in question by letting participants (that is, a citizens' panel) raise questions. In the process of pTA, it is the citizens' panel that poses the questions, while the role of experts is to answer them. In addition, it is important that the "lay citizens" be a diverse group of people, each with his or her own vocation, experience, knowledge, values, life history and vision for the future, based on which the panel can raise and discuss questions from diverse perspectives. Furthermore, it is quite common for some of these participants to be technical experts in fields other than the technology in question, in addition to some with expertise closely related to the issue. In any case, the citizens' panel can provide a valuable set of questions and insights whose scope is far wider and more comprehensive than the narrow perspectives of experts in the technology in question. In other words, they form a panel of "extended peer review" (Funtowicz and Ravetz, 1990, 1991, 1994).

Another merit of pTA is to include questions regarding the "utility" of a given technology. In the history of risk assessment of technology, questions of utility – such as, "Is the technology truly useful? If so, in what respects? For whom?" – have often been left outside the scope of enquiry. The claim that the technology or technological product is useful is left as an unspoken and unchallenged assumption in evaluating the impacts of technology. However, the claim of utility is often nothing more than what developers and promoters of the technology allege and is not necessarily shared by the public or supported by empirical evidence that it really is beneficial to society or the environment. In addition, the question of utility entails value judgements that are beyond the purview of technical reasoning. It is a question that must be addressed democratically. In order to develop and use the outcomes of science and technology appropriately for society and the environment, it is vital to ask the question of utility. In this regard, the DBT made the following declaration:

> Today the assessment and regulation of risks related to technology are carried out in a backward manner, and there is a need to turn the processes around. Rather than experts starting by analyzing risks, laymen should start by formulating questions to the experts. And rather than having the utility of a given technology as an unspoken precondition for risk analysis and assessment, the discussion of utility value should be tied in with the discussion about risk. This is the essence of a proposal for a new way of handling manmade risks, devised for the Danish Board of Technology by a broadly selected work group. (DBT, 1999)

Examples of methods of participatory technology assessment

Various methods of pTA have been invented and implemented in Denmark and other countries. They are largely classified into two types in terms of the attributes of the participants. One comprises methods in which the participants are members of the general public who do not have any immediate stakes in the development and use of the technology in question. Another type includes methods in which stakeholders such as policymakers, politicians, companies and affected citizens evaluate the technology. A typical example of the former is the consensus conference, whereas examples of the latter are the scenario workshop and the future search.

Steyaert and Lisoir have selected 13 participatory methods available for pTA and described their features (Steyaert and Lisoir, 2005). Table 4.6.1 is a comparative chart that describes the features of each method in terms of objectives, characteristics of topics, attributes of participants, time taken to convene the event and financial cost. Details of two examples of pTA methods presented in the table, the citizens' jury and the consensus conference, are explicated below.

Example 1: Citizens' jury

The citizens' jury is a method of obtaining informed citizen input into policy-making through citizen jurors' deliberations on and assessment of the social aspects of new technologies. It was originally developed by the Jefferson Center in the United States in 1974. In the United Kingdom, since being initially introduced in the medical field in 1996, citizens' juries have been conducted on many occasions. Generally, the method has been applied to a wide range of topics, including economic, environmental, social and political issues. In relation to science and technology, a typical issue to which this method was applied in the late 1990s was GM technology; one of the major issues of the 2000s is nanotechnology.

According to Steyaert and Lisoir (2005), the citizens' jury is particularly useful for building a bridge between the jury and the broader public, as well as for enabling participants to carry out thoughtful value-based discussions in developing policy recommendations. It is most applicable when one or more alternatives to a problem need to be selected and the various competing interests arbitrated.

A citizens' jury comprises 12–24 citizens selected in a demographically random manner. In the process of deliberation, the jurors form subgroups, each of which focuses on a different aspect of the issue, and they are informed of several perspectives on the issue by experts called "witnesses". Finally, the jurors produce a decision or provide recommendations in the form of a citizens' report. The sponsoring body (for example, a government department or local authority) is required to respond to the report either by acting on it or by explaining why it disagrees with it. Usually it takes four or five days over several weekends to complete the process of citizens' deliberation.

Table 4.6.1 Comparative chart of participatory methods

		Topic[a]					Time		Cost
Method	Objectives	Knowledge	Maturity	Complexity	Controversial	Participants	Event	Total	(1–4)
21st century town meeting	Engage thousands of people at a time (up to 5,000 per meeting) in deliberation about complex public policy issues	+	+/–	+	+/–	Anyone	1–3 days	A year	4
Charrette	Generate consensus among diverse groups of people and form an action plan	+/–	+/–	–	+/–	Average citizens or stakeholders. Others give input	1–5 days	2–3 months	3
Citizens' jury	A decision that is representative of average citizens who have been well informed on the issue	+/–	+/–	+/–	+	12–24 randomly selected citizens. Experts, stakeholders and politicians give input	3 days	4–5 months	4
Consensus conference	Consensus and a decision on a controversial topic	+	+/–	+	+	10–30 randomly selected citizens. Others give input	3 weekends	7–12 months	4

Deliberative polling©	Get both a representative and an informed (deliberative) view of what the public think and feel about an important public issue	–	+/–	–	+/–	A random and representative sample of the population	1 day	8 months	4
Delphi	Expose all opinions and options regarding a complex issue	–	–	+	+/–	Experts	Variable	Variable	1–3
Expert panel	Synthesize a variety of inputs on a specialized topic and produce recommendations	–	–	+	+/–	Experts	Variable	Variable	2
Focus group	Expose different groups' opinions on an issue and why these are held (reasoning)	+/–	–	m	+/–	Stakeholders and/or citizens	2 hours–1 day	1 month	1
PAME	Evaluating and learning	+/–	+/–	+/–	+/–	All stakeholders	Variable	Variable	Variable

Table 4.6.1 (cont.)

Method	Objectives	Topic[a]				Participants	Time		Cost
		Knowledge	Maturity	Complexity	Controversial		Event	Total	(1–4)
Planning cells	Citizens learn about and choose between multiple options regarding an urgent and important issue. Develop action plan	+/–	–	m	–	25 average citizens. Experts and stakeholders present positions	5 days	5 months	4
Scenario-building exercise	Planning and preparedness for uncertain future. Vision-building	–	–	+	+/–	Anyone	2–5 days	6 months	1–3
Technology festival	Provide a means for public debates about societal issues of science and technology	–	–	+/–	+/–	Anyone	1–2 days	6–12 months	4
The World Café	Generating and sharing ideas	+/–	–	–	+/–	Anyone	4 hours–1 day	1 month	1

Note: +/– means that the method can address subjects with either + or –. Cost: 1 = inexpensive; 2 = moderate; 3 = expensive; 4 = very expensive.

[a] Explanation of chart symbols:

Topic	+	m = medium	–
Knowledge	A lot of common knowledge exists		There is little common knowledge
Maturity	Most people have already formed opinions on the subject		The subject is new; people are still forming their opinions
Complexity	Highly complex or technical		Not very complex or technical
Controversial	Highly controversial		Not very controversial

Source: Steyaert and Lisoir (2005: 27).

One good example of a citizens' jury is the NanoJury UK jointly held in June and July 2005 by the Cambridge University Nanoscience Centre, Greenpeace UK, *The Guardian* newspaper and the Policy, Ethics and Life Sciences Research Centre (PEALS) of Newcastle University (Doubleday and Welland, 2007; Pidgeon and Rogers-Hayden, 2006). In this project, 15 randomly selected people from different backgrounds in Halifax, UK, heard evidence about nanotechnologies and their possible roles in the future. The result of the jurors' discussion was filed as a set of policy recommendations (NanoJury UK, 2005).

Example 2: Consensus conference

The consensus conference, including its variants, is one of the most popular methods of pTA in the world. It was originally developed in the United States, then refined for pTA in the late 1980s by the Danish Board of Technology. Since the mid-1990s and particularly during the early 2000s, with the emergence of controversial issues such as GM foods and crops, it has diffused throughout Europe and other parts of the world. The first series of consensus conferences in Japan, which was the first example of pTA conducted in Japan, was convened by a researchers' group on the themes of genetic therapy in 1998 and the advanced information society in 1999. In 2000, the Japanese Ministry of Agriculture, Forestry and Fisheries (MAFF) held a consensus conference on the theme of GM foods, the results of which were input to a research and development policy by MAFF. Since then, several consensus conferences and their variants have been held by researchers' groups and local governments in Japan.

The consensus conference is a public inquiry conducted by 10–30 lay citizens (a "citizens' panel"), who carry out an assessment of a socially controversial topic concerning technology. In the procedure of the conference (Figure 4.6.1), the panel of citizens propound a set of questions ("key questions") to a panel of experts (the "experts' panel"), evaluate the experts' answers, debate among themselves and finally produce a consensus statement.

The statement is filed in the form of a report, including the expectations, concerns and recommendations of the citizens' panel, and presented to policymakers, relevant expert communities and the general public. It takes three or four days, usually spread over weekends, to convene a consensus conference. All or a large part of the conference is open to the public.

The name notwithstanding, arriving at a consensus opinion is not a prerequisite for consensus conferences as practised in most countries, although in Denmark it is (Steyaert and Lisoir, 2005). The primary objective of the method is to broaden the perspective of debate on a topic by promoting the exchange of as many divergent views and opinions by

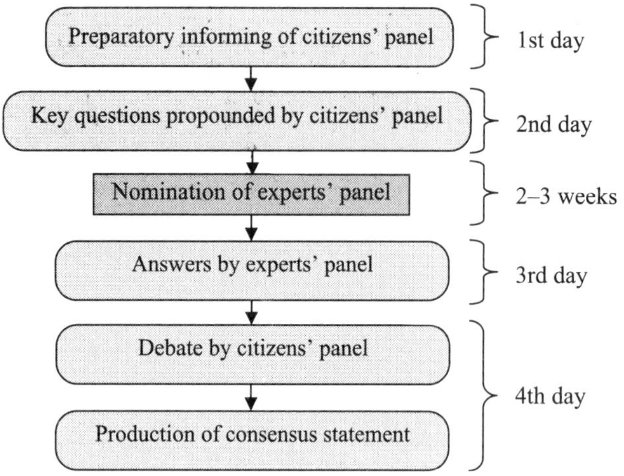

Figure 4.6.1 Key steps in the consensus conference.
Source: Steyaert and Lisoir (2005: 70).

non-expert citizens as possible. In this regard, the members of a citizens' panel are not required to be representative of a given population. In addition, the most beneficial effect of the method is to combine diverse forms of knowledge, including the local, traditional knowledge that lay citizens possess, and to utilize them with the technical knowledge of scientific experts regarding the technology in question. According to Steyaert and Lisoir (2005), this method is useful when all or most of the following criteria are present:

- citizen input is required for policies under review or development;
- issues are controversial, complex and/or technical;
- many diverse groups and individuals have concerns;
- ensuing decisions significantly and directly affect select groups or individuals;
- there is a need for increased public awareness and debate;
- there is citizen desire for a more formal involvement;
- the process of communicating information about the conference topic provides a strong educational component.

As mentioned above, the Japanese government held a participatory technology assessment of GM crops in 2000. The method employed was the consensus conference. The conference was conducted by the Society for Techno-innovation of Agriculture, Forestry and Fisheries (STAFF) under MAFF sponsorship from September to November 2000 with the aim of drawing up policy recommendations for research and development by MAFF, especially those associated with the safety of GM foods and crops. A panel of 18 citizens was selected from 479 applicants in a demographically random manner. From 2006 to 2007, a regional government, Hokkaido Prefecture, also convened a consensus conference on GM crops. The objective was to conduct public consultation about the possibility of the cultivation of GM crops in the prefecture. At both conferences, discussions reflected the diverse perspectives of citizen panellists on a broad range of issues such as food and environmental safety, the adequacy of regulatory policies and institutions, and the socioeconomic impacts of the commercial cultivation of GM crops.

4-6-4 Conclusion: Challenges for science and technology communication for sustainability

To conclude this section, some challenges for S&T communication in the pursuit of sustainability are presented. The process of S&T communication is still under development and its effectiveness needs to be tested empirically and theoretically on many occasions. In fact, the history of PEST-style communication practices is barely a decade long, so various experiments are still being conducted globally in order to improve PEST methods. Here, three major challenges for S&T communication are highlighted that have particular importance for Japan.

The first challenge for Japanese S&T communication activities is to enlarge the range of actors involved. So far, the target of Japanese S&T communication has been exclusively focused on the relationship between scientific and technological experts and the lay public. However, in order to develop and utilize science and technology in pursuing the sustainability of society, communication among all relevant actors is vital. S&T communication must be enhanced not only between experts and citizens but also among policymakers, experts in various disciplines, industries and civil society organizations.

The second challenge for S&T communication is to move "upstream" in the process of developing science and technology (see DEMOS, 2004). The traditional focus of S&T communication has been on the later phases of the innovation process, when the development of a new technology is almost completed or the technology has been put into practice in society. At that stage it is very difficult to change the applications or specifications of a technology by political will. In other words, public engagement in this later phase is in vain because there is little room for it to influence the course of innovation. Moreover, as discussed above, one of the virtues of pTA is that it addresses the question of utility, which is exclusively an upstream question. Finally, what is vital in pursuing sustainability is the strategic intention to build and rebuild comprehensive policy agendas integrating science and technology policy with other relevant social policies. This inevitably involves exchanges and interactions among diverse perspectives, values, interests and visions of possible futures, as well as the production and exchange of knowledge. In this regard, S&T communication among diverse actors needs to engage in formulating policy strategies and to move to the upstream phase of innovation, because strategy formulation is intrinsically an upstream activity.

The last challenge is to strengthen the incentives for academic experts to engage in S&T communication. Although their engagement is indispensable for effective communication, generally it is difficult for them to do so for several reasons. One is the lack of interest in S&T communication. Most scientists are so busy doing their own research that they are not interested in talking with outsiders. Secondly, if they have an interest in communicating with others, they tend to hesitate to do so because they have not been accustomed to speak before a general audience. Throughout their educational and professional career, their vocabulary and mindset have been so specialized in their narrow discipline that they have learned to be reluctant to communicate with outsiders. Lastly, the lack of academic credit for engaging in S&T communication makes researchers hesitate to do so. In the current academic evaluation system, contributing to communication with the public is not counted as an academic achievement. In any case, a range of institutional supports to scientists must be built in order to enhance their engagement in S&T communication.

REFERENCES

Bodmer, W. (1985) *The Public Understanding of Science*. London: Royal Society.

DBT [Danish Board of Technology] (1999) "Communication about Risk: Let Laymen Lay the Foundations", *Newsletter of the Danish Board of Technology*, No. 127, April. Available at <http://www.tekno.dk/subpage.php3?article=460&toppic=kategori10&language=uk> (accessed 1 June 2010).

DEMOS (2004) *See-through Science: Why Public Engagement Needs to Move Upstream*. London: DEMOS.

Doubleday, R. and M. Welland (2007) "NanoJury UK: Reflections from the Perspective of the IRC in Nanotechnology and FRONTIERS", FRONTIERS. Available at <http://www.frontiers-eu.org/archive/Nanojury%20final%20reflections%20Mar-07.pdf> (accessed 1 June 2010).

European Commission (2001) *Democratising Expertise and Establishing Scientific Reference Systems: White Paper on Governance*. Report of Working Group 1b. Brussels: European Commission.

Funtowicz, S. O. and J. R. Ravetz (1990) *Uncertainty and Quality in Science for Policy*. Dordrecht: Kluwer Academic Publishers.

Funtowicz, S. O. and J. R. Ravetz (1991) "A New Scientific Methodology for Global Environmental Issues", in Robert Costanza (ed.), *Ecological Economics: The Science and Management of Sustainability*. New York: Columbia University Press.

Funtowicz, S. O. and J. R. Ravetz (1994) "The Worth of a Songbird: Ecological Economics as a Post-normal Science", *Ecological Economics* 10: 197–207.

Grove-White, R. (2001) "New Wine, Old Bottles: Personal Reflections on the New Biotechnology Commissions", *Political Quarterly* 72: 466–472.

House of Lords Select Committee on Science and Technology (2000) *Science and Society: Third Report*. House of Lords, United Kingdom.

Irwin, A. and B. Wynne (1996) *Misunderstanding Science? The Public Reconstruction of Science and Technology*. Cambridge: Cambridge University Press.

Kasemir, B., J. Jäger, C. C. Jaeger and M. T. Gardner (2003) *Public Participation in Sustainability Science: A Handbook*. Cambridge: Cambridge University Press.

MAFF/DoH (1989) *Report of Working Party on Bovine Spongiform Encephalopathy*. London: Ministry of Agriculture, Fisheries and Food/Department of Health.

MEXT [Ministry of Education, Culture, Sports, Science and Technology] (2004) *White Paper on Science and Technology 2004*. Tokyo. Available at <http://www.mext.go.jp/english/news/2005/04/05051301.htm> (accessed 1 June 2010).

NanoJury UK (2005) "Our Provisional Recommendations", <http://www.nanojury.org.uk/pdfs/recommendations.pdf> (accessed 1 June 2010).

Parliamentary Office of Science and Technology (2001) *Open Channels: Public Dialogue in Science and Technology*. Parliamentary Office of Science and Technology (POST), House of Commons, United Kingdom.

Pidgeon, N. and T. Rogers-Hayden (2006) "Reflecting upon the UK's Citizens' Jury on Nanotechnologies: NanoJury UK", *Nanotechnology Law and Business* 2: 3.

Steyaert, S. and H. Lisoir (2005) *Participatory Methods Toolkit: A Practitioner's Manual*. King Baudouin Foundation and the Flemish Institute for Science and Technology Assessment (viWTA). Available at <http://www.eukn.org/eukn/themes/Urban_Policy/Social_inclusion_and_integration/Community_development/Capacity_building/participatory-manual_1003.html> (accessed 1 June 2010).

Wynne, B. (2002) "Risk and Environment as Legitimatory Discourses of Technology: Reflexivity Inside Out?", *Current Sociology* 50: 459–477.

4-7

Global governance

Makiko Matsuo, Hirotaka Matsuda and Hideaki Shiroyama

4-7-1 Introduction

The last several decades have witnessed serious challenges to global sustainability, such as climate change, biodiversity loss, fishery depletion, food insecurity and increasing poverty. These issues are complex in terms of policy areas and actors. Climate change is a good example of this complexity. The environmental effects of climate change can have a huge impact on socioeconomic systems. For example, extreme weather can reduce crop yields, which in turn can affect food security. Global warming may increase the spread of infectious diseases, which can have serious impacts on human health. A rise in sea level can put coastal communities in danger. Dealing with the effects of climate change requires actions across many policy areas by various relevant actors. And, despite the fact that global sustainability is recognized as a "common" challenge, the actors involved have different stakes and motivations. In addition, actions that either cause or mitigate global sustainability problems produce effects across territorial borders. Consequently these challenges cannot be solved by unilateral actions confined within national borders. A mechanism that guides concerted action at the global level is required.

The institutional arrangement of global governance as a tool is necessary to deal with these challenges. Global governance can serve as an effective tool in addressing global sustainability issues that require governance across national borders. The key function of global governance is *coordination* across various dimensions. Successful coordination can be realized through the concept of *doushouimu* (the Japanese word for a Chinese proverb, the literal meaning of which is "sharing the same bed, dreaming different dreams"), which implies coexistence among actors without the sacrifice of each actor's interests and concerns (see the discussion of *doushouimu* in Section 4-2 of this chapter).

Four dimensions of coordination are introduced in this section: (1) international coordination, (2) inter-regime coordination (including international organizations), (3) coordination with non-state actors, and (4) coordination of science and politics. The objective of this framework is to identify coordination issues in each of these dimensions as part of the design of global governance on sustainability issues. As a case study, the global governance of agri-food is examined, with particular attention to the problems of securing the quantity and quality of food. This subject was chosen because its multifaceted nature is illustrative of the complex characteristics of sustainability issues, and is illuminative in describing global governance. There are many international bodies that have interests related to agri-food issues, such as the Food and Agriculture Organization of the United Nations (FAO), the World Food Programme, the United Nations Environment Programme, the United Nations Development Programme (UNDP) and the World Bank. Even within the UN system, at least 30 bodies are said to have some special interest related to food (Shaw, 2009). Amongst various international bodies whose mandates are related to food, two international bodies have been selected as case studies, namely, the Consultative Group on International Agricultural Research (CGIAR) and the Codex Alimentarius Commission, because they are illustrative in highlighting various dimensions of coordination among divergent actors and issues presented in this section.

The framework of the four dimensions of coordination in global governance is described in subsection 4-7-2. Subsection 4-7-3 examines the nature of agri-food issues, and subsection 4-7-4 offers an analysis of global governance of agri-food, focusing on the four dimensions of coordination and using CGIAR and Codex as examples. Challenges for future studies of global governance towards the achievement of global sustainability are explored in the final section.

4-7-2 Coordination in global governance

The concept of global governance

Governance refers to "the process and institutions, both formal and informal, that guide and restrain the collective activities of a group" (Keohane and Nye, 2000: 12). As implied by Rosenau's famous "governance without government" (Rosenau, 1992), governance is not synonymous with government. It involves not only governmental activities but also activities of private companies, non-governmental organizations (NGOs), scientific communities and others. Hence, global governance can be understood as the governance of various types of global issues at the global level, including horizontal coordination among a variety of global actors not limited to nation-states.

Modes of global governance

Designs for the structures of global governance have been envisaged in the existing literature on global governance. Many studies have acknowledged the limits and weaknesses of the existing global governance framework in the management of sustainability concerns, and they contend that it should be reformed. The most ambitious proposal for global governance would be to set up a new authoritative single entity, exemplified by the constitutional approach described by Mitrany (1933). Others have suggested more modest alternatives. Haas (2004) presented a "network model of decentralized global governance", a form of global governance based on diffuse networks of diverse actors. Keohane and Nye (2000) suggested "network minimalism", a governance framework that requires extensive networked cooperation, with states remaining as the primary actors, supplemented by private entities and NGOs. In the context of current international relations, the creation of a new single organization seems rather naïve and ambitious, and the network model seems more realistic and compelling. How such a network-based mode of governance would work is still under consideration, however, and it would need to overcome many practical challenges in actual operation.

At this point, there seem to be no universal modes of global governance that can be applied to all cases. This section therefore works from the premise that there can be divergent modes of global governance. This turns attention to the functions and dimensions of coordination in global governance, rather than contemplation of the design of the overall governance architecture. Regardless of the mode of global governance, coordination among various actors and issues is critical.

Four dimensions of coordination

In an effort to simplify the complicated and interlinked issues associated with coordination, the following four dimensions of coordination are offered: (1) international coordination, (2) inter-regime coordination (including international organizations), (3) coordination with non-state actors, and (4) coordination of science and politics.[1] The importance of the first three dimensions of coordination has been acknowledged in previous studies of the analysis of dimensions of international administration (Shiroyama, 1997, 2001). However, the fourth dimension has been added to this framework because the relationship between scientific assessment and policy-making is an indispensable element in dealing with sustainability issues. As in the case of climate change, many of the measures taken for global governance are based on scientific assessment. The framework presented here can help identify various coordination dimensions in the consideration of global sustainability issues and reveal the challenges posed in each dimension. It also serves as a useful analytical tool for scholars of sustainability sciences as well as for practitioners.

International coordination

International coordination is coordination among states over national interests. The emergence of global challenges to sustainability makes it difficult for states to implement solutions unilaterally, which in turn leads states to engage in international cooperation. However, states are positioned to pursue national interests consisting of multiple objectives. They therefore have to engage in the so-called two-level game, dealing with both domestic and international politics (Putnam, 1988). The nature of international anarchy and sovereignty causes states to be concerned with relative gains even under the conditions of positive-sum games (Grieco, 1998). In an effort to maximize their national interests, states strategically choose to go into bilateral or multilateral negotiations, engage in forum-shopping to find rules that best suit their aims and form various types of coalition.

As Garrett Hardin's famous analogy of the "tragedy of the commons" predicts, "freedom in a commons brings ruin to all" (Hardin, 1968: 1244). For the promotion of sustainability, it is necessary to institutionalize rules, norms and agreements that restrict some of the freedom of self-interested states. It is not always a bad thing for states to restrict their own freedom through institutions, because international institutions among states can enhance the prospect of cooperation by decreasing transaction costs and uncertainties, avoiding moral hazard through monitoring, and raising the cost of deception (Keohane, 1984). A central question for global governance concerning the management of international coordination, therefore, is the degree of centralization: the extent to which states delegate their authority by creating informal or formal institutional structures, rules and procedures for decision-making, and voluntary or mandatory agreements.

Inter-regime coordination

Inter-regime coordination takes place between international regimes (often between international organizations as the main actors of international regimes) across the boundaries of policy fields. There has been much debate on the autonomy of international regimes and institutions (Keohane and Martin, 1995; Mearsheimer, 1994/95; Simmons and Martin, 2002). However, it is becoming increasingly evident that international regimes are an important element of global governance.

Historically, international regimes and organizations have evolved within specific functional issue-areas (Shiroyama, 1997). As a consequence, most existing regimes act to promote their own mandates within sector-based issue-areas, which makes global governance appear to be fragmented when faced with sustainability issues that require intersectoral action. For example, the World Trade Organization (WTO) has a primary mandate of trade liberalization, yet more cases that touch the domains of other regimes, such as the environment and labour, are now brought to the WTO for dispute settlement. Decisions in one regime inevitably influence behaviour in others.

Inter-regime coordination is imperative in this context. Without coordination, relations among regimes remain competitive and inconsistent owing to their conflicting objectives, and policies are rendered inefficient by overlapping or duplicated responsibilities. A central question is how to coordinate inter-regime relations by enhancing institutional interactions and synergies through the use of an integrated approach.

Coordination with non-state actors

Coordination with non-state actors involves coordination among actors over their own interests and missions. Non-state actors are gaining increased prominence at the global level; typical non-state actors are multinational corporations and NGOs. Firms possess technological knowledge and experts, and they use their strong ties with decision-makers or attempt to leverage standardization to enhance their commercial interests (Shiroyama, 2007). NGOs try to achieve their specific goals through advocacy. They are quick to respond to some issues on which governments often take slower action, and are also good at mobilizing public awareness. Multinational corporations and NGOs can have contentious relationships. However, there is also a movement towards industry–NGO partnerships. For example, the DuPont chemical company and an environmental NGO, Environmental Defense, jointly established a framework for the responsible development of nanotechnology in 2005. Another example of such a partnership is the US Climate Action Plan, which is a group of major business and environmental NGOs whose aim is to prompt the US government to take measures in reducing greenhouse gas emissions. These examples are of partnerships mainly in the domestic context, but partnerships can play important roles in global governance too.

Because these activities by non-state actors are driven by their own interests or missions, the legitimacy and accountability of their actions are often questioned. However, because public solutions are not the only way to tackle global issues, the complementary role of non-state actors in global governance should not be dismissed. Ruggie (2004: 519), for instance, used the example of the UN Global Compact to show how private voluntary efforts compensate for governance gaps, and argued for the emergence of a new global public domain where interactions between NGOs and transnational firms occur, in addition to those of traditional nation-states.

One of the challenges associated with global governance is to examine how to balance interests among actors, make use of voluntary initiatives and incorporate expertise and technical knowledge, as well as answer the legitimate concerns of non-state actors, particularly where public action is lacking or absent. In addition, ensuring the transparency of non-state actors is necessary as a measure towards resolving their inherent accountability issues.

Coordination of science and politics

The coordination of science and politics occurs at their interface in the process of global policy-making. Scientific assessment in policy-making is becoming an increasingly important element of global governance when addressing sustainability issues. Questions arise over how science should be reflected in policy measures in combination with various social values. In analysing the dynamic interface between science and politics, it is important not only to see the role of experts and the epistemic community in the decision-making process (Haas, 1992) but also to reveal the process of how and by whom knowledge and scientific assessments are framed, interpreted and produced (Litfin, 1994). Scientific assessment is not necessarily an objective activity, but is one that also involves framing subjects with political implications.

When introduced to society, scientific information needs to be transformed into a suitable form. It must be scientifically sound but at the same time politically acceptable (Agrawala, 1998). Even when the academic/scientific community is directed towards one solution, politics deals with people in a local context who often possess diverse values and different attitudes towards the scientific community.

At the heart of the global governance of sustainability, in light of the interaction between scientific assessment and policy-making, is the definition of the roles of scientific experts and policymakers, how they interact, and how science and social values are to be balanced and incorporated into policy.

4-7-3 Agri-food issues as an example of a sustainability issue

Securing the "quantity" and "quality" of food: CGIAR and Codex

The Rome Declaration, adopted at the 1996 World Food Summit, states the following with regard to world food security: "We ... reaffirm the right of everyone to have access to safe and nutritious food, consistent with the right to adequate food and the fundamental right of everyone to be free from hunger" (WFS, 1996). It is clear from this statement that there are two distinct aspects of food security – "quantity" and "quality". These two aspects must be taken into account when considering global agri-food governance.

CGIAR and Codex, two of the international bodies chosen for case studies in this section, have played critical roles in the governance of global agri-food issues by addressing these two major concerns. CGIAR has contributed to securing food quantity as well as quality by coordinating international efforts for technology development in agricultural research to increase agricultural productivity and improve food quality. Codex has contributed to food safety as well as quality by setting international food safety and quality standards with a view to improving human health and nutrition.

The multifaceted nature of agri-food issues

Both organizations are faced with common challenges in the global governance of sustainability. They both have to manage the whole array of interlinked and interconnected policy concerns with the various relevant actors.

In the early years, global governance of the quantity and quality of food seemed simpler. When CGIAR was successful in realizing the Green Revolution in its early years, its policy concerns were primarily limited to a few commodities, especially staple foods, and to the improvement of technology to increase agricultural productivity. However, as other issues, such as loss of genetic diversity and soil degradation, became increasingly viewed as relevant to food production, CGIAR had to expand its focus to include resource management (particularly of genetic resources), biodiversity and the entire social and economic infrastructure and system.

A similar policy expansion has occurred with Codex. Although consideration of fair trade practices and consumer protection was acknowledged from the beginning, it was not until the Codex standard became the formal WTO reference point that the socioeconomic aspects of food safety standards were so strongly highlighted. The issue of the relationship of free trade and food safety standards has been the subject of a good deal of discussion in the so-called "trading up and racing to the bottom" debate (Drezner, 2001; Kollman, 2003; Porter, 1999; Vogel, 1995; Vogel and Kagan, 2002; Wheeler, 2001). More recently, emerging issues exemplified by the use of hormones in beef and genetically modified (GM) foods have further drawn the concerns of other domains into food safety debates, including environmental, cultural and ethical issues.

The interlinked nature of agri-food issues has led to a broader involvement of stakeholders. Traditionally, nation-states and international organizations played dominant roles in securing both the quantity and the quality of food. Now, the growing role of private actors (agri-food business corporations and food companies) and NGOs (environmental and consumer groups) cannot be dismissed. Coordination among the interlinked issue-areas and actors has become an increasingly important matter for the governance of global agri-food issues.

4-7-4 Global governance of agri-food issues for sustainability in four dimensions

Case study 1: Global governance of agricultural production and food security

The Consultative Group on International Agricultural Research (CGIAR) was established in 1971. With the success of the Green Revolution, the importance of agricultural research and development in solving food shortages around the world was realized. Both the Rockefeller and Ford Foundations conducted and supported agricultural research to enhance agricultural productivity in developing countries, but they were faced with a lack of funding. A series of conferences (the Bellagio Conference) were held from 1969 to 1970 to support agriculture in developing countries and to discuss the practical aspects of the design of CGIAR. Sponsors included the FAO, the World Bank, UNDP, donor countries and other international institutions. This was the original movement for the creation of CGIAR.

The following paragraphs describe the current CGIAR system and also sketch the features of the planned new CGIAR system. (The description and the discussion in this section are mainly based on the CGIAR system as of December 2009. It is important to note that it was written in the midst of CGIAR reform.)

CGIAR is a strategic consortium composed of 15 CGIAR Centers, 64 members (including the Rockefeller and Ford Foundations) and four co-sponsors – the World Bank, the FAO, UNDP and the International Fund for Agricultural Development (IFAD) – supported by national agricultural research institutions and in partnership with NGOs and private companies. Each CGIAR Center is an independent institution with its own charter, administrative board, head of the secretariat and staff members. The term "CGIAR system" is used here to describe all of the above-mentioned CGIAR actors. The aims of the CGIAR system are to achieve sustainable food security and to reduce poverty in developing countries through scientific research and research-related activities in the fields of agriculture, forestry, fisheries, policy and the environment. In addition, maintaining an international gene bank to conserve plant genetic resources and keep them readily available is increasingly becoming a critical task at 11 of the CGIAR Centers. The chair of the CGIAR system is appointed by the president of the World Bank with the concurrence of the members.

The main bodies of the CGIAR system are the Annual General Meeting, the Executive Council, the Science Council and the Board of Trustees. The Annual General Meeting is held once a year to discuss strategic issues, make final decisions and evaluate executed projects. Agenda-setting and advice on major issues are conducted by the Executive Council twice a year.

The Technical Advisory Committee (TAC) was established as an independent committee in 1971 to provide scientific and technical advice to the CGIAR system and to CGIAR Centers. Its successor, the Science Council, was formed in 2003. The tasks of the Science Council are to monitor and evaluate, to set priorities and strategies for the CGIAR system and Centers, and to mobilize scientists and researchers. The Science Council is composed of six members and a chair and it has its own secretariat; its administrative office is at FAO headquarters in Rome.

CGIAR initiated a discussion on reform in 2007 and is currently undergoing reform to meet the challenges posed by increasing food demands and climate change, which require a global agricultural research response, and by the expanded scope of agricultural issues, and to reverse the trend of increasing restriction on funds from donors. The features of the planned new structure, according to the CGIAR reform proposal (Change Steering Team, 2008), are as follows. First, it has replaced some of the key founding principles – donor sovereignty with donor harmonization, centre autonomy with system coherence, and decision by consensus with results-oriented rules for prioritizing and managing research guided by independent advice that integrates science with the concerns of partners and development outcomes. Second, it will separate the "Doers" and the "Funders" by establishing a new legally constituted Consortium of Centers and a Fund. The Consortium will take the lead in the development and implementation of the Strategy and Results Framework and the Fund will coordinate donors to support research. The Consortium will become a single entry point for the Fund to contract the Centers. Third, with respect to scientific advice, an Independent Science and Partnership Council will be established to offer independent advice and expertise to the CGIAR system.

International coordination

The degree of centralization is a critical issue in terms of international coordination. In the CGIAR system, a high degree of centralization in terms of funding means that each donor country provides general-purpose funds to the CGIAR system and enables it to allocate the funds to the Centers based on priorities set by the CGIAR system. The degree of centralization has been decreasing, however.

From its inception, the structure of the CGIAR system can be characterized as a form of "dispersed governance" (CGIAR Independent Review Panel, 2008) and it has been considered to be a somewhat informal organization. Because there was no written charter, there were no clear definitions of the roles and responsibilities of the various actors until it was acknowledged that fundamental components of governance need to be made explicit and a charter was approved in 2004. Therefore, the degree of structural centralization in the CGIAR system has never been high. This relatively loose structure had its merit in giving flexibility and diversity in each Center's activity to respond to the local context, which was important in the development of agricultural research. Nonetheless, until the early 1980s, the system was substantially centralized in terms of the allocation of funding and was able to achieve its purposes by integrating the activities of individual Centers to promote international and regional public goods through a research plan set by the central CGIAR system. Funding was sufficient to follow the central system plan as laid out in the Technical Advisory Committee's initiatives, and most donor countries allowed funds to be spent without restrictions and in accordance with TAC's directives.

This centralized funding system has collapsed, however, because the proportion of unrestricted funds from donor countries has decreased. The World Bank faced financial crisis in the early 1990s and hence had to change the resource distribution scheme from donor of "last resort" to "matching grant" in respect of the CGIAR system. Because contributions from donors were becoming less related to the priorities set by TAC

(World Bank Independent Evaluation Group, 2003), the World Bank had long been faced with a big gap between the funds needed to perform the tasks prioritized by TAC and contributions from donors. Since stakeholders in donor countries are concerned with their national interests, donors selected a particular region, research centre, programme or domestic research institution that matched their interests (Ekboir, 2009; World Bank Independent Evaluation Group, 2003).

On the other hand, the number of undernourished people was not decreasing, and poverty reduction or alleviation in developing countries was not being achieved as expected. Some recent studies note that increasing the production of staple foods has not always contributed to poverty alleviation in developing countries, whereas integration into the global economy has contributed to poverty alleviation (IFAD, 2008; World Bank, 2005). In addition, the scope and number of issues addressed also expanded during this period to include such areas as natural resource management, biodiversity conservation, germplasm conservation, and environmental impact assessment. The specific issues vary widely among countries and regions, and donor countries therefore provide funds to National Agricultural Research Systems (NARS) in individual countries or regions and to CGIAR Centers on the basis of their direct national interests. The increased proportion of restricted funds relative to the total CGIAR system budget upset the centralized system, which had previously worked through TAC (or the Science Council).

Acknowledging the decline in unrestricted funds and the lack of donor coordination, the Consortium of Centers will become a single entry point for the Fund to contract the Centers, under the previously mentioned reform. How such institutional change can reverse the trends towards restricted funds is yet to be seen. However, this attempt towards more centralized governance is worth noting.

Inter-regime coordination

The primary way in which inter-regime coordination relates to CGIAR is in its relationships with its co-sponsors – the World Bank, the FAO, UNDP and IFAD – and particularly in the relationship between the FAO and the CGIAR system.

When the CGIAR system was founded, inter-regime coordination occurred primarily with the World Bank, whose objective is development in general, and the FAO, whose mandate is development in the agricultural sector. Whether the World Bank or the FAO should take the initiative as a large donor was discussed at the time of CGIAR's foundation by the World Bank's Board of Executive Directors, which decided that the World Bank should take the leadership role (Baum, 1986; Shaw, 2009). Finally, after formal and informal discussions with the main co-sponsors, the World Bank, led by then-president Robert S. McNamara, took the lead in terms of management. At the same time, TAC was established and the FAO took the lead on technical issues. The CGIAR system can be viewed as resulting from successful inter-regime coordination between the World Bank and the FAO.

Conflict between the FAO and the CGIAR system is the largest inter-regime coordination issue in the system's operation. According to Baum (1986), the FAO's interests included not only the interpretation and dissemination of information about agricultural technology but also more practical matters, such as testing new seed varieties and training researchers in agricultural disciplines. However, the generation of agricultural technology was not part of the FAO's mandate. The aim of the CGIAR system at the time of its establishment was to take action on food shortages in developing countries through research to increase production (Shaw, 2009). Thus the mandates of the FAO and CGIAR did not originally overlap. However, this has changed as the mandate of CGIAR has expanded. Issues regarding the conservation of genetic resources and strengthening national and extension programmes are causing major conflicts between the two (McCalla, 2007).

The main issue is whether the CGIAR should engage in non-agricultural research and development activities. The International Service for National Agricultural Research (ISNAR) was established in 1980 by the CGIAR system and later merged into the International Food Policy Research Institute in 2004. The purpose of ISNAR was to strengthen NARS in individual countries by conducting joint research, training researchers and distributing new knowledge. This goal was similar to that of the FAO, and the FAO initially objected to its establishment. However, it did finally agree to become a member of ISNAR's board. Donors realized the importance of ISNAR because they were able to use new plants and agricultural technology in the field and to collect genetic resources, so they wanted to fund ISNAR. This meant, however, that the CGIAR system was a competitor of the FAO in terms of donor funding. There was also a concern that the CGIAR system might not be able to handle the rapidly expanding scope and number of agricultural issues because it had grown too large. More specifically, the number of issues that should be addressed at the ISNAR level was expanding, but the CGIAR system was not able to provide sufficient services to address them.

Another issue between the FAO and CGIAR is the management of genetic resources. The International Board for Plant Genetic Resources (IBPGR) was established in 1974 to encourage, coordinate and support the conservation of genetic resources throughout the world. These activities are at the core of plant-breeding, including developing GM techniques and allowing access through CGIAR-sponsored programmes. However,

IBPGR was not an agricultural research centre; it was a service organization. It was anticipated that there would be considerable overlap between IBPGR and FAO activities. Originally, TAC requested that the FAO prepare a proposal for the IBPGR concept, and it was decided that IBPGR would be financed by the CGIAR system but housed in the FAO. As provider of the secretariat, the FAO influenced IBPGR activities even though IBPGR had an independent board of trustees, making it difficult for IBPGR to remain independent. Finally, IBPGR separated from the FAO in 1991 in terms of administration, but it remained based in Rome and the name was changed to the International Plant Genetic Resources Institute.

According to McCalla (2007), however, the FAO and CGIAR have recently enjoyed a more positive, synergistic and mature relationship with regard to genetic resources. Sovereignty over genetic resources, including the collections in CGIAR Centers, was discussed by the Convention on Biological Diversity. Simultaneously, the Agreement on Trade-Related Aspects of Intellectual Property Rights of the WTO has tried to codify the intellectual property rights of genetic resources in international trade agreements. Furthermore, mainly through the FAO's efforts, the International Treaty on Plant Genetic Resources for Food and Agriculture was established in 2001 and has succeeded in defining rules related to some genetic resources, including the CGIAR collections. In 1994, an In-Trust Agreement was signed between the CGIAR Centers and the FAO, which formally established that the CGIAR Center collections were on a trustee basis under the auspices of the FAO.

Coordination with non-state actors

The CGIAR as it is today was originally established through the efforts of the Rockefeller and Ford Foundations (that is, non-state actors) and intergovernmental organizations such as the World Bank, the FAO and UNDP. It can therefore be viewed as an achievement of coordination between intergovernmental organizations and non-state actors. Coordination with non-state actors is of increasing significance for the CGIAR system in the context of the current changes in the system.

Investment in research and development activities in the agricultural sector is on the rise from non-state actors, especially the private sector, including multinational agribusiness companies, whereas investment from the public sector has stagnated (Change Steering Team, 2008; World Bank Independent Evaluation Group, 2003). The influence of non-state actors has also increased in agricultural issues, and it has become almost impossible for the CGIAR system to conduct activities without the involvement of non-state actors. The introduction of biotechnology and GM organisms, as well as intellectual property rights, is an important issue in this context. CGIAR's Private Sector Committee (PSC) and NGO Committee (NGOC) coordinate the CGIAR system with non-state actors. The PSC defines the roles of CGIAR and the private sector and strengthens these partnerships. In the case of GM organisms and intellectual property rights, the PSC advocates the introduction of biotechnology, including GM organisms that are protected with intellectual property rights (CGIAR, 1998; Forum on Environment & Development, 1999). On the other hand, the NGOC offers critical evaluations of improvements in food security in developing countries, the environmental impacts of agricultural production, the management of natural resources, farmers' participation, and other issues. The partnership of actors in PSC and NGOC could be beneficial to CGIAR if the coordination mechanisms function in a well-balanced way, but this is yet to be achieved. Nonetheless, private companies have incentives to maintain and strengthen their relationships with the CGIAR system, including NGOs, even providing the output from technology without the execution of intellectual property rights, because they expect the CGIAR system and Centers to give them access to genetic resources and to serve as an intermediary actor in the creation of intellectual property rights in developing countries. Indeed, there are cases in which a public–private partnership in agriculture has been achieved. For instance, the African Agricultural Technology Foundation established a partnership with CIMMYT (the International Maize and Wheat Improvement Center), the Kenya Agricultural Research Institute, BASF and NGOs to provide herbicide technology to Kenyan farmers.

Coordination of science and politics

The Science Council sets priorities and strategies for the CGIAR system, according to which new scientific knowledge and technologies are diffused to farmers through the system, the Centers and NARS in each country and region. Because climate is a critical factor in applying agricultural technology in the field, NARS play a very important role in this process. The mechanism for distributing scientific knowledge and technology from the CGIAR system and Centers to farmers through NARS has been successful.

However, the situation is beginning to change for the previously discussed reasons – for example, stagnant unrestricted funding. The importance of conserving genetic resources and treating these resources as international or regional public goods is generally agreed upon in the global arena because genetic resources are absolutely necessary for breeding and developing GM techniques. Although countries understand the importance of international and regional public goods, as donors they tend to provide funds directly to CGIAR Centers or NARS in accordance with their own national interests.

As mentioned earlier, unrestricted funding has stagnated because donor countries have not always obtained the expected results for a number of reasons (Ekboir, 2009). Although circumstances have changed dramatically since the founding of CGIAR, a new theory or model of the role of agricultural technology in development has not yet been proposed (Rodrik, 2006).

As the mandate of CGIAR has changed, so has its scientific effectiveness. The goal of the CGIAR system now includes enhancing agricultural production in addition to alleviating poverty, in accordance with the United Nations' Millennium Development Goals. The global arena has increasingly recognized that connections to international markets, entering niche markets for agricultural products and remittances have much stronger immediate and apparent impacts in alleviating poverty than does enhancing the production of staple foods. As the scope of CGIAR's goals has expanded, so have the number of actors and issues that need to be considered. It has been difficult for the CGIAR system to respond to this increased number of actors and issues because the system already includes such a large number of actors and mandates. In addition, the role that public research should play in poverty alleviation is not clear. According to Ekboir (2009), the system sets priorities and strategies on the basis of a linear vision of science that is more suitable for enhancing agricultural production through plant-breeding than for alleviating poverty. Furthermore, CGIAR Centers do not have sufficient expertise in fields such as agricultural marketing that would enable farmers to earn higher incomes (Science Council Secretariat, 2006). Therefore, it is of great importance that CGIAR establish partnerships with actors in the private sector and with NGOs that have this type of expertise.

Case study 2: Global governance of food safety

The Codex Alimentarius Commission is an international, intergovernmental body jointly established by the FAO and the World Health Organization (WHO) in 1963 to set food standards. Its objectives are twofold: consumer health protection and fair trade practices. The organization is composed of the Codex Alimentarius Commission (CAC), an Executive Committee and subsidiary bodies. There are four types of subsidiary body: general subject committees, whose work applies across all commodities (for example, the Committee on General Principles); commodity committees, whose work is to develop specific food standards (for example, the Committee on Milk and Milk Products); ad hoc intergovernmental task forces, whose work has a limited mandate with a fixed period of time; and coordinating committees, whose work is to coordinate regional interests. Draft standards are elaborated in eight steps by the subsidiary bodies, which are serviced by host countries, for endorsement by CAC.

The standards developed at Codex have no binding effect; however, after becoming the reference point for the WTO's Sanitary and Phytosanitary (SPS) Agreement, they are regarded as a "yardstick" by food safety regulators (Dawson, 1995). Despite some criticism from consumer groups, Codex may well be said to represent a more "science-based" model of regulation compared with the approach taken at other international organizations (Huller and Maier, 2006: 269). For scientific advice, Codex relies on the Joint FAO/WHO expert consultations, which are institutionally independent of Codex. Their reports are considered to be of high quality and are well respected in general. Because Codex is an intergovernmental organization, its primary actors are nation-states (and one other entity, the European Commission or EC). It had 180 member countries and one member organization (the EC) as of May 2009. In addition, observers are actively involved and have relatively good access to the standards development process.

International coordination

In contrast to the diffuse network-based CGIAR, Codex is a formal intergovernmental organization mandated to elaborate food safety standards. Its governance structure, rules and procedures for decision-making are clearly stipulated in the *Procedural Manual* of the Codex Alimentarius Commission (CAC, 2008b). International coordination at Codex occurs through consensus-based decision-making, with the purpose of making standards legitimate and acceptable to all parties. Although there is a simple majority voting procedure, it has rarely been used except for the most controversial issues (for example, the cases of hormones in beef and natural mineral water; Codex Evaluation Team, 2002).

Historically, before Codex became a reference point as cited in the WTO's SPS Agreement, little attention was paid to its standard-setting activities. However, after becoming the reference point of the SPS Agreement, even though Codex standards were voluntary and had virtually no binding effect, the fact that its standards had the potential of being used in WTO dispute settlements made Codex of significant importance (Veggeland and Borgen, 2005; Victor, 2000). This status has conferred a "semi-binding" effect on Codex standards (Shiroyama, 2005; Veggeland and Borgen, 2005), raising the stakes for its members.

Various conflicts related to food safety and quality need to be coordinated internationally. Because there is a range of food-safety-related issues, the stakes of states are contingent on the agenda in question.

However, coordination issues largely fall into four categories. First, there are conflicts between importing and exporting countries over a food standard's impact on their food safety control and trade. Whereas importing countries tend to require stricter controls and argue that it is legitimate for them to do so to ensure consumer health, exporting countries generally demand a lesser administrative burden and perceive higher standards as disguised barriers to trade. Secondly, conflicts occur between developed and developing countries over feasibility issues, financial constraints and technical assistance. Although developed countries have the resources to respond to stricter standards, many developing countries are not able to comply if standards are set without regard for local capacity or technical assistance, which could influence the availability of foods in developing countries. Thirdly, there are different regulatory approaches to making food policies. For example, the European Union (EU) has institutionalized the "precautionary principle" in its governance of food safety and sees this principle as crucial in establishing global food standards (Poli, 2004). However, the United States does not adhere to this principle in the same manner. The last source of conflict is a diversity of values, such as the consumer's right to know, animal welfare and other ethical issues. For example, in the case of the labelling of GM foods, the EU and Japan sought a wider scope of information to be provided to consumers, including production methods (whether a product is GM food or not), but some countries, particularly the United States, objected and argued that information should focus only on the changed composition or nutritional value of a product (Matsuo, 2008b).

Decision-making based on consensus poses many challenges. It is time-consuming to share each position and reach consensus, and consensus-based decisions are prone to produce compromised standards. To be usable, standards need to have precise and clear definitions. However, agreements are often made in the areas where consensus can be reached, and matters are ignored in areas where consensus cannot be reached. For example, the issues of the "precautionary principle" and "other legitimate factors" (OLFs, discussed in detail below) are so contentious that they have been left ambiguously defined. As a consequence, they are discussed repeatedly in various Codex committees.

Because it is a standard-setting body, there is a demand for higher centralization of decision-making procedures at Codex in terms of timeliness and effectiveness, but the changed nature of Codex standards has induced stronger resistance from states defending their national interests. Hence the preference for a consensus-based approach is still emphasized at Codex. To overcome the time-consuming aspects of this approach, Codex has worked on procedural innovations such as critical reviews conducted by the Executive Committee and the establishment of time-bound ad hoc task forces.

Inter-regime coordination

Inter-regime coordination important to Codex includes coordination with (a) its parent bodies – the FAO and WHO – and the Joint FAO/WHO expert consultations (the latter are discussed in the subsection on coordination of science and politics); (b) the WTO; and (c) other international organizations.

Coordination issues with the FAO and WHO revolve around the degree of independence. Housed in the FAO, Codex follows the FAO's administrative rules (on budgets, job promotion, etc.). Its funding is provided by both organizations (almost 80 per cent from the FAO; Codex Evaluation Team, 2002). As a result, it is said that the Codex secretariat has less authority over the allocation of budgets. In addition, having two masters sometimes puts Codex in a difficult position. The policies and strategies of the two organizations are sometimes inconsistent because of their different mandates. These problems have led to discussions about whether Codex should be more independent (Codex Evaluation Team, 2002). At the same time, however, great trust and confidence are conferred on Codex by food safety regulators because the FAO and WHO are its parent organizations. Furthermore, the two organizations assist and complement Codex activities by providing informational resources, relevant expertise and technical assistance for developing countries.

The relationship with the trade regime (the WTO) is in the context of the SPS and Technical Barriers to Trade (TBT) agreements. Together with the World Organisation for Animal Health and the International Plant Protection Convention, Codex is one of the three "sister" organizations explicitly referenced as WTO standard-setting bodies. Various impacts of WTO activities on those of Codex can be observed. For example, as noted, the legal framework of the WTO has changed the incentives and expectations of Codex members (Shiroyama, 2005; Veggeland and Borgen, 2005), making their activities in Codex more politicized (Codex Evaluation Team, 2002; Victor, 2000). In addition, the systematic application of a risk analysis framework to each Codex committee can in part be attributed to the SPS Agreement (although the basic framework of risk analysis had already been developed at Codex even before the establishment of the WTO). Because the SPS Agreement requires that SPS measures be based on risk assessment, it is imperative for Codex to base every committee's conduct on a risk analysis framework in a systematic and consistent way. Discussions at the SPS Committee of the WTO have also stimulated actions, as in the case of drafting guidelines on judging the equivalence of sanitary measures at the Codex Committee on Food Import and Export Inspection and Certification Systems.

Relations with other international organizations depend on the issues in question. For example, at the seventh session of the Codex Ad Hoc Intergovernmental Task Force on Foods Derived from Biotechnology, an agreement was made to establish an information-sharing system on GM crops. In response to a request by Codex and its member states, it was agreed that the Organisation for Economic Co-operation and Development would provide its data on GM plants (from the BioTrack Product Database) to the FAO's International Portal on Food Safety, Animal and Plant Health website and establish an automated bidirectional data-sharing system (CAC, 2008a; Matsuo et al., 2008). Another important regime is the International Organization for Standardization (ISO). Interactions with the ISO are perceived to be of growing importance because ISO standards are now becoming de facto standards used by the dominant multinational corporations and in national regulations.

Coordination with non-state actors

The major non-state actors in the Codex process are food-related businesses and consumer and environmental NGOs. In this area, the greatest coordination challenge is maintaining a proper balance between business interests and legitimate concerns raised by the NGOs.

Although only states are entitled to voting power, the food industry has a long history as an observer at Codex and represents about 71 per cent of the membership of the observer organization. In contrast, consumer and environmental NGOs make up about 8 per cent of the observer members (Codex Evaluation Team, 2002). The food companies have long been criticized for exerting influence on Codex in an effort to promote their commercial interests at the expense of consumer protection. However, it is also acknowledged that these companies have the best experts and critical data needed for elaborating standards.

The question is how to reconcile these conflicting interests and to consider the experience and legitimate arguments of these non-state actors in the standard-setting process. In this regard, Codex has put in place a relatively broad and open system of stakeholder involvement. Observers do not have the right to vote, but they are entitled not only to "observe" meetings but also to submit comments and participate in the discussions during meetings.[2] Non-state actors therefore have good opportunities for consultation. The question of the accountability and legitimacy of non-state actors' involvement remains to be considered, but partnership with the non-state actors allows Codex to overcome the problem of limited public resources.

Coordination of science and politics

There are two major coordination issues in the process of scientific assessment and the use of scientific assessments in policy-making. The first is the clarification of the respective roles of risk assessment and risk management in the scientific assessment process, and the second is the combination of scientific assessment and so-called "other legitimate factors" (OLFs) in the risk management process.

During the scientific assessment process, the Joint FAO/WHO expert consultations are in charge of scientific and risk assessment, whereas Codex is in charge of policy-making and risk management. Codex makes an explicit distinction (that is, a functional separation) between risk assessment and risk management. At the same time, however, it has become evident that risk analysis is an iterative process, and interaction between risk managers and risk assessors is essential in practical applications.

Clarification of the respective roles of risk assessment and risk management in the scientific assessment process is critical. Because experts make scientific value judgements in the process of risk assessment, such as the use of default assumptions or scientific choices among a number of alternatives (FAO/WHO, 1997), the question has arisen of whether such judgements should be within the realm of risk management. Particularly in cases where assessment involves uncertainty, it is argued that how such scientific value judgements were undertaken must be clearly explained because they can affect the entire outcome. These were some of the focuses of discussions by several Codex committees (including the Codex Committee on Pesticide Residues, the Committee on Residues of Veterinary Drugs in Foods and the Committee on Food Additives) about the application of risk assessment policy,[3] which comprises documented guidelines on the choice of options and judgements at decision points in the risk assessment that are established by risk managers in consultation with risk assessors and the relevant stakeholders in advance of risk assessment.

The important point in the management of governance is to enhance transparency in the entire process so that the dynamic interactions between science and politics are made clear. The above discussion of risk assessment policy guidelines at Codex has contributed to a more transparent process of risk assessment and a clearer definition of the roles of both risk assessors and managers, which were hitherto ambiguous. In addition, in response to Codex's demand for clarification, the FAO and WHO initiated a workshop on a consultative process for scientific advice. In 2007, they produced the *FAO/WHO Framework for the Provision of Scientific Advice on Food Safety and Nutrition (to Codex and Member Countries)* (FAO/WHO, 2007). As a whole, these changes have provided clearer definitions of the principles of scientific advice and its practices and procedures in current operations.

The second coordination issue is the combination of scientific assessment and OLFs in the risk management process. The *Procedural Manual* of the Codex Alimentarius Commission states that "[w]hen elaborating and deciding upon food standards Codex Alimentarius will have regard, where appropriate, to other legitimate factors relevant for the health protection of consumers and for the promotion of fair practices in food trade" (CAC, 2008b). However, there is some disagreement over OLF criteria. Some consider OLFs to include all factors other than science, including economic (cost/benefit), social, cultural and ethical factors. By contrast, others, particularly the United States, view OLFs in a more restricted manner. As a result, this issue has repeatedly been discussed at various Codex committees and has been at the root of most recent controversial cases, such as the use of hormones in beef (Jukes, 2000) and GM foods (Matsuo, 2008a).

The issue involving OLFs is difficult to coordinate. In 2001, criteria for the consideration of "other factors" were adopted at the Codex Committee on General Principles. However, only a general description was provided, and there was little agreement on what precisely is meant by OLFs. Indeed, presenting universally applicable criteria is almost impossible because these are dependent on the particular situation, local values and location. As such, the OLF issue can be a repeated source of conflict.

4-7-5 Conclusion

As highlighted by the examples of agri-food issues examined in this section, sustainability issues are complicated and require action at the global level. It is difficult, however, to establish a single unified organization to deal with sustainability issues, and governance functions will have to be undertaken by many actors in collaborative ways. The importance has been emphasized of coordination across four dimensions: (1) international coordination, (2) inter-regime coordination (including international organizations), (3) coordination with non-state actors, and (4) coordination of science and politics. This discussion is intended to help identify issues and actors in the management of global governance for sustainability.

Future challenges for global governance

The analysis of coordination on agri-food issues in this section has mainly focused on issues related to research and development for food production, food quality and food safety. However, we now face even broader coordination challenges in the area of agri-food than those described in this section. For example, to manage biofuel issues, it is necessary to coordinate areas and actors related to both food security and energy. In addition, health issues such as obesity require the coordination of issues related to food safety and other areas such as healthcare, lifestyle and even culture.

Given the multifaceted nature of sustainability issues, an inclusive, comprehensive and integrated approach is necessary in global governance. However, in practical application, it appears impossible to include *all* relevant factors. The extent to which all relevant issues, values and actors can be included and where to set limits are important issues in global governance for sustainability.

Another important issue that has not been directly addressed here is the issue of power in global governance (Endo, 2008). Depending on how sustainability issues are framed, some actors get the benefits of sustainability whereas other actors face losses. The importance of coordination has been emphasized, but coordination sometimes involves the exclusion of certain actors. The dimension of power must not be forgotten in the discussion of global governance.

Notes

1. These approaches are not mutually exclusive or definitive.
2. The status of observers is defined in the Principles Concerning the Participation of International Non-governmental Organizations in the Work of the Codex Alimentarius Commission adopted in 1999.
3. The *Procedural Manual* of the Codex Alimentarius Commission defines risk assessment policy as "[d]ocumented guidelines on the choice of options and associated judgements for their application at appropriate decision points in the risk assessment such that the scientific integrity of the process is maintained" (CAC, 2008b: 73). It also states that: "Risk assessment policy should be established by risk managers in advance of risk assessment, in consultation with risk assessors and all other interested parties. This procedure aims at ensuring that the risk assessment is systematic, complete, unbiased and transparent" (CAC, 2008b: 69).

REFERENCES

Agrawala, S. (1998) "Structural and Process History of the Intergovernmental Panel on Climate Change", *Climatic Change* 39: 621–642.

Baum, W. C. (1986) *Partners Against Hunger*. Washington, DC: World Bank.

CAC [Codex Alimentarius Commission] (2008a) *Report of the Seventh Session of the Codex Ad Hoc Intergovernmental Task Force on Foods Derived from Biotechnology. Chiba, Japan, 24–28 September 2007*, ALINORM 08/31/34. Rome: FAO/WHO.

CAC [Codex Alimentarius Commission] (2008b) *Codex Alimentarius Commission: Procedural Manual*, 18th edn. Available at <ftp://ftp.fao.org/codex/Publications/ProcManuals/Manual_18e.pdf> (accessed 2 June 2010).

CGIAR [Consultative Group on International Agricultural Research] (1998) "Shaping the CGIAR's Future", Document No. ICW/98/16. CGIAR Genetic Resources Policy Committee, Report of the 8th Meeting. Available at <http://www.cgiar.org/corecollection/docs/icw9810c.pdf> (accessed 2 June 2010).

CGIAR [Consultative Group on International Agricultural Research] (2009) "Who We Are", <http://www.cgiar.org/who/index.html> (accessed 2 June 2010).

CGIAR Independent Review Panel (2008) *Bringing Together the Best of Science and the Best of Development*. Independent Review of the CGIAR System. Report to the Executive Council, Washington, DC. Available at <http://www.cgiar.org/pdf/agm08/agm08_cgiar_overview.pdf> (accessed 2 June 2010).

Change Steering Team (2008) "A Revitalized CGIAR – A New Way Forward: The Integrated Reform Proposal", 3 November. Available at <http://www.cgiar.org/pdf/agm08/agm08_reform_proposal.pdf> (accessed 2 June 2010).

Codex Evaluation Team (2002) *Report of the Evaluation of the Codex Alimentarius and Other FAO and WHO Food Standards Work*. Available at <http://www.fao.org/docrep/meeting/005/y7871e/y7871e00.htm> (accessed 2 June 2010).

Dawson, R. J. (1995) "The Role of the Codex Alimentarius Commission in Setting Food Standards and the SPS Agreement Implementation", *Food Control* 6(5): 264.

Drezner, D. W. (2001) "Globalization and Policy Convergence", *International Studies Review* 3: 53–78.

Ekboir, J. (2009) "The CGIAR at a Crossroads: Assessing the Role of International Agricultural Research in Poverty Alleviation from an Innovation Systems Perspective", ILAC Working Paper 9, Institutional Learning and Change Initiative, Rome.

Endo, K., ed. (2008) *Frontiers of Global Governance*. Tokyo: Toshindo (in Japanese).

FAO/WHO [Food and Agriculture Organization of the United Nations / World Health Organization] (1997) *Risk Management and Food Safety. Report of a Joint FAO/WHO Consultation, Rome, Italy, 27–31 January 1997*. FAO Food and Nutrition Paper No. 65. Rome: FAO.

FAO/WHO (2007) *FAO/WHO Framework for the Provision of Scientific Advice on Food Safety and Nutrition (to Codex and Member Countries)*. Rome/Geneva. Available at <http://www.fao.org/ag/agn/agns/files/Final_Draft_EnglishFramework.pdf> (accessed 2 June 2010).

Forum on Environment & Development (1999) *Public-Private Partnership for Global Food Security? The Cooperation between the CGIAR and the "Life Sciences" Industry*. Bonn: German NGO Forum on Environment & Development. Available at <http://www.forum-ue.de/fileadmin/userupload/publikationen/aglw_1999_ppp_engl.pdf> (accessed 2 June 2010).

Grieco, J. M. (1998) "Anarchy and the Limits of Cooperation: A Realist Critique of the Newest Liberal Institutionalism", *International Organization* 42(3): 485–507.

Haas, P. M. (1992) "Introduction: Epistemic Communities and International Policy Coordination", *International Organization* 46: 1.

Haas, P. M. (2004) "Addressing the Global Governance Deficit", *Global Environmental Politics* 4(4): 1–15.

Hardin, G. (1968) "The Tragedy of the Commons", *Science* 162: 1243–1248.

Huller, T. and M. L. Maier (2006) "Fixing the Codex? Global Food-Safety Governance under Review", in C. Joerges and E.-U. Petersmann (eds), *Constitutionalism, Multilevel Trade Governance and Social Regulation*. Oxford: Hart Publishing, pp. 267–300.

IFAD [International Fund for Agricultural Development] (2008) "Sending Money Home: Worldwide Remittance Flows to Developing Countries", <http://www.ifad.org/remittances/maps/index.htm> (accessed 2 June 2010).

Jukes, D. (2000) "The Role of Science in International Food Standards", *Food Control* 11(June): 181–194.

Keohane, R. O. (1984) *After Hegemony: Cooperation and Discord in the World Political Economy*. Princeton, NJ: Princeton University Press.

Keohane, R. O. and L. L. Martin (1995) "The Promise of Institutionalist Theory", *International Security* 20(1): 39–51.

Keohane, R. O. and J. S. Nye (2000) "Governance in a Globalizing World", in J. D. Donahue and J. S. Nye (eds), *Governance in a Globalizing World*. Washington, DC: Brookings Institution Press, pp. 1–41.

Kollman, K. L. (2003) "Biopolitics in the EU and the U.S.: A Race to the Bottom or Convergence to the Top?", *International Studies Quarterly* 47: 617–641.

Litfin, K. (1994) *Ozone Discourses: Science and Politics in Global Environmental Cooperation*. New York: Columbia University Press.

McCalla, A. F. (2007) "FAO, Research and the CGIAR", ARE Working Papers March 1/2007, University of California, Davis.

Matsuo, M. (2008a) "Dynamics of International Agreement on Food Safety: Case Study of Genetically Modified Foods", in H. Shiroyama (ed.), *Politics of Science and Technology*. Tokyo: University of Tokyo Press, pp. 191–224 (in Japanese).

Matsuo, M. (2008b) "The Global Politics of Labeling of Genetically Modified Foods", *Food Sanitation Research* 58(12): 15–24.

Matsuo, M., H. Shiroyama and T. Imamura (2008) "Low-Level Presence of Unauthorized GM Plant Material in Food: International Negotiation Process at the Codex Ad Hoc Intergovernmental Task Force on Foods Derived from Biotechnology (TFFBT)", *Food Sanitation Research* 58(1): 21–27.

Mearsheimer, J. J. (1994/95) "The False Promise of International Institutions", *International Security* 19(3): 5–49.

Mitrany, D. (1933) *The Progress of International Government*. London: Allen & Unwin.

Poli, S. (2004) "The European Community and the Adoption of International Food Standards within the Codex Alimentarius Commission", *European Law Journal* 10(5): 619–622.

Porter, G. (1999) "Trade Competition and Pollution Standards: 'Race to the Bottom' or 'Stuck at the Bottom'?", *Journal of Environment and Development* 8(2): 133–151.

Putnam, R. (1988) "Diplomacy and Domestic Politics: The Logic of Two-Level Games", *International Organization* 42(3): 427–460.

Rodrik, D. (2006) "Goodbye Washington Consensus, Hello Washington Confusion? A Review of the World Bank's *Economic Growth in the 1990s: Learning from a Decade of Reform*", *Journal of Economic Literature* 44(4): 973–987.

Rosenau, J. N. (1992) *Governance without Government: Order and Change in World Politics*. Cambridge Studies in International Relations. Cambridge: Cambridge University Press.

Ruggie, J. G. (2004) "Reconstituting the Global Public Domain – Issues, Actors, and Practices", *European Journal of International Relations* 10(4): 499–531.

Science Council Secretariat (2006) "Implementation of the CGIAR Performance Measurement System in 2005: Moving Forward", October. Available at <http://www.sciencecouncil.cgiar.org/fileadmin/user_upload/sciencecouncil/Performance_Measurement/PM_2005_moving_forward_Oct31.pdf> (accessed 2 June 2010).

Shaw, J. D. (2009) *Global Food and Agricultural Institutions*. London: Routledge Global Institutions.

Shiroyama, H. (1997) *Structure of International Administration*. Tokyo: University of Tokyo Press (in Japanese).

Shiroyama, H. (2001) "International Administration as an Indispensable Component of Global Governance", in A. Watanabe and J. Tsuchiyama (eds), *Global Governance*. Tokyo: University of Tokyo Press, pp. 146–167 (in Japanese).

Shiroyama, H. (2005) "Differentiation and Harmonization of Food Safety Regulations – Interplay of Scientific Knowledge, Economic Interests and Policy Judgment", in H. Shiroyama and R. Yamamoto (eds), *Melting Borders and Crossing Law 5: Environment and Life*. Tokyo: University of Tokyo Press, pp. 83–109 (in Japanese).

Shiroyama, H. (2007) "The Harmonization of Automobile Environmental Standards between Japan, the United States and Europe: The 'Depoliticizing Strategy' by Industry and the Dynamics between Firms and Governments in a Transnational Context", *The Pacific Review* 20(3): 351–370.

Simmons, B. and L. L. Martin (2002) "International Organizations and Institutions", in W. Carlsnaes, T. Risse and B. A. Simmons (eds), *Handbook of International Relations*. Thousand Oaks, CA: Sage Publications, pp. 192–211.

Veggeland, F. and S. O. Borgen (2005) "Negotiating International Food Standards: The World Trade Organization's Impact on the Codex Alimentarius Commission", *Governance* 18(4): 675–708.

Victor, D. (2000) "The Sanitary and Phytosanitary Agreement of the World Trade Organization: An Assessment after Five Years", *New York University Journal of International Law and Politics* 32: 865–937.

Vogel, D. (1995) *Trading up: Consumer and Environmental Regulation in a Global Economy*. Cambridge, MA: Harvard University Press.

Vogel, D. and R. A. Kagan, eds (2002) *National Regulations in a Global Economy*. GAIA Books, Global, Area, and International Archive, University of California. Available at <http://escholarship.org/uc/item/4qf1c74d> (accessed 2 June 2010).

WFS [World Food Summit] (1996) *Rome Declaration on Food Security*. Rome: FAO. Available at <http://www.fao.org/docrep/003//w3613e/w3613e00.htm> (accessed 2 June 2010).

Wheeler, D. (2001) "Racing to the Bottom? Foreign Investment and Air Pollution in Developing Countries", *Journal of Environment & Development* 10(3): 225–245.

World Bank (2005) *Pro-Poor Growth in the 1990s: Lessons and Insights from 14 Countries*. Washington, DC: World Bank.

World Bank Independent Evaluation Group (2003) "The CGIAR at 31: An Independent Meta-Evaluation of the Consultative Group on International Agricultural Research", <http://www.worldbank.org/ieg/cgiar/> (accessed 2 June 2010).

4-8

Conclusion

Hideaki Shiroyama

Sustainability science is "mode 2 science" that demonstrates its relevance through providing specific solutions to specific contexts, as discussed in Section 3-3. But sustainability science is more than just the collection of specific solutions to specific contexts. The knowledge used to provide specific solutions to specific contexts is sometimes called "tacit knowledge" or "prudence", which is not transparent. But sustainability science should provide the means to visualize ways of providing specific solutions through transparent tools and methods as analysed in this chapter.

To achieve sustainability in society, it is necessary to recognize that sustainability has many dimensions, including various aspects of environmental sustainability and various aspects of social sustainability. In addition, it is noteworthy that different actors within society hold different viewpoints on sustainability.

Hence it is important to understand the context, that is, the various frameworks within which perceptions of sustainability are framed by the many actors in society. Problem-structuring methods based on cognitive mapping introduced in Section 4-1 are useful for analysing actors' perceptions concerning sustainability. Actors' perceptions have to be summarized as an overall social assessment. Each actor's perceptions are different; but there then has to be a platform on which these multiple viewpoints are shared. One such assessment of the social implications of technology using stakeholder involvement was introduced in Section 4-2. The consensus-building process introduced in Section 4-4 also uses assessment of stakeholders' interests based on interviews.

Then there has to be coordination of interests and perceptions based on the assessment. It is not necessary for all the actors involved in decision-making on sustainability to share the same interests. The notion of "sharing the same bed, dreaming different dreams" (*doushouimu* in Japanese) introduced in Section 4-2 is an important approach for coordinating different interests. Different actors may support technologies such as nuclear energy and biomass energy for different reasons, such as the pursuit of energy security or the mitigation of global warming. Putting it the other way around, science and technology can facilitate the "solution" of sustainability issues by providing the physical infrastructure that enables *doushouimu*. Among the policy instruments discussed in Section 4-3, economic instruments can also be recognized as tools for *doushouimu* because the pursuit of economic interests and the pursuit of environmental protection can coexist. Consistency with other interests, such as promoting innovation and conformity with political values, also needs to be considered in the choice of policy instruments. The consensus-building process introduced in Section 4-4 is one way to achieve coordination. The concept in consensus-building of "living with" an agreement is similar to *doushouimu*.

However, it is not always possible to achieve *doushouimu*. Sometimes coexisting with someone might be a nightmare, which cannot be tolerated. In these cases, judgements involving trade-offs must be made. For example, risk trade-off refers to the fact that efforts to reduce specific risks end up increasing other risks as a result. A typical case might involve biofuels, where the benefits of energy security and possible CO_2 reductions are countered by the potential risk of food insecurity, especially in developing countries, as discussed in Section 4-2. There may be other cases where important issues concerning values such as individual rights, human dignity and religion have to be considered, whatever the other risks and benefits. Values frequently emerge as a key issue in the realms of the use of genetic engineering and the use of birth control as measures towards sustainability. In those cases, priority-setting on values is necessary. Because there is no inherent hierarchy of values, there can be multiple answers to problems related to sustainability involving values, based on the judgement of the priority of values by the actors involved. This kind of knowledge for judging trade-offs and values is an indispensable aspect that social science can contribute, and will be discussed further in Section 5-6. Another possibility for resolving values conflicts is the process of deliberation by mutual interaction, whereby actors' frameworks of values can change through learning, as discussed in Section 4-6.

To understand the perceptions and interests of the actors involved (that is, the context) and to facilitate judgements about trade-offs and values, the establishment of a communication mechanism is indispensable.

When science and technology are important for providing solutions to sustainability problems, science and technology communication is necessary, as discussed in Section 4-6. One important point is that science and technology communication is necessary not only between scientists/engineers and citizens and various stakeholders in society, but also between scientists and engineers, who tend to have fragmented perspectives. This is also illuminated in Chapter 1, which calls for the structuring of fragmented knowledge for sustainability.

In addition, because scientific assessment- and R&D-related sustainability issues cross national borders, the institutional framework of global governance is necessary to undertake those tasks. As analysed in Section 4-7, there are four dimensions of coordination concerning global governance: international coordination, inter-regime and cross-sectoral coordination, coordination with non-state actors, and coordination of science and politics.

The tools and methods introduced in this chapter – problem-structuring methods, technology governance (assessing the social implications of technologies, *doushouimu* and value judgements), policy instruments (including economic instruments), consensus-building (including assessment of stakeholders' interests and negotiation), public deliberation, science and technology communication, and global governance – are used in combination to manage the transition to a sustainable society. It is said that transition management utilizes innovative bottom-up developments in a more strategic way by coordinating different levels of governance and fostering self-organization through new types of interactions, generating cycles of learning and action for radical innovation that offer sustainability benefits (Kemp et al., 2007: 80). The tools and methods in this chapter, which play an important role in understanding the contexts and facilitating *doushouimu* and judgements about trade-offs and values, can also contribute useful instruments to transition management.

REFERENCES

Kemp, R., Loorback, D. and Rotmans, J. (2007) "Transition Management as a Model for Managing Processes of Co-evolution towards Sustainable Development", *International Journal of Sustainable Development & World Ecology* 14: 78–91.

5

The redefinition of existing sciences in light of sustainability science

5-1

Global change and the role of the natural sciences

Akimasa Sumi

5-1-1 Background

The basic purpose of the natural sciences is to understand the dynamism of nature by applying deduction based on physical and chemical laws and validating the results with observational and experimental data. Therefore, it is considered important to find the essential phenomenon by filtering the various noises that are included in the natural phenomena, and to analyse it. At the same time, quantitative expression is preferred to qualitative expression. Most discussion and deduction is based on countable quantities. Thus, the analytical method tends to become the dominant method because it applies logic to a target phenomenon and then explores its mechanism. Usually, mathematical expression is used to represent the logic. Of course, the analytical method is not the only method in the natural sciences. Other methods include observation and description, where the emphasis is on knowing nature itself and every detail is well documented. These primary data are organized on the basis of the topic's characteristics and the scientist's interest. Through this process, filtering is conducted.

However, a new problem has arisen in modern times and the situation of the natural sciences is changing. Most modern problems do not lie within the natural sciences. One example of an issue that cannot be handled within the traditional natural science framework is global warming. Tackling these issues then presents a challenge to the natural sciences. One line of attack is to deal with an element of the problem that can be handled within the framework of the traditional discipline. Another is to establish a new framework to handle the issue. The role of the natural sciences in relation to modern problems is discussed in connection with the global warming issue.

A global warming effect was discovered in the nineteenth century during the exploration of the mechanism that determines the Earth's climate. When the history of the Earth's climate is examined, great variability is noted. Why this kind of climate change has occurred is a fundamental question. It is well known that the Earth's climate is determined by solar energy and one problem is how to determine the distribution of the energy in the atmosphere over the Earth. Investigations were conducted by many researchers, including Jean Baptiste Joseph Fourier, John Tyndall and Svante August Arrhenius (Weart, 2003). At this stage, the research was conducted within the discipline of atmospheric radiation. A variety of knowledge was accumulated by the efforts of many researchers. It was compiled into the "radiative-convective equilibrium" model by Manabe and Wetherald (1967). All information about radiative energy transfer in the atmosphere is included in this model. It is a one-dimensional model, in which the temperature in the atmosphere is horizontally averaged and the vertical distribution of temperatures is taken into account.

However, this model demonstrates that the Earth's climate is determined not only by the radiative equilibrium but also by atmospheric motion; in other words, atmospheric convection is critical in the determination of the Earth's climate, especially in the troposphere. Thus, the distribution of energy in the atmosphere is influenced by atmospheric convection. In the radiative-convective equilibrium model, the effect of atmospheric convection is parameterized, where the lapse rate of atmospheric temperature in the model is automatically adjusted to be a criterion, that is, a mean value of observations (6.5°C per 1,000 metres), when the lapse rate becomes larger than the criterion while integrating the model. In the real atmosphere, this lapse rate is realized through various atmospheric phenomena such as cumulonimbus, cyclones and anti-cyclones. The lapse rate of the atmosphere is also computed by using a three-dimensional Atmospheric General Circulation Model (AGCM). An essential point of the three-dimensional AGCM is that the vertical distribution of the horizontally averaged temperature can be simulated by permitting these atmospheric phenomena.

Manabe's work clarified the basic mechanism of how the climate of a planet is determined, that is, the process is represented by a set of numerical equations based on physical laws and a solution is obtained by solving them numerically. With respect to obtaining solutions, advances in computing capability after World War II made a huge contribution. However, those equations are not sufficient for solving a real problem. For real problems, factors that are filtered out in order to uncover an essential aspect become important and it is necessary to understand the characteristics of these factors. In case of the Earth's climate, there are many factors to be investigated. One is the climate–aerosol interaction, which has been investigated since early times, and another is the effect of anthropogenic greenhouse gases on the Earth's climate, which has become important more recently. This suggests that one needs to change perspective. When discussing temperature, it is sufficient to pay attention to the physical aspect of the Earth's climate system but, when discussing the climate–aerosol interaction, one has to consider not only the physical aspect but also the chemical aspect of the Earth's climate system. Furthermore, it has been pointed out that the interaction between the biosphere and climate is very important. Therefore, the biosphere has to be taken into consideration as well. Now, human activity has to be included too. In other words, the components of the Earth's climate system are expanding: at the beginning it comprised the physical system; then the geochemical system and the biological process were added; finally it is expanded to include the humanosphere.

This trend of integrating various disciplines can be found in international collaboration on research projects. First, the World Climate Research Programme (WCRP) was established jointly by the International Council for Science (ICSU) and the World Meteorological Organization in 1980, and since 1993 has also been sponsored by the Intergovernmental Oceanography Commission. The WCRP's main research objective is the physical climate system. Then, in 1987, the International Geosphere-Biosphere Programme was started by ICSU. It studies the interactions between physical, chemical and biological processes and how they impact (and are impacted by) human systems. The International Human Dimensions Programme on Global Environmental Change was initiated in 1996. Its focus is social science research on global change. DIVERSITAS was established in 1991 as a partnership of inter-governmental and non-governmental organizations formed to promote, facilitate and catalyse scientific research on biodiversity. These four international programmes realized that their topics interact and jointly established the Earth System Science Partnership in 2001, a partnership for the integrated study of the Earth system, the ways that it is changing, and the implications for global and regional sustainability. However, it should be noted that the methodology is fundamentally the same although the objective has expanded.

5-1-2 The concept of a coupled system

In traditional climatology, field surveys and data analysis based on statics were the main research tools. The basic dynamics involved a linear concept. However, the real climate system is a highly non-linear and complex system. It should be noted that the chaos phenomenon was discovered during research into weather forecasting (Lorentz, 1963). As a result, intensive research on the non-linear system has been conducted; however, the research is being conducted within the conventional framework of physics and system dynamics. In order to understand the behaviour of the system, it is necessary to know the details of the components and their interactions. In general, these cannot be worked out analytically and the numerical expressions that are considered to represent the system are inevitably solved numerically. Computer simulation thus becomes a powerful tool and there has been continuous development in high-end super computers.

Various subsystems interact with each other in nature and these interactions can be represented as exchanges of physical quantities. However, the characteristic temporal and spatial scales are different in each subsystem and there are many issues in handling these interactions in a coupled system. The first example of a coupled system is the El Niño/Southern Oscillation (ENSO) phenomenon, where the global atmosphere and the ocean in the tropics interact with each other. The reason this coupled system was the first is because the characteristic timescale in both systems is of the same order and it is easy to treat both systems in one model (Philander, 1990). For example, the sea surface temperature (SST) in the mid-latitude does not change much in the course of a few days, so SST is assumed to be constant when one- or two-day predictions are being made. If the characteristic timescale of the subsystem is different, there is no need for a coupled system. The Earth's climate system consists of many subsystems whose characteristic spatial and temporal scales are different and, when formulating a coupled system, it is very important to specify the horizontal and temporal scales of the targeted phenomenon. The components in the coupled system vary depending on the temporal and spatial scales of the targeted phenomena.

As awareness of our environment becomes deeper, different components will be added to the coupled system and a new model is generated. For example, when the bio-geochemical cycle is added to the climate system, the concept of the Earth system emerges and the Earth system model is proposed. When simulating future climate change resulting from anthropogenic global warming, the Earth system and a socioeconomic model are combined, which produces a new concept, that is, the Earth–human system, and an impact assessment model is developed. Thus, as awareness of phenomena becomes deeper and wider, new problems emerge and the necessary knowledge is sought. A new science field will be generated by adding the newly acquired knowledge and a discipline. However, the methodology used is the same as in the traditional sciences, although the target changes.

5-1-3 The appearance of the anthropocene

The geological history of the Earth is classified according to fossils and materials in a bed. It has been proposed that the present era should be called the "anthropocene" because of the human influence on the Earth's climate (Crutzen and Stoermer, 2000). In the past, humankind was weak and dominated by nature. Now, the number of people on the Earth and human activities have expanded rapidly and are exerting a strong influence on nature. For example, air pollution indicates that human activity can modify air quality and deforestation shows that people can modify the condition of the land surface, which can have an impact on the local and global climate. In other words, a human system, or humanosphere, is incorporated into the Earth's climate system.

However, the inclusion of the human system in the Earth's climate system throws up different issues compared with the climate system or the Earth system without the humanosphere, where the object is nature and the subject is human beings, and there exists a clear distinction between the subject and the object. However, when human beings are included as actors in the system, they become both the subject and the object. As long as human beings can be treated as the object, a conventional scientific disciplinary approach may be applied. Numerical expressions can be defined to describe the development of the economy and the population. However, these formulations are based on assumptions and uncertainty is inevitable.

For the global warming issue a scenario approach is being used. Because the future development of human society is modelled using current data and expert judgements, there exist many scenarios of the future. Simulations are conducted on the basis of these scenarios, but there is no established method for integrating the results. When human beings are included as an active variable in the Earth's climate system, a new methodology is necessary and it has not yet been established how to investigate this issue.

5-1-4 Future development

Whenever a new target or a new issue appears, scientists take an interest and try to tackle the issue. Many attempts will be made, although most do not succeed. Clearly, it is not sufficient to amass knowledge from the various disciplines to resolve the issue. There remain essential differences between the humanities, the social sciences and the natural sciences, and no method has yet been found to integrate these disciplines.

However, even though the integration of knowledge and a holistic approach are essential, the deepening of knowledge in each discipline is also crucial. If one wants to integrate knowledge in a discipline, it is useless if it is out of date. The traditional approach in each discipline also needs to be brought up to date. Based on established and reliable knowledge in each discipline, there is the possibility of integrating the disciplines and creating new knowledge. Therefore, first, the issues that are considered to be components of a newly emerging problem need to be clarified, and a discipline found to resolve the problem. However, it is necessary to give up remaining complacently in the existing disciplinary atmosphere. New problems must be confronted. It is possible that this challenge will be met through the exchange of ideas, opinions and information between researchers. The Intergovernmental Panel on Climate Change (IPCC) was established to review and assess existing knowledge for policymakers, but it can also be considered as a forum where a range of information is collected and integrated. Stimulated by the success of the IPCC, many interdisciplinary meetings are held at many venues every year. This is the realistic way to find a new method while maintaining the opportunity for researchers in the humanities, the social sciences and the natural sciences to meet each other.

REFERENCES

Crutzen, P. J. and E. F. Stoermer (2000) "The 'Anthropocene'", *IGBP Newsletter* 41: 17–18.

Lorentz, E. N. (1963) "Deterministic Nonperiodic Flow", *Journal of the Atmospheric Sciences* 20: 130–141.

Manabe, S. and R. T. Wetherald (1967) "Thermal Equilibrium in the Atmosphere with a Convective Adjustment", *Journal of the Atmospheric Sciences* 24: 241–259.

Philander, S. G. (1990) *El Niño, La Niña, and the Southern Oscillation*. San Diego, CA: Academic Press.

Weart, S. R. (2003) *The Discovery of Global Warming*. Cambridge, MA: Harvard University Press.

5-2

Science and technology for society

Hiroyuki Yoshikawa

5-2-1 Introduction

Sustainability science presents a dilemma to scholars, decision-makers and practitioners worldwide. On the one hand, there is increasing recognition of the need for a new way to address complex contemporary problems that threaten the sustainability of planet Earth. On the other, there is confusion in scientific communities as to how to organize research to meet these threats head-on with the urgency that they demand (Kauffman, 2009). In recent years, activity has increased in support of the development of sustainability science as an academically established field, including the creation of academic posts, the development of curriculums, opportunities for scholarly publication in peer-reviewed journals, and the establishment of degree programmes in sustainability science.[1] A 2009 Special Feature edition of the journal *Sustainability Science* on "Education for Sustainable Development" illustrates the progress that is being made to prepare the next generation of scientists, engineers and decision-makers with the tools they need to address twenty-first-century challenges (Takeuchi, 2009). The addition to the *Proceedings of the National Academy of Sciences* (PNAS) of the United States of a section devoted specifically to sustainability science also points to this progress, as do efforts to build networks of scientists around the world to address sustainability issues.[2] These issues are complex problems that lie at the intersection of global, social and human systems and thus transcend disciplinary and geographical boundaries.

Yet, despite the proliferation of efforts to understand and address these issues, little progress has been made in the creation of new systems that will lead to global sustainability.[3] Sustainability science aims to overcome this weakness by creating knowledge for action, but the science is in its infancy. Criteria, approaches and even definitions of the science vary.[4] Although there is general agreement on three key concepts that underscore sustainability science (transdisciplinarity, integrative analysis, and the creation of knowledge for action), there is no established methodology, and the means employed to measure outcomes are inconsistent. Hence there is a need to clarify and elaborate the key concepts of sustainability science and to define how they can be implemented in research.

This section presents a perspective on the development of this emergent science and draws upon research conducted between 2001 and 2008 at the National Institute of Advanced Industrial Science and Technology (AIST), Japan, to develop and test an approach to sustainability science research. The main body of the section is divided into four parts: (1) discussion of the need for a new science; (2) elaboration of the key concepts of sustainability science and identification of the elements that set it apart from traditional scientific research; (3) an overview of the AIST initiative; and (4) the presentation of a new cyclical model and dynamic structure for conducting sustainability science research. The section concludes with consideration of the utility of this model to advance science in practice.

5-2-2 The need for sustainability science

The evolution of scientific methods

In order to fully understand the key concepts that underscore sustainability science and to develop a structure to support them, we must understand the limitations of the present discipline-based approach in addressing contemporary sustainability issues. This is, after all, an approach that has contributed to technological and economic progress and human welfare for centuries. The history of science is linked with social development, beginning with humankind's earliest struggles to bring order to chaos and gain control over the forces of nature. Over time, from Babylonian and Egyptian antiquity to the early development of scientific methods in ancient Greece, the struggle to overcome external threats to humankind evolved into formal methods of enquiry. What has come to be called "the scientific age" began in the seventeenth century with the articulation by Francis Bacon of a disciplined method for developing, testing and verifying theories using inductive reasoning to answer questions and understand natural phenomena and causation. Towards the end of that century, Isaac Newton revolutionized science and laid the foundation for modern science by developing rules for scientific reasoning, which he laid down in his "Mathematical Principles of Natural Philosophy", the *Principia* (1687). In this work Newton identified three laws of motion, set forth a new scientific philosophy establishing four rules for scientific reasoning, and demonstrated that analysis consists of making experiments and observations and drawing conclusions from them by induction. His methods applied universally to all branches of science, from the natural sciences to mathematics, social science, engineering and medicine, and they became the foundation for what we have come to think of as "traditional science", that is, enquiry based on hypothesis formulation, testing and validation through rigorous experimentation and observation in order to discover what is true. It is a tradition that has served humankind well, and in its universality is an approach that undergirds research across virtually all disciplines. But the problems that confront humankind today are of such complexity that they do not fall easily into one discipline and therefore require a new model or structure to address them.

The complexity of modern threats

By its nature, science is progressive in its identification and exploration of new phenomena and in its response to social problems. Although the aim of modern science is to add to humankind's understanding of the universe and our place in it, the expected outcome of scientific research is that the results of enquiry will contribute to the increased prosperity and security of humankind. As knowledge has advanced in separate disciplines, a new set of problems has arisen to confront modern humanity. These problems are vast in their scope and of very long duration. The threats they pose are not readily visible to modern science, in part because they are the result of unintended consequences of actions and artefacts that are meant to improve the quality of life. Moreover, we are often blind to their negative impacts, especially when these accrue slowly over time, be it decades, centuries or even millennia.

As the effects of these new problems come to light, researchers have begun to understand that they stem from the interconnectedness and degradation of the three systems that are crucial to the sustainability of the planet: global, social and human (Komiyama and Takeuchi, 2006). These problems include global warming, environmental degradation, the appearance of new diseases, the burgeoning world population coupled with growing inequities between rich and poor, terrorism, urban isolation, racial tensions and cyber-crime. These are problems that are more difficult to assess and address than were the direct threats faced by our ancestors. Since the roots of these problems are interconnected in the global, human and social systems, their specific causes are difficult to determine and therefore to address. And because their scope is potentially so vast, involving the planet as a whole, and of such long duration, they fall outside the scope of the present organization of scientific disciplines.

As scientists struggle to understand the complex problems that afflict humankind and the planet today, a fundamental question that must be addressed in organizing knowledge for sustainability is why, given the urgency of these issues, do researchers persist in developing a new science to address them when the aim of traditional science is, in fact, to understand everything in the universe? Given the apparent universality of the approach of traditional science, can one not assume that problems of sustainability would be adequately addressed if their study were to be subsumed within traditional science? The answer to this is no. It would be an incorrect assumption given that the fundamental difference between traditional science and sustainability science is that entirely different perspectives drive them. Traditional science is aimed at understanding, but the orientation of research in sustainability science is to action. Moreover, this action is aimed at achieving the sustainability of the Earth as a whole. Although it is fair to say that the traditional scientific method is meant to contribute to understanding everything in the universe, in practice the focus of enquiry is on individual components that comprise the universe.

The research results of traditional science that advance understanding of the components of the Earth are necessary but not sufficient to ensure the sustainability of the planet. Also necessary is a new generation of knowledge that stems from a high level of integration of the results of multidisciplinary research and the means to translate such knowledge into action. Without this holistic perspective, the scientific approach to both understanding and addressing sustainability issues in their full complexity will be inadequate. The new approach to addressing these complex modern problems is what we call "sustainability science".

5-2-3 Concepts that differentiate sustainability science from traditional science

Fundamental differences in perspective and orientation between traditional and sustainability science are the drivers for a number of other differences that affect the way research is organized and conducted under the two different approaches. Table 5.2.1 lays out these differences in six categories that will inform the model developed for the organization of

Table 5.2.1 Traditional science and sustainability science

	Traditional science	Sustainability science	Difference
Aim	To understand everything	To sustain the earth	separate/total
Object	Anything generally existing in the universe	Specific phenomena	open/bounded
Result of research	Knowledge for understanding	Knowledge for action	analysis/synthesis
Mode of change	Additive	Non-additive	linear/non-linear
Measure	Unchangeable (any change can be deduced from existence)	Slowly changing	stable/unstable
Expected practical results	Prosperity and safety of human beings	Sustainability of the Earth	prosperity/ sustainability

research in sustainability science. The categories of differences are: (i) the aim and (ii) object of research (discussed above); (iii) the methodologies used to interpret the results of the research; (iv) the mode of change or development as the research progresses; (v) the focus of measurement (what is being observed and how it is measured); and (vi) the expected practical results (what the researchers wish to achieve).

Here, these differences will be discussed in the context of the three concepts that underscore sustainability science: transdisciplinarity, integrated analysis and creating knowledge for action. Given the broad spectrum of differences between traditional and sustainability science, it is necessary to consider how the organization of research may be affected by them. Once this has been done, it will be possible to use this understanding of the differences to develop a model for sustainability science that supports the three fundamental concepts of the new science and contributes to its goal of safeguarding the planet.

A transdisciplinary approach

In a broad sense, the origins of academic disciplines can be traced back to the need to understand and gain control over untamed and chaotic forces. The problems addressed were seen not as of humanity's own making but, rather, as stemming from natural causes over which human beings had no dominion, such as storms, drought, floods, earthquakes, disease and pestilence. As science evolved to address these problems, disciplines were created within academic institutions to study them: meteorology to study storms, seismology for earthquakes, microbiology to better understand disease and plagues. Physics, one of the oldest academic disciplines, covers a wide range of phenomena from subatomic particles to galaxies and is motivated primarily by curiosity – the desire to understand natural phenomena. The social sciences and the humanities, including such fields as ethics, economics and logic, were created to increase understanding of human behaviour and to establish rules to guide societal development through, for example, maintaining order, structuring economic development or providing means for civic participation and debate.

Today, in most universities around the world, departments and sub-departments have been established to study specific phenomena, from astronomy to zoology, with numerous subcategories of ever-increasing specialization. The accumulation of knowledge over generations in each of these specialized fields has been significant and their contributions to technological and economic development substantial. Humankind has made enormous progress using science and technology to overcome diverse problems that threaten human welfare, from the conquering of diseases to improvements in standards of living, the provision of clean water and shelter, more efficient food production, and stable social organization for many of the Earth's inhabitants.

Academic disciplines typically build knowledge in a distinct sphere. They are independent of each other in vocabulary and rarely function as a unity in the creation of knowledge. Inevitably, there are inconsistencies both in approach and in the manner in which results are interpreted. Such disparities can have positive benefits if they lead to more robust interpretations of results. But this requires communication and consistency across the disciplines, which is usually lacking. Rather than leading to more coherent results, the diversity and division of the disciplines may simply exacerbate confusion in the interpretation of results by society, resulting in a fragmented response to problems that is ultimately detrimental to systems that support the sustainability of the planet.

Increased knowledge specialization at an abstract scientific level through the proliferation of disciplines has been accompanied by the training of ever more specialized professionals in those fields and a concomitant division of labour. Consider, for example, the subdivision of the discipline of engineering science into mechanical, electrical and chemical engineering. Each of these sub-disciplines trains its own professional practitioners, each with a distinct specialization aimed at producing artefacts that contribute to human welfare and economic prosperity. This is not unique to the engineering sciences; a similar pattern of subdivision and the training of specialized professionals accompanies the evolution of many other academic disciplines in the preparation of medical professionals, economists, politicians, lawyers, artists and managers, to name but a few. These professionals are trained to delve deeply into their subject or field with little cross-fertilization of ideas and methods across the professions. As a result, the solutions and artefacts of social development that accrue to address needs or problems in one area may be inconsistent with, and even harmful to, those in another area.

Ironically, the growth of diversified disciplines and specialization of knowledge that has helped humankind to overcome earlier threats and has contributed so much to human progress has also hindered the ability to recognize the emergence of new threats to the sustainability of the planet. By focusing on components of the "Earth system" (that is, anything in the universe) rather than on the planet as a whole, science has largely ignored how the interconnectedness of the global, natural and human systems can result in perverse outcomes within the systems that support the sustainability of the planet. In traditional science, elements of the universe, for example atoms, are investigated in isolation within specific disciplines. Unfortunately, little thought has been given to the coherence or compatibility of knowledge between different scientific disciplines. From the sustainability perspective, the diversified evolution of disciplines and concomitant growth in specialization and contradictory artefacts have already had a detrimental effect on the planet, endangering the environment through excessive, localized and uncoordinated human actions.

Consider two very simple examples by way of analogy. Doctors specializing in internal medicine will examine a patient's complaint from the perspective of their specialty and recommend a prescriptive therapy based on their training. But if the patient is suffering from an ailment outside the realm of internal medicine, the doctors will be relatively powerless to address it. Similarly, mechanical engineers have produced the automobile, whereas electrical engineers have produced the mobile phone. Each of these inventions, used separately, facilitates an individual's life and is greatly appreciated. Used together, however, they may result in highly undesirable and unintended consequences such as financial and legal penalties or physical harm. And, as in the case of the contemporary sustainability issues discussed above, the threat itself is partially invisible in these cases because it derives from the collision (that is, interconnectedness) of perceived benefits. Although these examples are simplistic compared with problems on a global scale, they illustrate how presumed benefits from the fruits of research can be cancelled out by inconsistency, insufficient integration of knowledge and, hence, suboptimal use of that knowledge.

Integrated analysis

Because traditional science focuses on the accrual of knowledge in specific disciplines, it is ill-equipped to deal with the inconsistencies and incoherence on a larger scale that result from this narrow vision. This has led to the present crisis of seemingly uncontrollable global system degradation that only now is beginning to be recognized and questioned.

Through the separation of disciplines and their inconsistent approaches to the advancement of knowledge, traditional science has developed in an asymmetrical manner, separating the advancement of knowledge for "fact" or "truth" from knowledge for use. Over time, as the demand for specialized knowledge has increased, the concepts of fact-oriented and use-oriented knowledge have been divided, separated and abstracted. This asymmetry is illustrated in Figure 5.2.1. In the physical and natural sciences, theories about what is real or true are developed through hypothesis, observation and testing, then subjected to verification and re-evaluation through the application of deductive and inductive reasoning. The laws that emerge from this systematic approach are assumed to be true ("fact") to the extent that they are arrived at through the scientific method. The knowledge or "natural laws" that derive from this practice are then applied in a less systematized way in other disciplines relying on abductive[5] reasoning to create or construct social artefacts that range from technological innovations to public policy and works of art (see

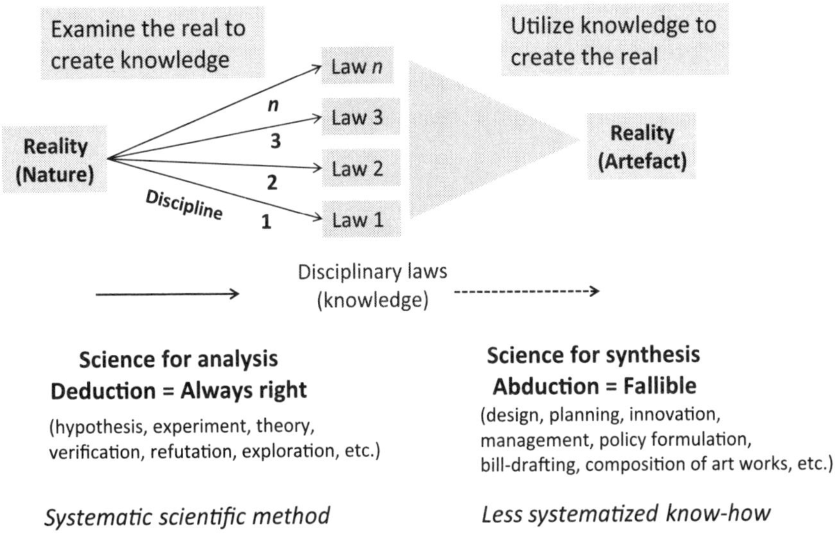

Figure 5.2.1 Analysis/synthesis: Asymmetry of human thought.

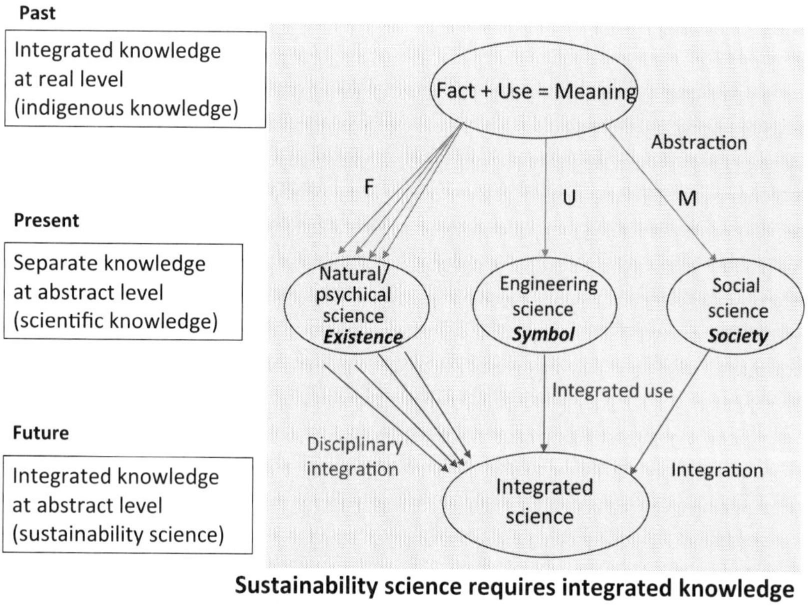

Figure 5.2.2 Two levels of integration of knowledge necessary for sustainability.

Burch, 2009). The separation of these approaches to reasoning for knowledge generation may have been necessary in the past, but sustainability requires a robust integration of knowledge at several levels.

Creating knowledge for action

Because it is aimed at action, sustainability science requires a realignment of knowledge generated for "fact" or truth with knowledge generated for use. There is a need for this integration of knowledge within the three major classes of science (natural, engineering and social), as well as across those classes, as illustrated in Figure 5.2.2. This integration might be expressed through a simple formula that has been used to describe indigenous knowledge: Fact + Use = Meaning. Here "fact" is knowledge generated through the psychical (or human) and natural sciences to understand reality, "use" is operational knowledge generated through engineering science, and "meaning" is the result of knowledge generated by the social sciences for understanding and influencing society's response.[6]

The sophisticated level of integration that is required for sustainability science presents two significant and related challenges to the scientific community. First, there is a need to strive for greater consistency across disciplines at a time when they are moving towards ever-increasing independent specialization. Second, there is a need to develop methodologies to realign and integrate the advancement of knowledge for fact and for use. The absence of a methodology that supports the optimal use of the vast amount of knowledge emanating from the many scientific disciplines perpetuates asymmetry in the application of science for the creation of artefacts. At present, this level of integration towards unified knowledge and its application seems impossible. Indeed, one might refer to the sophisticated integration of research in multiple domains as a "nightmare" phase as opposed to "dream" research in which knowledge is generated for its own sake without necessitating a reality check of the societal implications of that knowledge. Perhaps mathematics may offer a viable solution to these twin problems, devising a system of logic to explain the differences between disciplines and to prescribe a methodology for integration. But since there is at present no such solution proposed, it may be beneficial to consider a practical approach that addresses the problem of asymmetry through the creation of research teams that bring together researchers working in different domains.

5-2-4 The AIST initiative

Between 2001 and 2008, the National Institute of Advanced Industrial Science and Technology (AIST) in Japan undertook an initiative to develop and study sustainability science with the aim of creating useful knowledge for sustainability in the context of industrial product development. From this experiment, which brought together over 3,000 researchers in 56 separate units with numerous scientific disciplines represented in each unit, it is possible to begin to create a model that will be helpful in overcoming the inconsistencies that derive from more traditional approaches.

Closing the gap between basic and applied research

Scientific research is often broadly organized into two major categories, basic and applied, the terms generally attributed to the purposes and organization of the research effort. Basic research builds on existing knowledge to increase understanding, whereas applied research is geared towards the resolution of a problem or question. Typically, a sharp distinction is drawn between the two. The former is considered "blue-sky" or curiosity-driven research that aims to advance the understanding of natural phenomena. Applied research has as its aim the development or improvement of an artefact or technology that contributes in some way (great or small) to human prosperity. To the extent that there may be a relationship between the two domains, the passage of knowledge from one to the other occurs in a linear fashion, with knowledge moving from a basic understanding of conditions to the identification of the social value or need for a product or service. Integration between the two rarely occurs.

Although necessary to the advancement of knowledge for sustainability science, neither basic nor applied research in the traditional sense is sufficient, for two reasons. The first is the organization of both domains into disciplines that are inadequate for sustainability science research. The second is the difference in their aims. Sustainability science is centred on "use-inspired basic research" (Clark, 2007), which is different from both the basic and applied sciences.

Creating new knowledge for use through synthesis

In the AIST experiment, with many teams of diverse researchers working together, the gap between basic and applied research could be addressed by reformulating traditional basic and applied research into units that included what were dubbed Type I and Type II basic researchers along with product designers. The AIST units covered bio- and nano-science and bio- and nano-technology, manufacturing science and technology, and robotics, as well as energy and the environment. In this model, Type I basic researchers focused on generating new scientific knowledge. Type II basic researchers aimed at creating new values for society by subsuming ongoing applied research with a view to creating deeper understanding of their impacts for the purpose of application. The goal of product designers working with Type I and II basic researchers was to create products for society.

The head of each unit supported the integration and cross-fertilization of ideas and knowledge among the three groups, maintaining coherence and concurrence in the work and encouraging the researchers to move freely across the three categories. One of the lessons learned from this experience is that the unit head must be a philosophical thinker. In order to put theory into practice, the head of the unit must understand the concepts of "dream" and "nightmare" research, be willing to experiment with the new forms of Type I and II basic research, and be ready to facilitate extensive collaboration among researchers accustomed to working in more narrow domains.

By bridging the gap between basic and applied research to create new knowledge, bringing scientists together in the research effort and integrating results from many disciplines, it is possible to overcome some of the obstacles to sustainability that are frequently present in product development and that blind researchers to potential impacts. But this model is not complete with regard to ensuring sustainability. A solid structure for sustainability science must incorporate another dimension, which is a form of social technology. In the following subsection a structure is proposed that adds this dimension and incorporates the application of synthesis based on abductive reasoning.

5-2-5 A cyclical model and structure for sustainability science

A robust structure for sustainability science for product development and applicability must be flexible, take account of change and accommodate modifications. One of the reasons humankind finds itself facing a sustainability crisis today is that modern science has tended to focus heavily on stable physical matter through such disciplines as solid state physics, elementary particle physics and bioscience, rather than on subtle changes occurring in a more holistic manner, as in such fields as geology, archaeology and palaeontology. Although both streams of enquiry are important, researchers have, perhaps unwittingly, come to focus on stable phenomena at the expense of a full appreciation of the subtle ways in which the Earth changes.

In the context of sustainability science, where change, not stability, is the crux of enquiry, synthesis of knowledge from both streams will contribute to the creation of new knowledge for action and to the development of artefacts that are not detrimental to Earth-supporting systems. Today, there are advanced tools that facilitate researchers' ability to create this new knowledge and computer simulations that increase the ability to predict the future. Such tools are crucial to sustainability science research. However, when sustainability science is applied to the development of artefacts for society, the tools of social technology also need to be employed.

Consistent collaboration between science and society in sustainability science is essential as a means to test and measure the effects and impacts of actions taken in the light of scientific knowledge and thus to make corrections over time. Such collaboration also provides a means to interpret the effects of change that scientists observe through the use of increasingly complex measurement tools.

In order to ensure that the output of sustainability science contributes to safeguarding the planet, social technology must be added to the structure or model that is created for sustainability science. The incorporation of social technology transforms the model from a linear process to an evolutionary or cyclical process. Figure 5.2.3 illustrates this process, beginning with Type I basic research and continuing through Type II basic

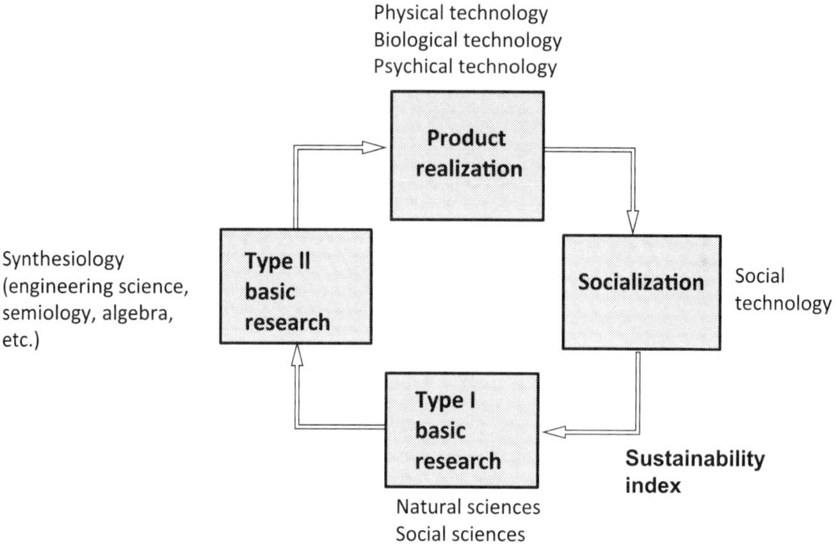

Figure 5.2.3 Social technology for sustainability.
Note: This figure is a primitive representation of sustainability science.

research integration as applied in the AIST model, then moving on to actual design and advice through collaboration between academic scientists and engineers with actors in industry for the testing and design of psychical technology, followed by exposure to social technology, which allows for normative issues to be included in the development of artefacts. Through social technology, the research community can maintain consistent collaboration with actors in civil society, for example decision-makers, opinion leaders and the public at large. The synthesized results of research, analytical information on the proposed artefact and proposals for application can be shared using both direct and indirect means of communication with the public, such as Web-based dialogues, roundtables and public surveys. Through these means, the results of the application of a proposed technology can be observed and the information fed back to the research community. If, in this phase, the product is found to be detrimental to sustainability, then the information gained through the cycle can be reassessed and reassigned to further research in a new cycle of sustainability science. Hence, the model adds greater flexibility to the development and deployment of products.

As the Austrian philosopher Karl Popper noted in his work *The Poverty of Historicism*, "we make progress if, and only if, we are prepared to learn from our mistakes, to recognize our errors and to utilize them critically instead of persevering in them dogmatically" (1957: 80). Popper criticized historicist attempts to foretell the future. The course of human history, he argued, is strongly influenced by the growth of knowledge, a growth that cannot be foretold. Popper developed a theory of "piecemeal social engineering" that is relevant to the development of the proposed cyclical model for sustainability science. Drawing on his concept of the scientific method, Popper argued that the only form of social engineering that may be rationally justified is engineering that is small-scale, incremental and continuously amended in the light of experience. Thus, what Popper called "piecemeal social engineering" and what might be considered relevant to the model for sustainability science is the integration of scientific methods with planning and politics.

5-2-6 Conclusion

Sustainability science, although nascent, is necessary to confront contemporary problems that threaten the future of the planet and humankind. It is a science that is fundamentally different from traditional science in its perspective, aims, organization of research and desired outcomes. Although there is growing activity within the international scientific community to advance the fundamental concepts of sustainability science, little progress has been made in applying it to safeguarding the planet. Understanding how the two approaches to science differ and what steps may be taken to resolve these differences may free science from this impasse. A cyclical model for sustainability science that includes social technology to support consistent collaboration between scientists and society offers a cohesive, holistic and democratic approach to scientific research, an approach that is necessary if we are to achieve the aims of sustainability science to create knowledge for action that will contribute to the well-being of the planet. The model includes multiple actors from the three classes of natural, engineering and social sciences in a process that refines information with each cycle to produce actionable knowledge for sustaining humankind and the Earth. It is one concrete step in the journey to develop a science with the power to transform humans' existence for the better.

Acknowledgements

The author wishes to thank Dr Joanne Kauffman for her invaluable help in preparing this section.

Notes

1. A number of networks now provide gateways to information on sustainability science on the Web. Examples include the IR3S Program of The University of Tokyo at <http://www.ir3s.u-tokyo.ac.jp>; the Forum on Science and Innovation for Sustainable Development at <http://www.sustainabilityscience.org>; Proceedings of the National Academy of Sciences at <http://www.pnas.org/site/misc/sustainability.shtml>; the website of the Global System for Sustainable Development at <http://gssd.mit.edu/GSSD/gssden.nsf>; and the website for the Network of Networks in sustainability science at <http://nns-u.org/>.
2. See, for example, G8 University Summit (2008) and (2009) for declarations by research university presidents on the need to develop scientific networks for sustainability.
3. See, for example, Kajikawa (2008) for an overview of achievements in sustainability science and various approaches to its research core and framework.
4. Definitions of sustainability science vary, but all incorporate three basic assumptions: that it is transdisciplinary, provides integrative analysis, and is aimed at action that contributes to solving complex global problems that lie at the intersection of environmental, social and economic issues. The various definitions may be explored through the online gateways listed in note 1 above.
5. Abduction is a method of hypothetical reasoning that transcends both induction and deduction. The philosopher Charles Sanders Peirce (1839–1914) arrived at his argument for abductive reasoning by forming a new argument from interchanging a conclusion (a result) with the minor premise used in a previous argument (or case). Unlike inductive or deductive reasoning, abduction is a form of probable argument. It has the air of conjecture or "educated guess" about it. An important contribution that Peirce's argument makes to sustainability science is his development of ways to integrate the three reasoning forms into his view of the scientific method (Burch, 2009).
6. Charles Sanders Peirce described the psychical sciences as consisting of three sciences: nomological psychics or psychology, classificatory psychics or ethnology, and history. For a historical and philosophical explanation of these terms, see the classification of the sciences in Peirce (1931).

REFERENCES

Burch, R. (2009) "Charles Sanders Peirce", in Edward N. Zalta (ed.), *Stanford Encyclopedia of Philosophy*. Available at <http://plato.stanford.edu/archives/sum2010/entries/peirce/> (accessed 4 June 2010).

Clark, W. C. (2007) "Sustainability Science: A Room of Its Own", *Proceedings of the National Academy of Sciences USA* 104: 1737–1738.

G8 University Summit (2008) "Sapporo Sustainability Declaration (SSD)", <http://g8u-summit.jp/english/index.html> (accessed 4 June 2010).

G8 University Summit (2009) "Torino Declaration on Education and Research for Sustainable Development (Turin Declaration)", <http://www.g8university.com> (accessed 4 June 2010).

Kajikawa, Y. (2008) "Research Core and Framework of Sustainability Science", *Sustainability Science* 3(2): 215–239.

Kauffman, J. M. (2009) "Advancing Sustainability Science: Report on the International Conference on Sustainability Science (ICSS) 2009", *Sustainability Science* 4(2): 233–242.

Komiyama, H. and K. Takeuchi (2006) "Sustainability Science: Building a New Discipline", *Sustainability Science* 1(1): 1–6.

Newton, I. ([1687] 1972) *Philosophia Naturalis Principia Mathematica: Vol. 1.* Cambridge MA: Harvard University Press.

Peirce, C. S. (1931) "An Outline Classification of the Sciences", in *The Collected Papers Vol. I: Principles of Philosophy. Book II: The Classification of the Sciences*. Available at <http://www.textlog.de/4257.html> (accessed 2 August 2010).

Popper, K. R. (1957) *The Poverty of Historicism*. London: Routledge & Kegan Paul.

Takeuchi, K., ed. (2009) "Special Feature: Education for Sustainable Development", *Sustainability Science* 4(1): 1–59.

5-3

Science for sustainable agriculture

Mitsuru Osaki

5-3-1 Introduction

Modern agriculture, which underwent revolutionary progress to achieve its twentieth-century form, can be defined in a single phrase as "petroleum-dependent agriculture". Inexpensive fossil fuels are used to operate large machinery, which enables large-scale cultivation, mono-cropping, standardization and high-volume transport; and the synthesis of fertilizers and agricultural chemicals permits the improvement of soil fertility and simplified management of ecosystems. Until the nineteenth century, agriculturalists were basically forced to engage in stable sustainable agriculture that took maximum advantage of nature's functions. But during the twentieth century, by taking full advantage of the energy sources provided by inexpensive fossil fuels, "using natural functions" gave way to "applying technology to transform nature". This form of agriculture is referred to as the modern agricultural revolution because its basic technologies were established at the start of the twentieth century, permitting a great leap in production. The modern agricultural revolution is characterized by (1) the mechanization and increased scale of agriculture, with high-volume transport made possible by motorization using fossil fuels (petroleum) to power internal combustion engines; (2) the management of soil fertility and ecology using chemical fertilizers and agricultural chemicals; (3) the development of high-yield varieties; (4) advanced water management in some regions; and (5) remarkable increases in labour productivity as a result of the first four characteristics. During the late nineteenth and early twentieth century, the basic technologies for motorization through the use of gasoline to power internal combustion engines were established. Also, Fritz Haber and Carl Bosch developed the technology to produce ammonia by fixing atmospheric nitrogen gas. The Haber–Bosch process is a method of producing ammonia using an iron oxide catalyst to trigger a reaction of nitrogen gas and hydrogen gas under supercritical conditions, 300–550°C and 15–25 MPa. Thanks to this process, it is now possible to manufacture the nitrogen fertilizers that play a crucial role in agriculture. In this way, the fundamental technologies for modern agriculture were almost entirely established in the late nineteenth and early twentieth century. One more important factor in the development of petroleum-dependent agriculture was crop yield improvement technologies based on large-scale fertilization and the breeding of extremely high-yielding varieties. These varieties included Norin 10 wheat and Yukara rice, which were bred by Japan in the 1950s and successfully contributed to heavier yields of tropical wheat and rice, an event called the Green Revolution.

It is true to say that, in this way, a series of technological revolutions during the twentieth century transformed agricultural production systems, resulting in the twentieth-century agricultural revolution. Because this was partly a product of revolutions in engineering technology and biotechnology, it can also be described as the industrialization of agriculture, which subsequently promoted the twentieth-century agricultural revolution in the world's most industrialized nations. Because this revolution involves petroleum-dependent agriculture, flat topographical conditions have made an extremely important contribution to its progress.

However, petroleum-dependent agriculture reached a major turning point during the latter part of the twentieth century and early years of the twenty-first century. This section will point out the problems with twentieth-century agriculture (petroleum-dependent agriculture) and present a sustainable design for twenty-first-century agriculture.

5-3-2 Categorization of the world's agriculture

Categorization of the world's agriculture on the basis of structural policy

Kimio Noda (2006) found the salient characteristic of the twentieth-century agricultural revolution to be structural policy, and classified world agriculture in the twentieth century into four categories according to its adaptability to structural policy. Here, structural policy refers to: selecting a number of very small enterprises, replacing them with partial large-scale management, and concentrating policy on the small number of enterprises so created in order to entrust agriculture as a productive industry to the industrial management bodies formed in this way. The following categories are created with reference to Noda's categorization (2006).

Category I (regions where structural policy is unnecessary). These are frontier agricultural regions in North America, South America, Australia and South Africa that were settled by West Europeans and where structural policy is almost entirely unnecessary. This process occurred against a background of the use of fossil fuels to carry out large-scale improvement of the ecology and the establishment of ecological management technologies. The causes and special characteristics of the establishment of agriculture in this category can be listed (1) geographically, in terms of flat continental topography, low soil fertility and relatively light rainfall, or (2) from the agricultural technology perspective, in terms of large-scale mono-crop agriculture made possible by large machinery powered by inexpensive fossil fuel, the breeding of high-yield varieties, supplementing soil fertility with chemical fertilizers, managing the ecology with agricultural chemicals, and partially managing water through irrigation or subsurface irrigation.

Category II (regions where structural policy has been achieved). In the old agricultural regions in Western Europe, bold structural policies were realized through the European agricultural revolution in the last half of the eighteenth century, the European agricultural crisis at the end of the nineteenth century and in the early years of the twentieth century, and other processes culminating in the switchover to modern agriculture.

Category III (regions where structural policy is impossible). These are Asian agricultural zones, including Northeast and Southeast Asia. As a result of the natural environment and agricultural methods, and under the weight of history, structural policy has not advanced because of the impossibility of overcoming such problems as (1) agricultural systems based on the dispersion of working land units and extremely small fields, and (2) high population pressure, mixed agricultural and urban populations, and farmers taking non-agricultural employment.

Category IV (regions where structural policy has not been undertaken). These occur in Africa as well as in parts of Asia and South America. They include cases where, as a result of state and capital interests, commercial crops are grown partly under duress, including plantation farms. The need for structural policy has been given almost no consideration in these regions.

Categorization based on the relationship between climate and agricultural method

Jiro Iinuma (1982) has proposed a categorization of agricultural methods on the basis of annual rainfall (heavy or light) and the rainfall period (rain does or does not fall in the summer agricultural season) (Table 5.3.1). The aridity index (I) of de Martonne, which distinguishes dry land and wet land meteorologically, is represented by the formula:

$$I = R/(T + 10),$$

where R is cumulative rainfall (mm) and T is average air temperature (°C).

Here the land is classified as wet land if the annual aridity index is 20 or higher, as dry land if it is 20 or less, and as desert if the index is 10 or lower. For agriculture, the summer aridity index is more important than the annual index, so a calculation covering only the period from June to August defines a region as the summer rain type if the index is 5 or higher and as the winter rain type if it is below 5. Based on the annual

Table 5.3.1 Categorization based on the relationship between climate and agricultural method

		Annual aridity index	
		Dry	**Wet**
Summer aridity index	**Dry**	**Region I:** Southwest Asia, Mediterranean (South) and Russia (part of the South) **(fallow-period water retention work)**	**Region II:** Mediterranean (North) and Russia (part of the South) **(fallow-period weeding work)**
	Wet	**Region III:** Punjab and Northern China **(intertillage water retention work)**	**Region IV:** Northern Europe, Siberia, Southeast Asia and East Asia **(intertillage weeding work)**

Source: Iinuma (1982).
Notes: Region I: annual aridity index is 20 or less, and summer aridity index is 5 or less; Region II: annual aridity index is 20 or more, and summer aridity index is 5 or less; Region III: annual aridity index is 20 or less, and summer aridity index is 5 or more; Region IV: annual aridity index is 20 or more, and summer aridity index is 5 or more.

aridity index and the summer aridity index, agricultural regions around the world are classified into four categories as explained below (see Table 5.3.1).

Region I (region where the annual aridity index is 20 or less, and the summer aridity index is 5 or less) – includes Southwest Asia, the Mediterranean (South) and Russia (part of the South). These are dry regions where dry farming is performed: crops strongly resistant to dryness are cultivated, and shallow ploughing and compaction of the ground surface are performed periodically, preventing the evaporation of moisture from the ground surface (water retention work during the fallow period). Because these are winter rain zones, the land is fallow from spring to autumn. Water retention work is done during the fallow period, and, counting on subterranean water that has been held in this way, winter crops (mainly wheat) are sown in October and, after germination, depend on winter rainfall for growth. Land is divided into fallow land and winter cropland, which are alternated every year, an agricultural method that is called the two-field system.

Region II (region where the annual aridity index is 20 or more, and the summer aridity index is 5 or less) – includes the Mediterranean (North) and Russia (part of the South). Weeding work is done in the fallow period, as in Region I, but, because the annual aridity index is above 20, the stability of crop production is far higher than it is in Region I.

Region III (region where the annual aridity index is 20 or less, and the summer aridity index is 5 or more) – includes the Punjab and Northern China. Like Region I, the annual aridity index is 20 or less, but summer crops can be cultivated, and water retention work is done repeatedly using spades during periods of heavy rainfall (water retention work during intertillage).

Region IV (region where the annual aridity index is 20 or more, and the summer aridity index is 5 or more) – includes Northern Europe, Siberia, Southeast Asia and East Asia. These are the wettest of the four types of region. Summer crops can be grown but, because weeds flourish, weeding is an indispensable part of agricultural work. Weeding is done in two ways. In Northern Europe, weeding is done mechanically by deep ploughing and turning over the ground during the fallow season of the three-field system of agriculture: alternating winter crops, summer crops and fallow in a three-year cycle (weeding work in the fallow period). In Southeast Asia and East Asia, weeding is done occasionally by hand during summer crop cultivation (weeding during cultivation work). Northern Europe and Southeast and East Asia are regions with a summer aridity index of 5 or more, but the summer aridity index has a distribution of 5 to 11 in Northern Europe and of 9 to 108 in Southeast and East Asia. The ways that weeds flourish vary. In Northern Europe, cultivation can be done without weeding for two years; in the third year the land is left fallow and then may be deep-ploughed and turned over. In Southeast and East Asia, in contrast, weeding must be done frequently every year. Moreover, on the continent of Europe, glaciers have reduced the topography, creating flat landforms, destroying fertile ground and lowering its crop productivity. This makes large farm operations feasible, and grazing, which is suited to land with low productivity, has become the foundation of the three-field system.

Categorization based on the fertile components of the soil and population density

Wakatsuki and Miwa (1993) have studied food productivity from the perspective of rainfall and the fertile components of soil, with population density as an index. They recognize huge differences in the distribution of population densities around the world, and analyse which factors determine population density distribution viewed macroscopically. In Figure 5.3.1, which shows the world's population densities and annual rainfall distribution, the regions where black dots are concentrated are regions with concentrated populations. This clearly shows that population density is restricted by rainfall (or the supply of water by rivers). In temperate zone regions with low evaporation (except for Egypt on the Nile Delta), areas of high population density exist only in regions with annual rainfall of 500–1,000 mm or more; in regions with high evaporation, areas of high population density exist only in regions with annual rainfall of 1,000–2,000 mm or more. However, there are regions with an uneven distribution of high population densities, and, regardless of rainfall, there are also areas with extremely low population density. Consequently, it is assumed that the fertile components of the soil make a significant contribution as a factor other than rainfall. Geological fertilization, which increases the fertility of the soil, can be classified mainly into the following four categories:

(1) *Transport and alluviation action of rivers*. Floods that occur repeatedly on a timescale of several years to several decades form fertile meadow soil (Inceptisol). In tropical Asia, the Himalayas and the monsoons cause particularly conspicuous formation of deltas. The Nile Delta is extremely fertile, a benefit of fertile volcanic ash soil distributed on the Ethiopian Plateau and around Lake Victoria in the region surrounding the upstream Nile River.

(2) *Supply of volcanic ash and lava by volcanic activity*. Volcanic ash supplied on a timescale of several hundred to several thousand years rejuvenates soil, forming fertile soil that is rich in nutrients (Andosol). The Ethiopian highlands, the region surrounding Lake Victoria and

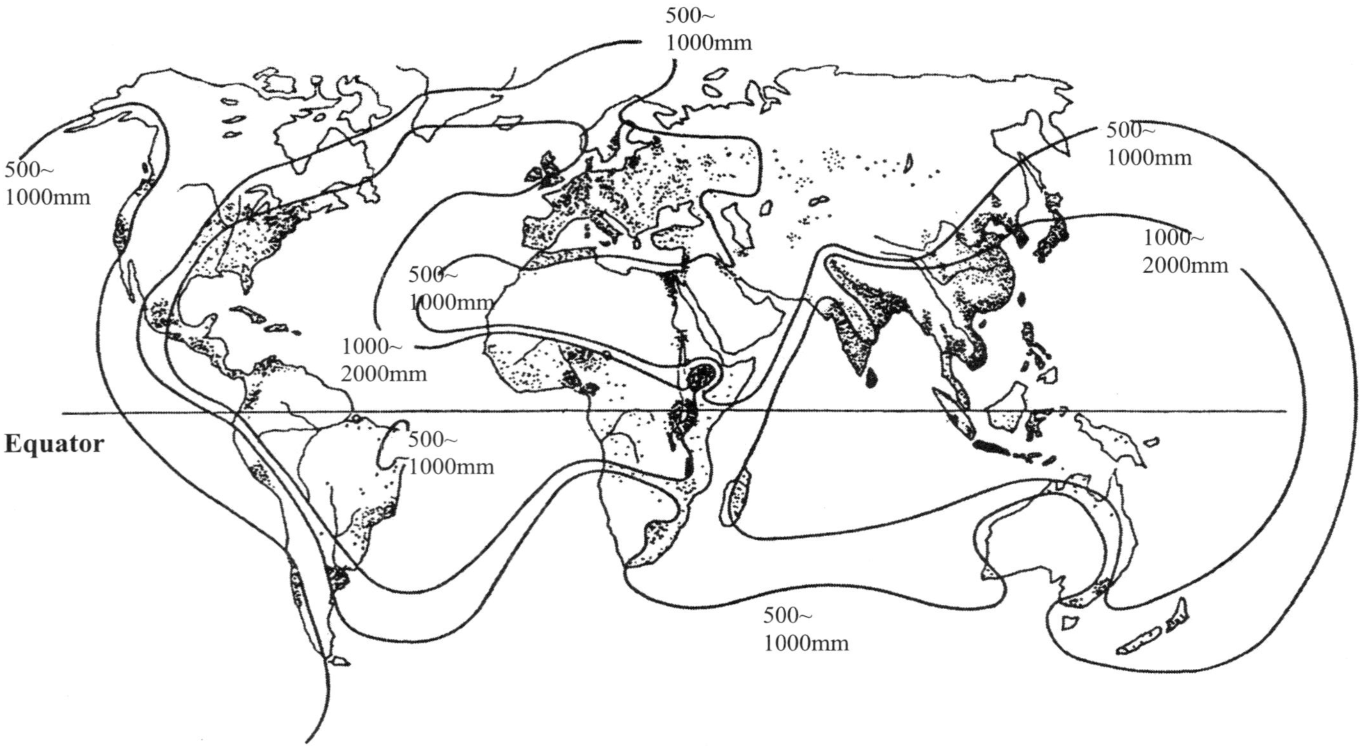

Figure 5.3.1 Population density based on the fertile components of the soil and precipitation.
Source: Wakatsuki (1994).
Notes: Lines and dots indicate precipitation and population, respectively.

the circum-Pacific volcanic belt all benefit from volcanoes. The Mayan and Incan civilizations were also established thanks to the high agricultural productivity of fertile volcanic ash soil.

(3) *Supply of loess by the wind.* The loess of the Sahara Desert is fertile, benefiting southern Nigeria. Eastern China is a fertile region thanks to the loess supplied from the Gobi, the Ocher Plateau, etc.

(4) *Rejuvenation action of soil caused by a suitable degree of erosion.* On the Deccan Plateau in India, which is a lava plateau made up of basalt with an age of several tens of thousands of years, fertile black cotton soil (Vertisol) is formed with this basalt as its base material. The age of the Vertisol is estimated as less than 10,000 years, and this Vertisol is fertile because a balance between soil erosion and soil creation is maintained. Excessive erosion removes the topsoil, causing desertification; but, where soil erosion is much less than soil creation, over the long term it eluviates and dissipates soil nutrients, forming aged soils (Oxisols).

5-3-3 The present state of the world's agriculture and of its future in the twenty-first century

Category I agriculture (regions where structural policy is unnecessary): Twentieth-century petroleum-dependent agriculture, its limits and shortcomings

Because Category I is typified by agriculture developed in the United States and is dependent on inexpensive petroleum, it can also be called petroleum-dependent agriculture. Under the rainfall categorization, it is primarily part of Region IV (annual aridity index of 20 or more and summer aridity index of 5 or more), and has made the development of North European-type agriculture possible. Geological fertilization action is high along the Mississippi River and in the Rocky Mountains, but, generally, glaciers removed the surface soil and dry regions expanded into inland districts, resulting in little fertile soil. Consequently, its natural conditions with respect to agriculture are similar to those in Northern Europe.

According to the categorization of Noda (2006), structural policy in the United States is classic Category I. Here, there is almost no need for a structural policy for historical reasons. Prior to the switch to petroleum-dependent agriculture after the beginning of the twentieth century, plantation-type agriculture dependent on slaves as a labour force had already appeared. The switch from slave labour to the use of internal combustion engines powered by inexpensive fossil fuel led to an explosion in productivity in North America, a region with few social and geographical restrictions. This established petroleum-dependent agriculture, the typical twentieth-century form of agriculture, in North America. This type of agriculture has, from the beginning, displayed strong features of capitalistic acquisition: selling food products in the same manner as industrial products to acquire capital, as opposed to producing food in support of people's daily lives. But limitations on American-style petroleum-dependent agriculture have appeared since the beginning of the twenty-first century. These limiting factors are described in detail below.

Restrictions on the use of inexpensive fossil fuels

This type of agriculture is becoming increasingly difficult because the use of inexpensive fossil fuels has been restricted by (1) the oil peak (various predictions assert that the peak will be reached between 2010 and 2025) (Rojey, 2009); (2) the sharp decline in the energy profit ratio (output energy/input energy) for petroleum production; and (3) international carbon dioxide emission restrictions, with the goal of cutting greenhouse gas emissions by nearly 80 per cent by 2050.

The inefficiency of bioethanol production

In the United States, the Energy Policy Act of 2005 incorporated a Renewable Fuel Standard (RFS), which increases the quantity of bioethanol that must be utilized to 28.4 million kilolitres, more than double the 2004 level, by 2012. The result has been competition between food and biofuel.

The net energy balance (NEB) ratio (energy output/energy input) of biofuel for the entire production process is 1.25 for corn grain ethanol and 1.93 for soybean biodiesel; for biofuel only (that is, after excluding co-product energy credits and energy allocated to co-product production) it is 1.25 for corn grain ethanol and 3.67 for soybean biodiesel (Hill et al., 2006), indicating that the NEB ratio of corn grain ethanol has quite a low value. David Pimentel et al. (2009) concluded that environmental problems, including water pollution from fertilizers and pesticides, global warming, soil erosion and air pollution, are intensified with biofuel production. Including environmental factors, most conversions of biomass into ethanol and biodiesel result in a negative energy return based on careful up-to-date analysis of all the fossil energy inputs, such as corn ethanol at –46 per cent, switchgrass at –68 per cent, soybean biodiesel at –63 per cent and rapeseed at –58 per cent. Even palm oil production in Thailand results in a –8 per cent net energy return when the methanol requirement for transesterification is considered in the equation. Pimentel et al. also claim that publications promoting biofuels have used incomplete or insufficient data to support claims of net energy provided by cellulosic ethanol, and that such claims have not been experimentally verified because most of the calculations are theoretical. Using corn for ethanol increases the price of US beef, chicken, pork, eggs, breads, cereals and milk by more than 10–30 per cent, which exacerbates food and fuel shortages and raises major nutritional and ethical concerns around the world.

The production of greenhouse gases (GHGs) by domestic livestock

In a report on the environmental impact of livestock production, Steinfeld et al. (2006) found that the production of meat is currently contributing between 4.6 and 7.1 billion tonnes of GHGs each year to the atmosphere, which represents 15–24 per cent of total current GHG production. Much of this effect is the result of deforestation for grazing and the processes that many countries are still using to produce meat, which require the animals to live longer than do other, more economically efficient processes. Nathan Fiala (2008) reported that beef production accounts for the majority of CO_2 production and is increasing, though pig products also have a large aggregate impact owing to their high use. Total potential GHG emissions, if all meat were produced by the same method as the US Concentrated Animal Feeding Operation system and there was no deforestation, would have been 1.3 billion tonnes of CO_2 equivalent in 2000. This number increases by 17 per cent to 1.5 billion tonnes in 2010, 33 per cent to 1.7 billion tonnes in 2020, and 47 per cent to 1.9 billion tonnes in 2030. In 2007, the total CO_2 output was approximately 30 billion tonnes of CO_2 equivalent (Fiala, 2008). If future CO_2 production stays at the current level, meat production will account for 5.0 per cent of total production in 2010, 5.7 per cent in 2020 and 6.3 per cent in 2030.

The impact of soil erosion

Reduction of soil depth can impair the land's productivity, and the transport of sediments can degrade streams, lakes and estuaries. As a consequence of conservation efforts associated with explicit US government policies, total soil erosion between 1982 and 1992 was reduced by 32 per cent and the sheet and rill erosion rate fell from an average of 4.1 tons per acre per year in 1982 to 3.1 tons in 1992, while the wind erosion rate fell from an average of 3.3 tons to 2.4 tons per acre per year over the same period (Uri, 2001). However, large amounts of soil erosion continue to occur. In the results of simulations of potential changes in erosion rates in the Midwestern United States, in 10 of 11 regions of the study area runoff increased from +10 per cent to +310 per cent and soil loss increased from +33 per cent to +274 per cent in 2040–2059 relative to 1990–1999 (O'Neal et al., 2005). Thus, it is predicted that soil erosion will increase even more with climate change, which will in turn cause a significant decrease in crop production.

It is predicted that, under US-style petroleum-dependent agriculture, there will be a marked deterioration in soil ecology in the future. Basic factors that will cause this to occur are: (1) the decline and destruction of the soil aggregate structure formation capability of micro-organisms and the decrease in fertile soil components under the impact of the impoverishment of microbiota and the decline in the quantity of organic material caused by the use of agricultural chemicals and chemical fertilizers; (2) a decline in air permeability and destruction of the soil aggregate structure under the effects of soil compaction by large machinery; (3) acceleration of soil erosion by the prolongation of the period when soil is not mulched by mono-cropping; and (4) drying as a result of the decline in water-retention capacity owing to the removal of topsoil.

Destabilization and depletion of water resources

According to a prediction for rainfall between 2070 and 2099 under global warming based on a scenario from the Intergovernmental Panel on Climate Change, rainfall will decline slightly in the southeast and southwest regions, but in other regions rainfall will tend to increase. It is assumed that in this way average rainfall will increase slightly, boosting agricultural production. However, under the impact of future global warming, El Niño and La Niña in particular will occur more frequently, and changes in rainfall patterns, including droughts and torrential rainfall, will occur with greater frequency (Rosenzweig and Hillel, 2008). When heavy rainfall occurs during sowing and harvesting, large machinery cannot enter the fields for long periods of time, growth is delayed and severe loss of harvests occurs. In fact, an examination of yield fluctuations between 1950 and 2000 (yield as a percentage of that in the previous year) has shown that, since around the 1970s and early 1980s, yields of corn, soybeans and cotton, which grow in the summer, have fluctuated greatly, whereas yields of wheat, which grows in a different season, have fluctuated relatively little (Rosenzweig and Hillel, 2008). From this it can be surmised that, although water issues were not necessarily the sole cause, the production of major grains in the United States began to destabilize in the 1970s and early 1980s, when global warming was not very pronounced, and that this trend will become even more pronounced in the future.

Another water issue concerns a water shortage in the High Plains (Ogallala) Aquifer. This aquifer underlies 111.4 million acres (174,000 square miles) in parts of eight US states: Colorado, Kansas, Nebraska, New Mexico, Oklahoma, South Dakota, Texas and Wyoming. The area overlying the aquifer is one of the major agricultural regions in the world, and represents 20 per cent of cultivated land in the United States (McGuire, 2007). Water-level declines began in parts of the High Plains Aquifer soon after the beginning of extensive groundwater irrigation. By 1980, water levels in the aquifer in parts of Texas, Oklahoma and south-western Kansas had declined more than 100 feet (McGuire, 2007). Also, because water pumped up from the aquifer contains sodium ions, salinity has become a serious issue, causing decreases in crop productivity.

Category II agriculture (regions where structural policy has been achieved): The prospects for twenty-first-century agriculture in the old agricultural regions of Western Europe

Under the rainfall categorization, Category II agriculture corresponds to Region IV (annual aridity index of 20 or more and summer aridity index of 5 or more) where summer crops can be grown but, because weeds flourish, it is a type of agriculture that must include weeding. The geological fertilization effect is almost absent; during the ice ages the topsoil was scraped off, reducing the fertility of the soil. In order to undertake agriculture in Northern Europe, continuous cropping of wheat had to be avoided because (1) it was necessary to supply barnyard manure, etc. to increase the fertility of the soil, (2) the soil was heavy clay, requiring tillage using a deep-tillage plough pulled by animal power, and (3) there were problems of disease and insect damage. In Europe, the three-field system of agriculture (for example, winter crop – summer crop – fallow, repeated in three-year cycles) developed as a way to overcome these conditions (Iinuma, 1982). Oats (spring cropping), turnip (winter feed) and clover (increasing nitrogen nutrients using nitrogen-fixing bacteria by cultivating forage legumes on fallow ground) were grown to provide feed for draught animals, and animal-powered tillage, supplying barnyard manure, etc., permitted the production of food with high energy value. This new system required close cooperative work by farm families, transforming formerly dispersed rural hamlets into centralized villages. One of the benefits of this three-field system was that, during the fallow stage, the use of ploughs to perform deep ploughing and overturn the soil removed weeds (fallow-period weeding work). Also, the soil could be conserved because this system prolonged the period when the soil was covered by crops. The introduction of livestock and legume forage also improved the fertility of the soil.

In any case, in the European Union, food self-sufficiency has been almost completely achieved, natural renewable energies have been extracted from agricultural resources, and new initiatives to achieve material recycling have been undertaken. An example cited as a reference is a new agricultural system that combines compound agricultural and renewable energy and is introduced in the Bioenergy Village Project in Juehnde, a small town in Germany, and in Denmark, where 20 per cent of all energy produced in 2009 was renewable energy (for details see Osaki, Braimoh and Nakagami, eds, 2011, *Designing Our Future: Local Perspectives on Bioproduction, Ecosystems and Humanity*, Sections 4-4 and 4-5, also part of this series). In the village of Juehnde, a population of 770 people in 200 households farms 1,200 hectares, with 9 households practising dairy farming and raising about 400 head of cattle. The core of this concept is (1) supplying electric power and heat produced through cogeneration by biogas facilities, with the fuel obtained from energy crops cultivated on fallow land and night soil from livestock in the village, and (2) providing a regional heating resource by supplying woody biomass mainly for use as a supply of heat in the winter, with the fuel for this process obtained as thinned wood and pruned branches collected in the village. Another important point is that fermentation liquor produced by fermentation is returned to the dry fields as liquid fertilizer, permitting organic agriculture and cyclical agriculture. This system can be implemented in Germany because, under the Renewable Energy Law, the electricity power supply company operating a power plant nearest to a facility that produces electric power from renewable energy sources (for example, solar, wind, biomass) is obligated to purchase the power produced by that facility. This permits such plants to sell their electric power for a good price to earn profits. In Germany, the concept of "energy towns" that combine various kinds of renewable energy has also been established (El Bassam and Maegaard, 2004).

Category III agriculture (regions where structural policy is impossible): The prospects for local infrastructure-oriented agriculture in the Asian monsoon region

Under the rainfall categorization, this category corresponds to Region IV (annual aridity index of 20 or more and summer aridity index of 5 or more), but the summer aridity index distribution ranges from 5 to 11 in Northern Europe and from 9 to 108 in Southeast and East Asia, resulting in extremely heavy rainfall. This rainfall is particularly heavy when the Asian monsoon arrives. Moreover, glaciers in the Himalayas and the Tibetan Plateau act like huge dams, supplying vast quantities of water when the glaciers melt in the summer, feeding the Indus, Ganges, Brahmaputra, Salween, Mekong, Yangtze and Yellow Rivers. It is therefore an ecosystem where weeds grow profusely, so that, if soil management is neglected, the land rapidly deteriorates. There is also vigorous geological fertilization action: (1) rivers in the Himalayas and Tibetan Plateau supply vast quantities of clay; (2) in the circum-Pacific volcanic belt, which includes Indonesia, the Philippines and Japan, volcanic activity is frequent, producing volcanic ash and lava that rejuvenate the soil; (3) the supply of loess by the wind fertilizes land in Eastern China; and (4) the soil rejuvenation action of appropriate erosion forms black fertile soil (Vertisol) with basalt as its base material on the Deccan Plateau in India. Thus, in the Asian monsoon regions in particular, the monsoon climate and the glaciers on the Tibetan Plateau form a system that supplies adequate water during the summer, when crops grow vigorously, and the soil is extremely fertile thanks to various kinds of geological fertilization action. Consequently, the Asian monsoon region is occupied by about 40 per cent of the world's population. Inhabitants are supported by a rich productive infrastructure, enabling them to live full lives in a limited area where they have created diverse cultures, peoples and languages. Linguistic diversity is high in the Asian monsoon region (South India, Southeast Asia, Southern China); linguistic diversity is likewise established in Central Africa and Central America, where the growing seasons are long and rainfall is plentiful (Nettle, 1999). In monsoon Asia, there are many mountains and islands where adequate food supplies can be guaranteed, so there is no need to make strenuous efforts to expand one's territory, a fact that promotes seclusion, preventing expansionist ideas from spreading horizontally, and directing people's attention to enriching their lives in their homelands, developing richly diverse nationalities and cultures.

Ecosystems and biota are more diverse in the Asian monsoon region than anywhere else on Earth, because it is a region with high biological productivity and the biota themselves are geographical and historical conditions (Nakashizuka, 1998). The most unique feature of this region is that, from the tropical area near the Tropic of Capricorn to latitude 60° North in the Arctic zone, there is a continuous moist climate and fertile soil. In many other regions of the world, dry zones often spread and forests are divided by deserts and grassy plains, because there are convergence zones of trade winds near subtropical regions, constantly forming high atmospheric pressure. From the Asian monsoon region to New Zealand in Oceania, particularly along coastlines, forest vegetation is linked, forming a so-called green belt.

In the Asian monsoon region there are many mountains, and rivers often flow rapidly because of the effects of the Himalayas, the Tibetan Plateau and volcanoes. Thanks to the region's adequate rainfall and fertile soil, weeds flourish, so weeding plays an extremely important role as an agricultural technology. In regions with topography and climate of this kind, neglect of land management and environmental destruction cause serious disasters, so meticulous conservation of nature is crucial. The wisdom of coexisting with nature has therefore advanced in these regions. This relationship between human society (defined in Japanese as *sato*) and nature, particularly mountainous regions (defined in Japanese as *yama*), finds expression in the Japanese concept of *satoyama* (for details, see Section 5-3 in Osaki, Braimoh and Nakagami, eds, 2011, *Designing Our Future: Local Perspectives on Bioproduction, Ecosystems and Humanity*).

Exporting surplus foods was a basic strategy of the United States in the twentieth century; this disrupted the world's economy, but the future will bring a return to a value that humanity has long maintained:

that food is the foundation of human life. When this occurs, the Asian monsoon region must again aim to develop a type of agriculture that offers a symbiosis between humans and nature, which is representative of the concept of *satoyama*, and to ensure a rich life for all by assessing the multifaceted functions of agriculture. Considering the limited ability of industrial methods to deal with this diverse ecosystem, conservation of the ecosystem of the Asian monsoon region is possible only if it is based on the coexistence of people and nature, as in the *satoyama* system. This ecosystem is home to almost 40 per cent of the world's population so, if it is destroyed, the harm will be irreparable.

The *satoyama* concept is a search for relationships between ecosystems – mountain and town, or forest, town and ocean, for example – or for a way to stabilize ecosystems through cyclical actions or coexistence. An examination of relationships between humans and nature shows that city–town and city–nature relationships are also important, and that, without links with urban residents, it will be difficult to maintain the *satoyama* system. Until now, greater economic efficiency has been pursued through the one-way concentration of materials and people in cities, but it is now clear that it will be difficult to preserve the ecosystems of habitable regions in perpetuity. It is extremely difficult to construct material cycling systems between cities and towns, and between cities and nature. In contemplating how to incorporate urban systems into the concept of *satoyama*, it is vital to construct a cyclical system of spiritual values as well, and to create a location conducive to a compound form of *satoyama*: one encompassing (1) tourism, including recreational ecotourism, green tourism, agritourism and sustainable tourism; (2) educational proposals for environmental education, nature education, conservation of nature, intergenerational exchanges and lifestyles; (3) long-term residency, rehabilitation and animal therapy in natural curative-based environments; and (4) the administration and construction by volunteers of partnerships between diverse groups, including urban residents, and participation in environmental conservation projects (Osaki, 2007). The *satoyama* concept, which comprehensively considers the diverse links between people and nature in an effort to construct new social infrastructures and foundations for human life, may be expected to emerge as a vital concept for human culture and society in the twenty-first century.

Category IV agriculture (regions where structural policy has not been undertaken): Food production dependent on natural conditions and immature infrastructure in Africa

Among the regions of Africa, rainfall is heavy in tropical Africa but the subtropical northern and southern parts are dry and include the Sahara Desert and the Namib Desert (see Figure 5.3.1). In tropical Africa, 65 per cent of the land area is covered with Oxisol or Psamment, which are senescent leached soils, or Aridisol, which has almost no effective moisture content, making this region totally unsuited for agricultural use. Such soil is almost completely nonexistent in tropical Asia, and the distribution of Oxisol is almost 45 per cent centred in the Amazon in tropical South America. Consequently it is extremely difficult to develop agriculture in tropical Africa (Wakatsuki, 1994). The only effective candidate as a method of improving these deteriorated soils is biochar, which is charcoal created by the pyrolysis of biomass. Biochar can be used as a soil conditioner to increase plant growth yield (Lehmann et al., 2003), improve water quality, reduce soil emissions of GHGs, reduce leaching of nutrients, reduce soil acidity and increase micro-organism activity (see Section 2-4 in Osaki, Braimoh and Nakagami, eds, 2011, *Designing Our Future: Local Perspectives on Bioproduction, Ecosystems and Humanity*).

5-3-4 The prospects for twenty-first-century world agriculture

The Millennium Ecosystem Assessment has developed four global scenarios exploring plausible future changes in drivers, ecosystems, ecosystem services and human well-being (see Figure 5.3.2). These scenarios

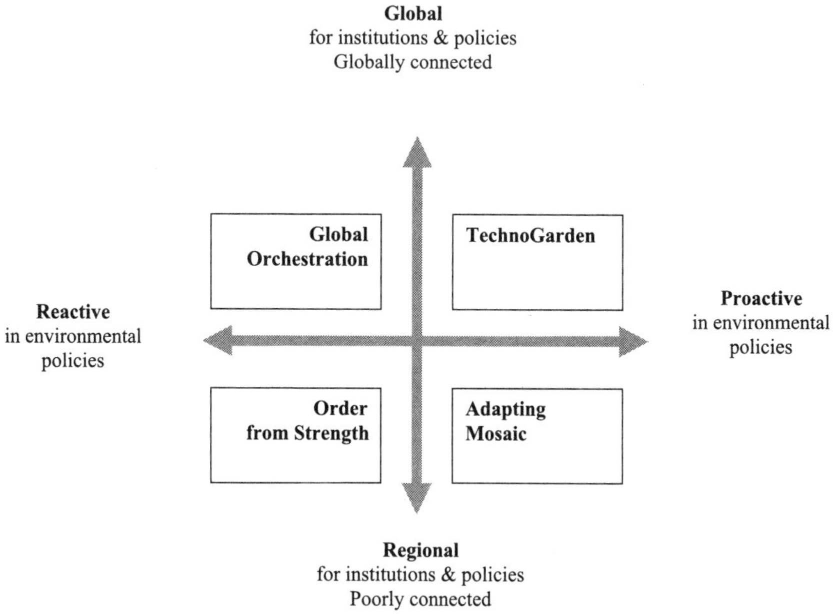

Figure 5.3.2 Four global scenarios developed by the Millennium Ecosystem Assessment.
Source: Millennium Ecosystem Assessment (2005).

are: (1) the Global Orchestration scenario, which "depicts a globally connected society in which policy reforms that focus on global trade and economic liberalization are used to reshape economies and governance"; (2) the TechnoGarden scenario, which "depicts a globally connected world relying strongly on technology and highly managed, often engineered ecosystems to deliver ecosystem services"; (3) the Order from Strength scenario, which "represents a regionalized and fragmented world that is concerned with security and protection, emphasizes primarily regional markets and pays little attention to common goods"; and (4) the Adapting Mosaic scenario, in which "regional watershed-scale ecosystems are the focus of political and economic activity" (Millennium Ecosystem Assessment, 2005: 72–73). The Adapting Mosaic scenario "sees the rise of local ecosystem management strategies and the strengthening of local institutions. Investments in human and social capital are geared towards improving knowledge about ecosystem functioning and management, which results in a better understanding of resilience, fragility, and local flexibility of ecosystems" (Millennium Ecosystem Assessment, 2005: 72).

Twentieth-century agriculture was petroleum-dependent, and, being almost completely unconcerned with ecosystems and intended solely to extract profits from land, it has led to the spread of competitive production. The Millennium Ecosystem Assessment portrays four ecosystem management scenarios according to the degree that environmental policies are connected globally, but it has a structure that is conceptually extremely similar to the structure of agriculture. Figure 5.3.3 shows the results of categorizing twentieth-century agriculture according to the factors of environmental capacity and structural policy. It can be concluded that twentieth-century agriculture sought the form of agriculture (Category I) that most efficiently seeks profits without concern for environmental conservation. In brief, this form was established in a fragile crop production environment where it was possible to ensure high productivity by supplementing this fragile environment with chemical fertilizers. Then, as a result of the need to export surplus products, globalization strategies – eliminating customs tariffs and promoting free trade – were adopted. Category II regions (where structural policy has been achieved) have somehow achieved food self-sufficiency through trade disputes and protecting regional agriculture on the basis of the Common Agricultural Policy. Northeast and Southeast Asia, which are Category III regions (where structural policy is impossible) and unable to compete with petroleum-dependent agriculture, include countries whose self-sufficiency has fallen abruptly, and it is difficult for them to support their national economies with agriculture.

However, petroleum-dependent agriculture does not display concern for the environment, leading to conspicuous environmental degradation.

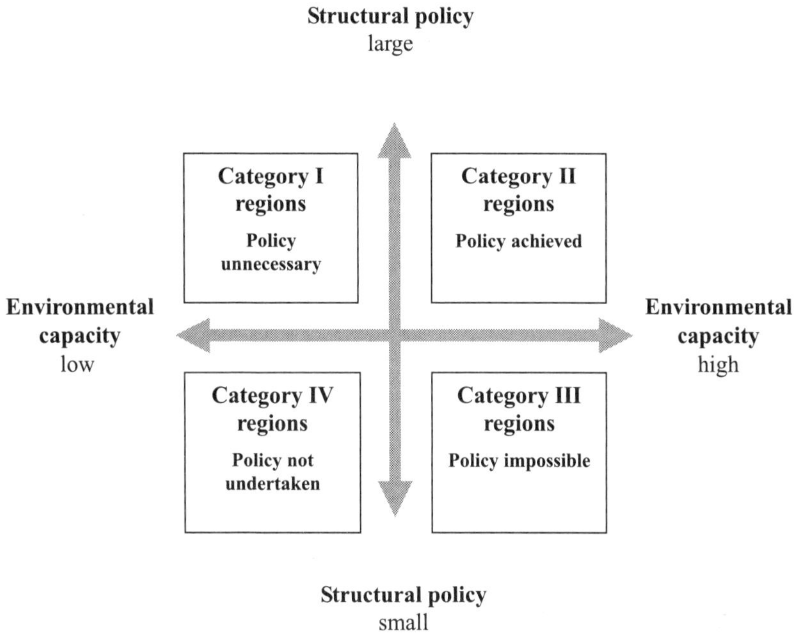

Figure 5.3.3 Twentieth-century agricultural categorization using environmental capacity and structural policy as criteria.

Petroleum-dependent agriculture has reached a major turning point owing to the exhaustion of and restrictions on fossil fuels and environmental deterioration. At the beginning of the twenty-first century, the need to consider restrictions on inexpensive fossil fuels and environmental conservation will make it extremely difficult to maintain Category I region agriculture (where structural policy is unnecessary), that is, petroleum-dependent agriculture. Given that agriculture is an industry that captures broad but weak levels of solar energy, it is clear that it is difficult to establish large-scale agriculture without inexpensive fossil fuel.

In studying forms of agriculture for the twenty-first century, the form to aim for will be made clearer if, instead of using the criteria of environmental capacity and structural policy applied to the categorization of twentieth-century agriculture (Figure 5.3.3), environmental policy is incorporated into the criteria and the forms of twenty-first-century agriculture are categorized according to environmental policy and structural policy (Figure 5.3.4), as in the Millennium Ecosystem Assessment scenarios. Substantially transforming structural policy alone will be difficult considering various social, historical, cultural and geographical conditions. Figure 5.3.4 shows four categories of twenty-first-century agriculture:

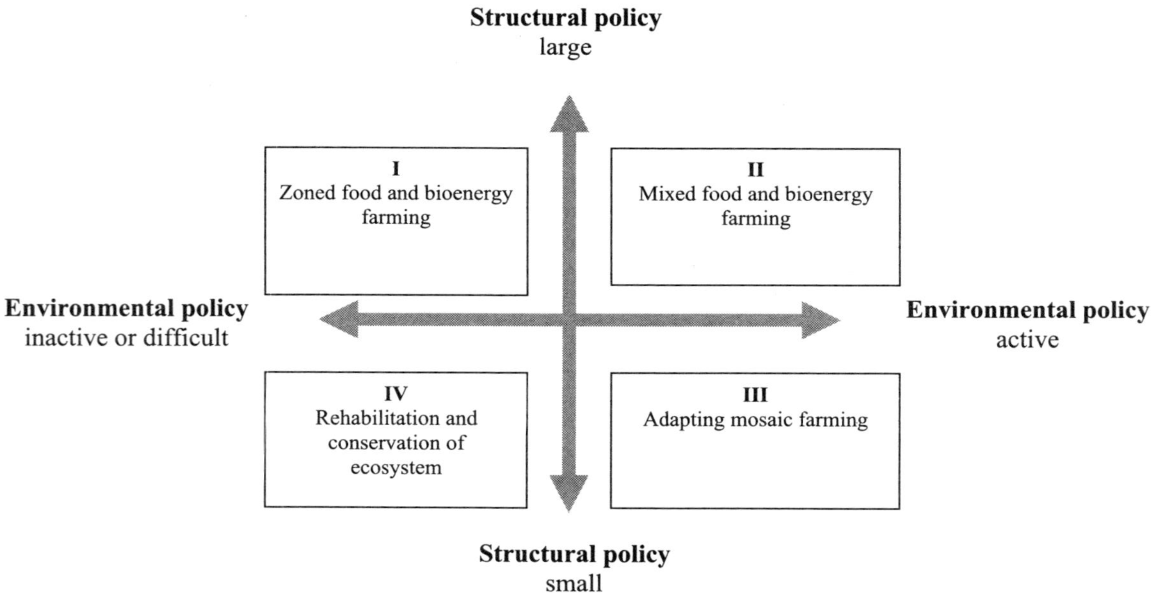

Figure 5.3.4 Twenty-first-century agricultural categorization using environmental policy and structural policy as criteria.

- *Type I (zoned food and bioenergy farming).* This category comprises large-scale agriculture like that in the United States. It is difficult to perform as mixed agriculture, but, because it requires energy obtained from biomass, it can produce grains on the most fertile land if high-yield technologies are implemented. On low-fertility marginal land, the zoning of food and bioenergy will progress as types of vegetation are planted that produce large quantities of biomass and do not require fertilization.
- *Type II (mixed food and bioenergy farming).* This is combined farming, exemplified by mixed farming in the European Union, and is performed on a regional scale with reference to the conventional three-field method. In addition, bioenergy production and physical cycling (of fertilizer constituents, organic materials, etc.) promote agriculture that aims to achieve both food and energy self-sufficiency.
- *Type III (adapting mosaic farming).* This applies to regions where geographical and cultural conditions make it difficult to develop mixed farming as in the European Union and force the adoption of distributed farming, but where rainfall is plentiful and the soil is extremely rich. In such regions, agricultural development maximizing the use of natural ecological services is predicted. Although basically of the distributed type, the construction of small-scale mixed agricultural units permits mosaic agriculture adapted to complex topographies and ecosystems.
- *Type IV (rehabilitation and conservation of ecosystems).* This applies to regions where soil fertility is extremely low and it is difficult to introduce existing agricultural technologies. Here a long-term perspective – boosting the soil's fertility, introducing technologies such as biochar, retaining carbon in the soil, and restoring forests – is necessary. It is important to provide an economic system that will sustain the lives of the inhabitants and conserve the environment and ecosystems by performing ecology management, contributing to the global environment and providing new financial mechanisms such as carbon credits.

The twentieth century was dominated by petroleum-dependent agriculture that made effective use of petroleum resources but severely disrupted the environmental and economic systems. In particular, it threatened local infrastructure-oriented agriculture and forced a reduction in food self-sufficiency, dramatically lowering the value of local infrastructure-oriented agriculture. The twenty-first century will require petroleum-independent agriculture but, if environmental and ecosystem conservation are established as core values, the value of local infrastructure-oriented agriculture will probably rise, permitting sustainable conservation of the environment and ecosystems. However, an economic model adapted to a new agricultural system of this kind has not yet been established, so it is still difficult to design a specific new agricultural system. The model presented here is still too broad, so there is an urgent need to construct specific models adapted to individual regions and to conduct model corroboration research, which must be evaluated through international networks and used to create international models.

REFERENCES

El Bassam, N. and P. Maegaard (2004) *Integrated Renewable Energy for Rural Communities: Planning Guidelines, Technologies and Applications*. Amsterdam: Elsevier.

Fiala, N. (2008) "An Estimation of Potential Future Greenhouse Gas Emissions from Meat Production", *Ecological Economics* 67(3): 412–419.

Hill, J., E. Nelson, D. Tilman, S. Polasky and D. Tiffany (2006) "Environmental, Economic, and Energetic Costs and Benefits of Biodiesel and Ethanol Biofuels", *Proceedings of the National Academy of Sciences* 103: 11206–11210.

Iinuma, J. (1982) *Research on Agricultural Revolutions*. Tokyo: Rural Culture Association Japan (in Japanese).

Lehmann, J., J. Pereira da Silva Jr, C. Steiner, T. Nehls, W. Zech and B. Glaser (2003) "Nutrient Availability and Leaching in an Archaeological Anthrosol and a Ferralsol of the Central Amazon Basin: Fertilizer, Manure and Charcoal Amendments", *Plant and Soil* 249: 343–357.

McGuire, V. L. (2007) "Water-Level Changes in the High Plains Aquifer, Predevelopment to 2005 and 2003 to 2005", US Geological Survey Scientific Investigations Report 2006-5324 <http://pubs.usgs.gov/sir/2006/5324/pdf/SIR20065324.pdf> (accessed 4 June 2010).

Millennium Ecosystem Assessment (2005) *Ecosystems and Human Well-Being: Synthesis*. Washington, DC: Island Press.

Nakashizuka, T. (1998) "Biodiversity in Monsoon Asia", in T. Inoue and E. Wada (eds), *Iwanami Koza Global Environmental Science 5: Biodiversity and Its Conservation*. Tokyo: Iwanami Shoten (in Japanese).

Nettle, D. (1999) *Linguistic Diversity*. Oxford: Oxford University Press.

Noda, K. (2006) "World Agricultural Categorization and Japanese Agriculture: Agriculture and Farm Village Autonomy in Small Farm Society", *at quarterly* 6 (in Japanese).

O'Neal, M. R., M. A. Nearing, R. C. Vining, J. Southworth and R. A. Pfeifer (2005) "Climate Change Impacts on Soil Erosion in Midwest United States with Changes in Crop Management", *Catena* 61: 165–184.

Osaki, M. (2007) "The Satoyama Concept Which Integrates Biological Production Ecology and Regional Societies", in H. Komiyama (ed.), *Challenge of Sustainability Studies*, Iwanami Science Library 137. Tokyo: Iwanami Shoten (in Japanese).

Pimentel, D., A. Marklein, M. A. Toth, M. N. Karpoff, G. S. Paul, R. McCormack, J. Kyriazis and T. Krueger (2009) "Food Versus Biofuels: Environmental and Economic Costs", *Human Ecology* 37: 1–12.

Rojey, A. (2009) *Energy & Climate: How to Achieve a Successful Energy Transition*. Chichester: Wiley.

Rosenzweig, C. and D. Hillel (2008) *Climate Variability and the Global Harvest: Impacts of El Niño and Other Oscillations on Agroecosystems*. New York: Oxford University Press.

Steinfeld, H., P. Gerber, T. Wassenaar, V. Castel, M. Rosales and C. De Haan (2006) *Livestock's Long Shadow: Environmental Issues and Options*. Rome: Food and Agriculture Organization of the United Nations.

Uri, N. D. (2001) "The Environmental Implications of Soil Erosion in the United States", *Environmental Monitoring and Assessment* 66: 293–312.

Wakatsuki, T. (1994) "Soils, People, and Sustainable Agriculture in Tropics: Potential of Sawah Based Agriculture in Tropical Africa", *Tropics* 3(1): 3–7.

Wakatsuki, T. and E. Miwa (1993) "Population Distribution and Soil Fertility in Jomon Period", *Prehistory and Archaeology Research* 4 (in Japanese).

5-4

Defining the sustainable use of fishery resources

Gakushi Ishimura and Megan Bailey

5-4-1 Introduction

Sustainable use of a fishery resource is an important goal for many management agencies worldwide. Sustainability in world capture fisheries can provide two major benefits to society, namely food and income security from both direct (harvesting) and indirect (for example, processing) industries associated with fishing activities. Through both wild capture fisheries and aquaculture, fish offer a major source of protein to much of the world's population and can impart substantial economic returns, either in the short term, or in the long term if managed in a sustainable manner. The scientific evidence today, however, indicates failures in the sustainable use and management of fisheries resources, with researchers predicting a 90 per cent removal of predatory fish (Myers and Worm, 2003) and warning that shortfalls in the supply of fish could have devastating consequences for human populations. What we see today are many fisheries suffering from too many boats fishing too few fish (Pauly et al., 2002), resulting in fewer catches globally and even full stock collapses. Although a limited number of these collapses may have been caused or exacerbated by natural phenomena (for example, climate variability), human activities and overfishing – essentially non-sustainable management – are the primary culprits (Pauly et al., 2002).

There are two fundamental elements of a fishery: (1) fish are a renewable resource, that is, they are replaceable through natural processes (such as geothermal power, fresh water and forests); and (2) the market provides the catch (food source) to consumers and income to fishers for that catch. It is important to understand that fish constitute not only biomass in the ocean but a potential economic input into society. Together, these two elements imply that fisheries are an economic activity by which fishers can catch fish resources in perpetuity if they are managed sustainably. A sustainable fisheries management regime is one that aims to ensure a flow of benefits from fisheries resources to society through the regulation of current and future fishing activities. This section attempts to identify the ideas that constitute our concepts of the sustainable use of fisheries resources through biological and economic tools. One tool that can provide information on the potential consequences of various fisheries management decisions is bio-economic analysis. The bio-economic basis for sustainable fisheries management is explored in this section.

This section will focus exclusively on the sustainability of wild capture fisheries, so aquaculture is not addressed here. First, the section will define what is meant by the sustainable use of fisheries resources. Second, it will introduce a bio-economic model of a fishery to explain how overfishing – essentially, depletion of the fish stock at too high a rate – occurs under open access. Finally, it will demonstrate the importance of combining biological and economic indicators to promote a fisheries management regime that fits with the concepts of sustainability science.

5-4-2 Definition of sustainable fisheries

For the purposes of this section, a sustainable fishery is one that provides substantial stable returns to society over time by means of a regulated fishing sector. In mathematical terms, a sustainable fishery is defined here as one whose catch (in biomass) is equal to the growth (in biomass) of the target stock. This will be explained below through the use of bio-economic modelling.

The bio-economic fisheries model, which merges population dynamics of the fish stock with economic components of the system, is the primary economic approach to estimating the economic and biological consequences of decisions made by fishery managers. This simple yet insightful approach consists of a fish population dynamics model as the biological component and a market model for catch as the economic component, with a production model of fishing effort bringing the two together. The response of the fish stock to human activities and the subsequent economic performance of the fishery can be examined through simulations with the bio-economic model of the fishery.

The bio-economic approach to fisheries is not new, having been initiated by a Canadian economist, H. S. Gordon, and described in his monumental 1954 paper (Gordon, 1954). During the 1970s and 1980s, C. W. Clark and G. R. Munro (Clark and Munro, 1978; Munro, 1979; Munro and Scott, 1985; Clark et al., 1985; Clark, 2006) extended the idea of fisheries bio-economics by introducing financial and economic theories to fisheries science (that is, game, capital and investment theory). The number of bio-economic applications for practical management, though growing, is limited. This is generally because most fisheries management policies rely on estimated catch levels based on biological criteria rather than on economic consequences. The result is both biological and economic waste in many of the world's fisheries. Economists often refer to this as the failure to capture economic rent. The theoretical background of bio-economic analysis has been extensively reviewed by Hannesson (1993) and Clark (1990).

In this section, a surplus production model, the simplest of biological models, is applied to analyse the bio-economic equilibrium of a fishery. The biomass of a stock X, in time $t + 1$, is given by

$$X_{t+1} = X_t + g(X_t) - h_t, \tag{1}$$

where $g(\ )$ is the growth function of the fish stock, which is dependent on the biomass at time t, and h is the catch. Note that growth in the surplus production model is the combination of individual growth and reproduction. The operational definition of the sustainable use of a fishery resource is that catch should be equal to growth in a given time unit, as in equation (2).

$$h_t = g(X_t). \tag{2}$$

This also implies that the size of the biomass remains the same over time.

$$X_{t+1} = X_t. \tag{3}$$

This mathematical definition of sustainable use of a fishery resource is now extended and completed. The term "sustainable" indicates continuousness and recursive modes of use of fishery resources. This implies that the size of catch and biomass today should not be detrimental to the potential catch and biomass in the future. Thus the size of the catch and biomass must be the same over time in a sustainably managed fishery. The time step term (t) is therefore dropped in the discussion to follow.

Catch is defined as a function of effort (e.g., hours of operation, number of fishing vessels participating in the fishery, number of hooks set, etc.), E, and the level of the fish stock.

$$h = q \cdot E \cdot X, \tag{4}$$

where q is the catchability coefficient, which is essentially the proportion of the stock taken by one unit of effort.

In this model a logistic growth function is assumed:

$$g(X) = X \cdot r \cdot \left(1 - \frac{X}{K}\right), \tag{5}$$

where K is the species- and environment-specific carrying capacity.

As economic incentives drive fishing activities, the price of fish, which is defined in the market by demand and supply, is one of the key elements to consider in support of sustainable fisheries. With the integrated global market in seafood, the identification of such demand and supply is challenging.

For convenience, it is assumed here that the unit price of fish (p) is perfectly elastic; that is, no matter what quantity is supplied on the market, the price is constant. Cost (c) is restricted here to mean the variable cost, and is proportional to effort. Variable costs include fuel and boat maintenance, as well as the opportunity cost of labour. This is essentially the lost wages the fisher is giving up in order to fish; it is a measure of his or her next-best opportunity. In this model, fixed costs (capital investment to purchase boats, insurance, etc.) are ignored, because they are assumed to be "sunk" costs.

$$c = \mu \cdot E, \tag{6}$$

where μ is a constant. Profit (π) from fishing is calculated as:

$$\pi = p \cdot h - c. \tag{7}$$

A key feature of the surplus production model is to keep the stock size in a sustainable state, that is, one where the catch equals the growth of the stock in each time step (equations (2) and (3)). In other words, under the assumption of the surplus production model, catching the surplus induces a sustainable fishery. Furthermore, from the above equations, and based on the assumptions of sustainable catch and the perpetually stable size of the fish stock, one can define a unique effort level corresponding to this sustainable fish stock size as:

$$E = \frac{h}{qX}. \tag{8}$$

Throughout this section the premise is held that a sustainable fishery is one where catch equals growth for each time step. This enables a static

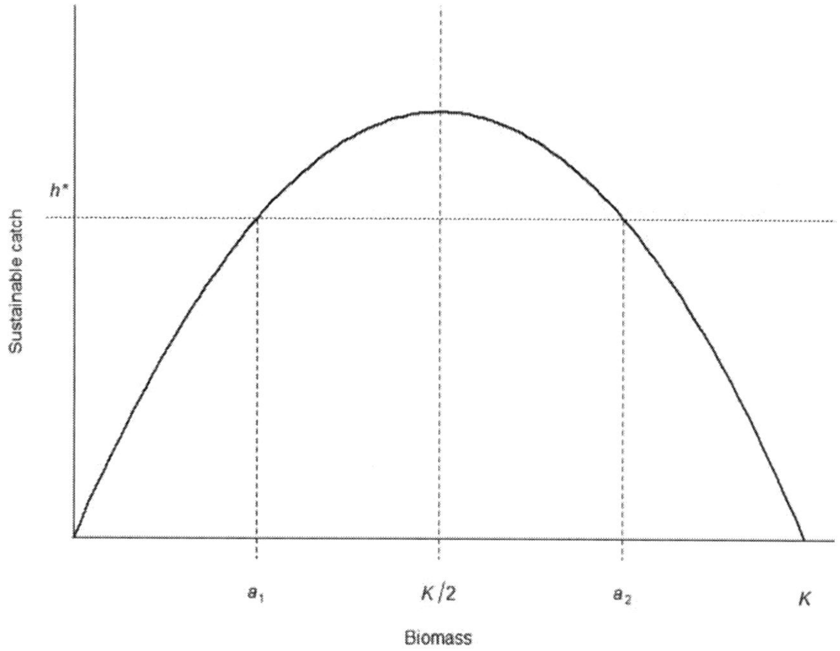

Figure 5.4.1 Sustainable catch with the surplus production model.

picture of a sustainable fishery to be drawn (Figure 5.4.1). Note that this static picture expresses the equilibrium status of the fish stock level as well as the catch level. This implies that the domain of the fish stock level is implicitly constrained at less than or equal to the carrying capacity (that is, $0 \leq X \leq K$).

Figure 5.4.1 shows catch as a function of the fish stock level. As the stock increases from 0, catch is increased and maximized at $K/2$, a level called the maximum sustainable yield (MSY). Managers have often tried to target this stock size in order to maintain the most productive stock level. As Clark (2006) discusses, MSY is not a stable equilibrium point, and is therefore a dangerous target because deviations from MSY can be extremely detrimental to the stock and hence the fishery. As the fish stock increases to more than $K/2$, the catch level actually decreases as a result of a less productive stock, owing to natural mortality factors such as density effects (for example, food availability, cannibalism). The growth of the stock, and hence of the catch, reaches 0 at K, the carrying capacity. Now biomass size is compared at the same catch level, h^*, given by a_1 and a_2. Note that the biomass at a_1 and a_2 results in exactly the same yield, h^*, although a_2 is a much higher fish stock level. Thus, if it is the case that catch is the only performance measure for a fishery, a_1 and a_2

are equally preferable. From a biological point of view, however, the fish stock is affected differently: a_2 maintains a much higher biomass level than a_1.

5-4-3 Overfishing and open access

The main feature of the fishery system is the nexus between fishery resources and society through fishing activities driven by economic motivations, whereby overfishing is the result of failure to control the economic motivations for fishing that destroys the renewable nature of the fishery resource. In this section, the bio-economic framework is again used to illustrate why the open access condition leads to overfishing.

In the bio-economic model from Gordon (1954), the revenue curve (R) is a function of the unit price of fish and the catch, and is equal to the catch curve (given constant prices, as explained above) (Figure 5.4.2).

$$R = p \cdot h = p \cdot g(X). \tag{9}$$

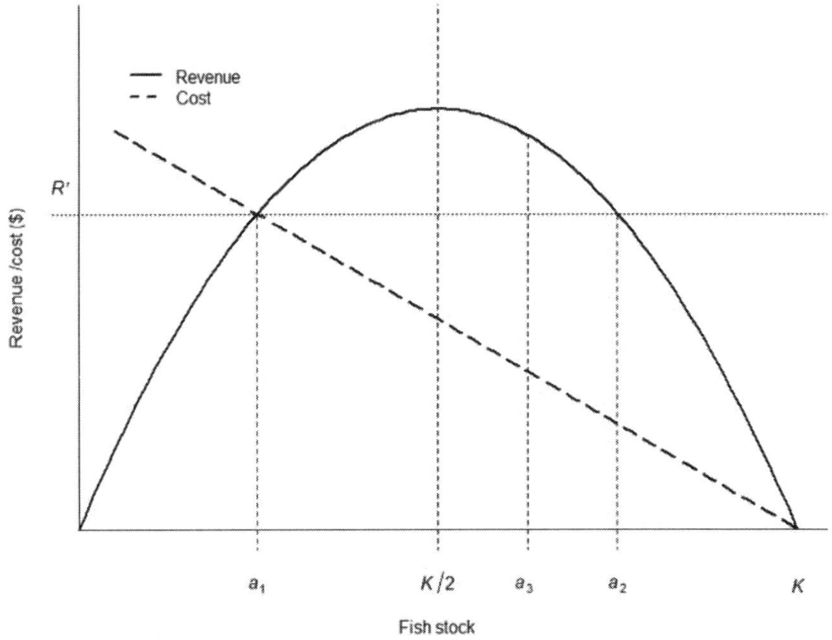

Figure 5.4.2 Revenue and cost in a sustainable fishery with the surplus production model.

It is assumed that the cost function of the fishery is linear, and inversely proportional to the fish stock level. Given the assumptions regarding the sustainable use of the fishery resource and surplus production, the cost of a sustainable catch is expressed as:

$$c = \frac{\mu \cdot r}{q}\left(-\frac{1}{K}X + 1\right). \tag{10}$$

The cost per harvest is:

$$\frac{c}{h} = \frac{\mu}{qX}. \tag{11}$$

This analysis suggests that the cost per unit of harvest decreases as the biomass increases. This is intuitive: it is generally more costly to fish fewer fish, because the stock density has been reduced.

As mentioned above, the opportunity cost of labour is a crucial component of the cost function in a bio-economic model. Opportunity costs reveal the earning potential by fishers if their labour is used in alternative projects other than the fishery. Although one cannot ignore the opportunity cost of labour in this analysis, the opportunity cost of fishing capital is generally ignored. This is the cost of having that capital invested in the fishery instead of in some other endeavour. It is ignored on the assumption of the irreversibility (non-malleability) of fishing capital, which is a unique characteristic of fishery industries and is discussed by Clark et al. (1979). Generally, capital investments for fishing, such as the purchase of a boat, are designed for specific fisheries and usually are not transferable to other fisheries or other industries without the loss of the original value. This irreversibility of the investments could lead to destructive results in the sustainable management of fishery resources. One therefore can assume that, once the investment in fishing capital is made, fishing must go on until the business owner receives negative profits or eliminates the fish stock. In this study, these assumptions are maintained in the following analysis.

The difference between revenue and cost is often referred to as the economic or resource rent of the fishery. Here it is assumed that the resource rent is equal to the profit (π), as in equation (7), for the definition of, and argument for, resource rent in a fishery (Stoneham et al., 2005). Although the catch, and hence revenue, at a_1 and a_2 are the same as described above, they yield very dissimilar profits. Whereas the profit at a_1 is equal to zero, a_2 yields positive profits (Figure 5.4.2). If economic rent is the performance measure of the fishery, instead of simply catch, a manager would most probably prefer to maintain the fish stock level at a_2

rather than a_1. The maximum sustainable profit is at a_3, where the first-order condition of the revenue curve (that is, the marginal change in revenue) is equal to the slope of the cost curve, and stays between $K/2$ and K.

The marginal change in the revenue (the first-order condition of the revenue curve), per equation (9), is given by:

$$\frac{\partial R}{\partial X} = p \cdot \left(r - \frac{2X}{K} \right). \tag{12}$$

The biomass size at the marginal change in the revenue is equal to the slope of the cost curve ($-\mu r/qK$), and is given by:

$$X_{MEY} = \frac{r}{2}\left(\frac{\mu}{pq} + K \right). \tag{13}$$

The sustainable catch at this biomass size is often called the maximum economic yield (MEY). Thus, comparing the biological (catch) and economic (rent) performance measures, it appears that the economic argument is actually more biologically conservative for a_1.

The point where revenue and costs are equal, at a_1, is called the bionomic equilibrium, where all resource rents are essentially dissipated. This is often the condition for stocks to be both overfished and overcapitalized (that is, a high amount of fishing capacity is necessary to maintain a sizeable catch). The Food and Agriculture Organization of the United Nations (FAO) suggests that depletion of a stock occurs when demand for the fish product outstrips the biological capacity of that stock for sustaining itself (FAO, 2005). In economic terms, this is essentially what is called overfishing. Overfishing is almost always the end result of an open access system, one where access to the fishery for the purpose of catching fish is unrestricted (for example, no exclusive rights for fishing). It is increasingly recognized, and rarely contested, that restrictions on the open access condition are a necessary, but of course probably not sufficient, means of moving a system away from overfishing. The simple bioeconomic model in this section can help one understand why overfishing occurs under open access.

In Figure 5.4.2, one can see that, for any biomass size greater than a_1, the economic rent from the fishery is positive. Current and potential fishers compare their expected profits from the fishery with their next-best alternative and, if rents in the fishery are positive, generally effort will continue moving into the fishery. As effort moves in, and the fishery heads towards overcapitalization, rents are slowly dissipated until a_1 is reached, where the marginal profit is zero and there is no incentive to enter or to leave the fishery. The majority of management measures put in place to control catch (output controls) or effort (input controls) may initially succeed in rebuilding the stock to a larger size, but these attempts will eventually fail because, as soon as rents return to the fishery, effort will move back in. This is the perverse nature of overfishing, and is entirely predictable given the simple Gordon bio-economic model.

An overfished resource is often accompanied by an overcapitalized fishery. Overcapitalization simply means that too many boats, too many nets, too many hooks or the like are being used to catch the fish. As seen in Figure 5.4.3, the same catch is brought in at stock sizes a_1 and a_2, but more effort is being used at stock size a_1 (owing to greater costs at a_1 than a_2). From society's point of view, this is wasteful, because that labour (and capital) could be contributing to society in some other way. Overcapitalized fisheries are often the result of government subsidies (Sumaila et al., 2006). Subsidies can take many forms (Clark et al., 2007), but some of the more biologically harmful subsidies include those that reduce fishers' costs, including fuel and vessel subsidies.

As can be seen in Figure 5.4.3, subsidies decrease the cost of fishing. This results in a bionomic equilibrium at a lower biomass size than one would otherwise find (b_1). Fishers keep fishing and effort continues to

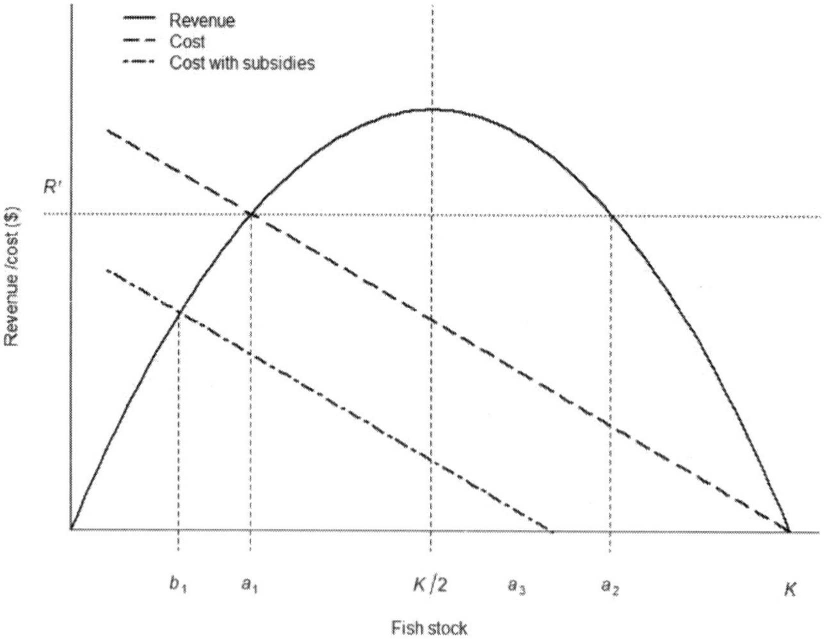

Figure 5.4.3 Cost and cost with subsidies in a sustainable fishery with the surplus production model.

move into the fishery at higher levels because the economic indicator that rents have dissipated is falsely nonexistent. Many scientists have called for the elimination of harmful subsidies (Pauly et al., 2003; Sumaila et al., 2006).

5-4-4 Conclusion: Integrating biological and economic indicators towards sustainable use of fishery resources

It has often been said that managing fish is really about managing people. If one sees a fishery not just as biomass in the ocean (or lake) but as the economic activity of fishers, one can understand that the lack of consideration for economic motivations and incentives can result, and has resulted, in failures of fishery management institutions. Even with a simple view of a sustainable fishery, it has been observed that using only biological or only economic performance indicators would result in dissimilar optimal stock levels. The conclusion here is that, to be considered sustainability science, fisheries science should also consider economic rent generated by the fishery resource, rather than just catch. That is, fisheries science needs to embrace interdisciplinary perspectives and methods. As long as economic motivations for catch exist without regulation of fishing activities, the depletion of biomass, which endangers the sustainable use of the fishery resource, and the failure to capture economic rent cannot be avoided.

Many attempts have been made to alter the incentives of fishers in order to remove effort from fisheries and to rebuild stocks. Some of these have been biological measures, such as closing areas to fishing (sometimes called marine protected areas). Others have been predominantly economic in their approach, for example allocation schemes such as individual transferable quotas. It is increasingly evident to scientists and managers, however, that biological and economic measures need to be integrated to ensure sustainable fisheries – those that catch only the surplus (or interest) from the fish population, allowing the biomass (as capital) to replenish itself in perpetuity. Market-based incentives, such as campaigns aimed at increasing consumer awareness and altering consumer demand, have also been promoted in an attempt to encourage sustainable fisheries. Such fisheries can be capable of providing both food and income benefits to society over time.

REFERENCES

Clark, C. W. (1990) *Mathematical Bioeconomics: The Optimal Management of Renewable Resources*. New York: John Wiley & Sons.

Clark, C. W. (2006) *The Worldwide Crisis in Fisheries: Economic Models and Human Behavior*. Cambridge: Cambridge University Press.

Clark, C. W. and G. R. Munro (1978) "Renewable Resource Management and Extinction", *Journal of Environmental Economics and Management* 5(2): 198–205.

Clark, C. W., F. H. Clarke and G. R. Munro (1979) "The Optimal Exploitation of Renewable Resource Stocks: Problems of Irreversible Investment", *Econometrica* 47(1): 25–47.

Clark, C. W., G. R. Munro and A. T. Charles (1985) "Fisheries, Dynamics, and Uncertainty", in A. Scott (ed.), *Progress in Natural Resource Economics: Essays in Resource Analysis by Members of the Programme in Natural Resource Economics (PNRE) at the University of British Columbia*. Oxford: Clarendon Press, pp. 99–120.

Clark, C. W., G. R. Munro and U. R. Sumaila (2007) "Buyback Subsidies, the Time Consistency Problem, and the ITQ Alternative", *Land Economics* 83(1): 50–58.

FAO [Food and Agriculture Organization of the United Nations] (2005) "Governance of Capture Fisheries", <http://www.fao.org/fishery/topic/2005/en> (accessed 15 June 2010).

Gordon, H. S. (1954) "The Economic Theory of a Common-Property Resource: The Fishery", *Journal of Political Economy* 62(2): 124–142.

Hannesson, R. (1993) *Bioeconomic Analysis of Fisheries*. New York: Halsted Press.

Munro, G. R. (1979) "The Optimal Management of Transboundary Renewable Resources", *Canadian Journal of Economics* 12(3): 355–376.

Munro, G. R. and A. D. Scott (1985) "The Economics of Fisheries Management", in A. V. Kneese and J. L. Sweeney (eds), *Handbook of Natural Resource and Energy Economics, Volume II*. Amsterdam: North-Holland, pp. 623–676.

Myers, R. A. and B. Worm (2003) "Rapid Worldwide Depletion of Predatory Fish Communities", *Nature* 423: 280–283.

Pauly, D., J. Alder, E. Bennett, V. Christensen, P. Tyedmers and R. Watson (2003) "The Future for Fisheries", *Science* 302: 1359–1361.

Pauly, D., V. Christensen, S. Guénette, T. J. Pitcher, U. R. Sumaila, C. J. Walters, R. Watson and D. Zeller (2002) "Toward Sustainability in World Fisheries", *Nature* 418: 689–695.

Stoneham, G., N. Lansdell, A. Cole and L. Strappazzon (2005) "Reforming Resource Rent Policy: An Information Economics Perspective", *Marine Policy* 29(4): 331–338.

Sumaila, U. R., A. Khan, L. Teh, R. Watson, P. Tyedmers and D. Pauly (2006) "Subsidies to High Seas Bottom Trawl Fleets and the Sustainability of Deep Sea Benthic Fish Stocks", in U. R. Sumaila and D. Pauly (eds), *Catching More Bait: A Bottom-up Re-estimation of Global Fisheries Subsidies*. Fisheries Centre Research Report 14(6), revised version, 2007, pp. 49–53. Fisheries Centre, University of British Columbia, Vancouver.

5-5

The market economy and the environment

Takamitsu Sawa

5-5-1 Market versus government

Adam Smith's thesis

It was in the 1980s that market fundamentalism – the faith in the market's omnipotence – became extremely fashionable in Europe and the United States. In 1979, Margaret Thatcher took office as prime minister of the United Kingdom and, in 1981, Ronald Reagan became president of the United States. In Japan, Yasuhiro Nakasone became prime minister in 1982. Each of these three politicians was second to none in terms of their vitality and leadership, and they resolutely pushed free-market reforms forward in their respective countries. Based on their conviction that the best policy is to entrust all economic activity to the market, these three administrations loosened or rescinded legal regulations, privatized state-owned industries, liberalized financial markets and relaxed or abolished protectionism.

Compared with the UK and US reforms implemented under Thatcher and Reagan, however, the Japanese reforms ended up being done in a rather half-baked way. As a result, not long afterwards, when the administration of Junichiro Koizumi came to power in the first year of the twenty-first century, it felt compelled to forcefully declare a new round of free-market reforms under the label of "structural reform". In that sense, free-market reforms in Japan – a country that is not very good at reforms

– were delayed by about 20 years in comparison with those in the United Kingdom and the United States.

The pioneer of market fundamentalism was Adam Smith (1723–1790), who taught moral philosophy at the University of Glasgow in the United Kingdom. In his classic work *The Wealth of Nations* (1776), Smith wrote: "By pursuing his own interest [an individual] frequently promotes that of the society more effectually than when he really intends to promote it" (Book IV, chapter II, para. IX).

These remarks, which declare that the pursuit of one's selfish desires is connected to the "public interest", imply that government policies that seek to "promote the interests of society" tend to be ineffective, and that acts by individuals that might at first glance seem to be the "pursuit of self-interest" and therefore contrary to the public interest in fact unintentionally promote the interests of society. Adam Smith's thesis forms the basis of the admonition that is heard from market fundamentalists today, at the start of the twenty-first century, to "leave everything to the private sector", that is, to the market.

An example illustrates how the pursuit of selfish interests and desires is connected to the "public interest". Let us say that there is a single road that connects two villages that were previously isolated, and that, thanks to this road, the people of the two villages travel back and forth on a daily basis, trading goods. The exchange of people and the trading of goods unquestionably contribute to the welfare of the people of both villages. The reader is probably thinking that it was undoubtedly through planning by some unknown person, a wise person blessed with foresight, or perhaps through consultations between leaders from both villages, that this single road was built. But people are not clever enough to foresee that "trade" would benefit the people of both villages.

Realistically speaking, the road more likely came about as follows. A man decided that he wanted to have the chicken whose crows he heard each morning from the neighbouring village, and so he walked through the thick brush, snuck into the neighbouring village, stole one of the chickens under cover of darkness, and fled home. Several days later, a man from that neighbouring village came to the first man's village and stole some liquor. He undoubtedly walked along the same path as the first man, since that man would have beaten down the brush as he walked and it would be easier to walk along that route. If the second man returned the way he came, then the tracks where the two men had walked would start to look like a path. The next man who went to steal something would probably walk the same way.

And so, as a result of many men walking over the same path, the public good known as a road was formed. The essence of Adam Smith's thesis can be found in the fact that not a single person had the "intention" to build the road. And not only did nobody do it intentionally, but it was scoundrels who were trying to steal to satisfy their own interests and desires who unintentionally built this public good called a road. The result was that the welfare of both villages certainly improved.

The place where corporations and consumers in pursuit of self-interest meet, and the place where goods and services are traded, is none other than the *market*. In a market in which an infinite number of sellers and buyers participate, the *equilibrium price* for goods and services is set in such a way that demand and supply balance out (that is, excesses or deficiencies in supply and demand disappear). If a fluctuation in demand or supply occurs, the price or transaction volume will be revised so that the equilibrium between supply and demand is restored. For example, the outbreak of mad cow disease caused a rapid drop in the demand for beef, while the demand for chicken and pork rose. As a result, to varying degrees the price of beef went down and the price of chicken and pork increased. That type of dynamic behaviour by the market is called the *market mechanism*.

The end of laissez-faire

From the 1840s through the 1870s, the classic free-market theory known as "laissez-faire" held sway, particularly in the United Kingdom. However, the age of laissez-faire economics did not last long. John Maynard Keynes, who in 1926 wrote *The End of Laissez-Faire*, explained why the concept rose and fell. He gave three reasons for the sweeping conquest of nineteenth-century Europe by the idea of "a divine harmony between private advantage and the public good", that is, laissez-faire thinking. First, the corruption and ineptitude of eighteenth-century government had "strongly prejudiced the practical man in favour of laissez-faire" (Keynes, [1926] 2004: 19). Second, the material progress made between the mid-eighteenth and mid-nineteenth century had been the product of individual initiative, and there was little recognition of governmental contributions. Third, Darwinism, which explained that "free competition had built Man", resonated with the theories of economists who explained that "free competition built London" ([1926] 2004: 20).

Against the backdrop of these ideological tides after the middle of the nineteenth century, "the ground was fertile for a doctrine that ... State Action should be narrowly confined and economic life left, unregulated so far as may be, to the skill and good sense of individual citizens actuated by the admirable motive of trying to get on in the world" ([1926] 2004: 20). Moreover, not only was laissez-faire influenced by the prevalent political doctrines of the day, but it "[conformed] with the needs and wishes of the business world of the day" ([1926] 2004: 34).

It was in the late 1970s, more than 200 years after the publication of *The Wealth of Nations*, that laissez-faire thinking made its comeback under the new name of market fundamentalism. Exactly 10 years after he had declared the "end" of laissez-faire in 1926, Keynes wrote *The General Theory of Employment, Interest and Money* (1936), which formed the cornerstone of the Keynesian economics that would go on to shape the economic policies of advanced capitalist nations for more than 40 years. When the stock market crash hit New York's Wall Street in October 1929 and the world was plunged into the Great Depression, President Roosevelt's New Deal policies (entailing large-scale public works projects such as the construction of dams) pulled the US economy out of the depression. At first glance, it would appear that these New Deal policies intentionally followed the teachings of Keynes, but in fact that was not the case. When the Great Depression hit, President Hoover thought that reducing the nation's budget deficit would be the panacea for the economy and thus shifted to a money-tightening policy. However, the economy continued to worsen and there was not even the slightest indication that it might improve. Hoover's successor, Roosevelt, believed that the cause was insufficient domestic demand, and so he launched large-scale public works programmes without worrying about the budget deficit. The resulting New Deal policy turned out to be a tremendous success. It was Keynes' theories, however, that logically explain why the New Deal policies were effective.

If one sums up the essence of Keynesian economics in a few lines, it would be as follows. The market is imperfect. Accordingly, government use of fiscal and monetary policy to intervene in the market in order to stabilize the national economy and correct inequities such as unemployment is both necessary and desirable.

The imperfection of the market refers to the following types of issues. There are many things, such as the price of labour – that is, wages – for which price is inelastic. Since friction in the market is difficult to avoid, it takes a long time to shift from the current equilibrium to the next. Future projections by companies and households are prone to egregious errors. As a result, the market mechanism functions imperfectly and imbalances such as unemployment, for example, are not resolved. We are perpetually shadowed by the instability of economic fluctuations.

The perfectly competitive market is "efficient"

Keynes believed that government intervention in the market is necessary because the market is "imperfect", and he targeted his criticism at the classical economists' hypothesis of a *perfectly competitive market*.

A perfectly competitive market must meet the following four conditions:

1. The goods and services exchanged in the market must be of exactly the same quality. If you take cell phones as an example, the functions, design, customer service and so on must be equivalent, or, to put it another way, there can be no product differentiation among brands.
2. There must be an "infinite" number of producers and consumers so that the volume of goods and services purchased by an individual consumer or supplied by an individual producer is like a drop in a vast ocean when compared with the volume being exchanged in the market as a whole. Otherwise, the behaviour of one producer or one consumer is liable to increase or decrease the market price. All producers and consumers must be price takers who obediently accept the price set by the market.

   Only when these two conditions are met does the law of one price become viable. It was stated above that "there can be no product differentiation", but it is possible to look at differentiated products as different goods. However, if one does so, then the number of producers of those different goods becomes "finite", so it is difficult to meet the second condition.
3. All producers and consumers are equally well informed regarding "information" on the quality and other aspects of the goods and services being traded on the market, and they can completely foresee the future.
4. Free entry and withdrawal from the market must be guaranteed. There can be no regulations or barriers.

   A market that meets these four conditions is a perfectly competitive market but, for the perfectly competitive market to be *efficient*, it must also meet five additional, rather technical conditions.
5. There must be a market for the exchange of all goods and services that might be considered.
6. A consumer's *utility function* is a function only of the volume of goods and services that the consumer in question purchases by him- or herself, while a company's *production function* is a function only of the volume of goods and services that the company in question invests in and produces itself.
7. The scale of companies cannot expand beyond a fixed limit, and one needs assumptions such as a diminishing marginal rate of substitution and diminishing marginal costs in order to achieve perfect competition.
8. The factors of production (capital and labour) must move freely between companies.
9. There must be a stable competitive equilibrium.

Assuming that these nine conditions are met (there are a number of other technical conditions as well, but they are in a sense trivial, so they are omitted here), the thesis that "the perfectly competitive market economy is efficient" can be mathematically deduced.

When used in everyday conversation, newspapers and magazines, government publications and elsewhere, the term "efficient" generally means "good in terms of cost/benefit considerations": in other words, a specific effect is produced with minimal expenditure, or the maximum effect is achieved with a given amount of expense. However, when one says that a perfectly competitive market is "efficient", the term has a different meaning.

A situation in which one cannot make any person better off without making someone else worse off is called *Pareto optimal*. Conversely, the situation is *Pareto non-optimal* if one can make someone better off without making anyone else worse off. To put this in simpler terms, even if every individual who is pursuing his or her own interests and desires has an equal right of veto, then, as long as envy or jealousy is not a factor, there is room for improvement under a Pareto non-optimal situation (that is, there is not a single person who would be worse off as a result of a change, so nobody uses their right of veto). On the other hand, as long as individuals have the power of veto, no changes can be made to a Pareto-optimal situation because the result of a change would be that at least one person would be worse off, and that person would then exercise his or her right to veto. The theorem that "perfect competition results in Pareto optimality and, conversely, any Pareto-optimal situation is achieved through perfect competition" is the fundamental theorem of welfare economics. The substance of this argument is a rephrasing of the statement that "perfectly competitive markets are efficient" or "perfectly competitive markets optimally distribute resources". What this thesis aims towards is none other than proof of the existence of what Adam Smith figuratively referred to in *The Wealth of Nations* as the "invisible hand".

Supplementing the "imperfect" market

One would have to say that all four conditions listed above as requirements for a perfectly competitive market are "unrealistic". Be that as it may, it would be an oversimplification to say that, because actual markets differ to some degree from the perfectly competitive market, the basic assumption of welfare economics is unrealistic. Even when deriving some kind of theory in physics, the hypothesis upon which it is premised is more or less unrealistic.

What must be questioned is how robust the theorem is that indicates that "a perfectly competitive market is efficient" (even if the market is slightly imperfect, to what extent is the market economy pseudo-efficient?). In other words, how accurately does the perfectly competitive market as an "ideal type" predict actual movements in the market economy? Unfortunately, today's economics has not yet reached the point where it can definitively answer that question.

As noted above, Keynesian economics assumes that, since the market is imperfect, governmental intervention in the market is indispensable to avoid disequilibrium as well as instability. During the period from the 1950s to the 1970s, everyone was convinced that it was possible, and indeed essential, for the government to "control" the macroeconomy. Furthermore, it was thought that discretionary fiscal and monetary policies (quantitative adjustments to government spending, tax rates and the official discount rate) were the best means to get the job done.

At the time, there was an overwhelming tendency to worship science and technology as omnipotent. The natural sciences exist to "control" nature. The accumulation of scientific knowledge has produced useful technologies for controlling and simulating nature, has overcome disasters and disease and has provided us with various kinds of machinery powered by electricity and oil products, as well as the means for faster and broader travel and communication. In the same way, it was felt, the social sciences must come up with useful technologies for "controlling" society. Keynesian economics, which at least superficially concurs with social engineering concepts, was consistent with the social trend towards scientific omnipotence.

Keynes' recognition of the economic reality that "markets are imperfect" became the common wisdom throughout the economics field around the late 1960s to mid-1970s. At that time, the phrase "market failure" was commonly heard. Obstruction for one reason or another of the market's function of "granting an efficient allocation of resources" was referred to as market failure. If one asks why markets fail, it is because the hypothesis on which the fundamental theorem of welfare economics is premised does not hold true.

Around the same time, attacks on the immorality of the market economy flourished. In short, whether or not a market is perfect, the market economy widens income disparity and is filled with such examples of immorality as pollution and degradation of the environment by corporations. Therefore, if the government does not take the necessary steps to control the market, a market economy is likely to lead to a gruesome "law of the jungle" conclusion. Those who were inclined to endorse this anti-free-market view styled themselves the school of radical economics, and they had a considerable impact on the economics field in the first half of the 1970s.

To put this another way, the market economy might be efficient but it is not fair. From the perspective of the normative value of fairness, the market economy is filled with flaws. In order to achieve an efficient and fair society, government must play a supplementary role in the market. Such arguments by these critics of free-market economics unquestionably played a role in making government bigger.

The return of market fundamentalism

The "oil shock" of 1973 closed the book on a period of rapid economic growth. The growth rates of all advanced industrial nations were reduced by half, and the expansion of tax revenues slowed. In order to avoid budget deficits, it was necessary to re-examine the steady expansion of government expenditures. In Japan, as well as elsewhere, it is extremely difficult to reduce government spending once it has expanded. Who, then, was to blame for this expanded government spending? The search for a scapegoat began. The one who stood falsely accused was none other than John Maynard Keynes. From the late 1970s to the early 1980s, Keynes was like a fallen idol, bearing the brunt of thunderous criticism as the culprit who had increased budget deficits.

In the field of economics, essays started to appear in academic journals that demonstrated (that is, established a hypothesis and then deductively drew a conclusion) that the discretionary fiscal and monetary policy continually advocated by Keynesian economics was "ineffective" and "harmful", although there were many schools of thought on the issue. At one point, the rational expectations school, supply-side economics, monetarism and other schools of anti-Keynesian economics became the mainstream factions in the field. Even the economist on the street began to use such phrases as "Keynes is dead", "small government", "government inefficiency", "private over public" and "deregulation".

Certainly, Keynes was extremely influential. Governments in the latter half of the twentieth century came to perceive it as their duty to intervene actively in markets with the intention of eliminating the instability and inequity of macroeconomics. In other words, governments began to claim for themselves the role of supplementing the imperfect market. In this way, Keynesian economics unquestionably formed the basis of economic policy-making in industrialized nations from the 1950s to the 1970s.

The fiscal policy proposed by Keynes aimed at stabilizing the economy and correcting disequilibrium (for instance, unemployment) by freely expanding and shrinking government spending, just like a toy balloon. But the actual practice of Keynesian fiscal policy in Japan, for example, was reduced to nothing more than a weapon used by politicians to steer funds towards their own constituencies, and there can be no denying that this was the main cause of progressive increases in the budget deficit. However, the real fault lay with the politicians who, once government spending had ballooned, would not allow it to be shrunk again, and so the accusations levelled at Keynes were completely without merit.

The free-market economy is a product of social engineering

According to British political philosopher John Gray, the free-market economy is not "natural", but rather is something designed through social engineering by those in power and by the state. In short, if the free-market economy were truly "natural" and the government did not do anything to create one, it would materialize on its own. The current situation in Japan, in which the objective of structural reform is to "create a free-market economy", substantiates Gray's argument. Gray (2002: 17) states:

> The free market is not, as New Right thinkers have imagined or claimed, a gift of social evolution. It is an end-product of social engineering and unyielding political will. It was feasible in nineteenth-century England only because, and for so long as, functioning democratic institutions were lacking.
>
> The implications of these truths for the project of constructing a worldwide free market in an age of democratic government are profound. They are that the rules of the game of the market must be insulated from democratic deliberation and political amendment. Democracy and the free market are rivals, not allies.

Critics of market fundamentalism argue that those who reap the benefits of free-market reforms are, relatively speaking, the minority with power, whereas the weak majority are nothing more than the victims of those reforms. Free-market proponents rebut that argument by saying that free-market reforms raise the growth rate of the economy, so the benefits trickle down to the impoverished class. Certainly during Japan's period of rapid economic growth, when manufacturing industry propelled the national economy, the trickle-down theory held true. However, as recent statistics indicate, that theory has lost its relationship to reality.

Under a majority-rule democracy, as long as democratic debate and political revisions are unimpeded, free-market reforms are unlikely to move forward. Isolating reforms from democratic debate requires a theoretical weapon that demonstrates that the free-market economy is desirable. In that case, the most persuasive strategy is to circulate and popularize the theory that, whereas the government is inefficient and foolish, the market (i.e. private business) is efficient and wise.

To prove this theory logically is not a simple task. Even if one employs the "fundamental theorem of welfare economics" noted above to demonstrate the efficiency of the market economy, it does not necessarily follow that "government is inefficient and foolish". However, the quotation at the start of this section from Adam Smith's *The Wealth of Nations* (1776) makes precisely that argument.

Looking back at post-war Japan's economic growth, the role played by the government, for better or for worse, was certainly not minor. The priority production method used immediately after the war, the industrial policy, the administrative guidance, the convoy method, the enormous sums spent on public investment and so on are often mistakenly described as state socialism, but the government must be given high marks for maximizing economic growth rates and providing the country with balanced growth. However, the downside of this approach was the interdependence of government, bureaucracy and business, the influence of ruling party politicians in public works, the excessive concentration of power, and other factors that impeded the functioning of a free-market economy. There is also no question that this model was flawed in that it perpetuated social injustices.

With the end of the rapid-growth period, Japan's economy headed towards maturation and, as it did so, these flawed elements of the Japanese model of administration became increasingly conspicuous. As a result, the argument that government is foolish and inefficient suddenly gained credibility.

Market forces can easily become violent

In the 1980s, based on an understanding that "the actual market is imperfect", Prime Minister Thatcher and President Reagan tried to convert the imperfect market into a perfect one. Why is government intervention in the market necessary? Why does the government have to be large? The answer was "because markets are imperfect". What can be logically drawn from that answer is the proposition that, "in order to make government smaller, you need to make the market more nearly perfect".

One by one, Thatcher privatized Britain's state-owned businesses. And, one by one, she rescinded regulations that had been imposed on the communications, electric power, finance and other industries. Income taxes were lowered, the welfare budget was cut and fat was resolutely trimmed from the overgrown government. In this way, the British market gradually became more like a free competition market. As a result, in contrast to the 1970s, when the term "the English disease" was frequently heard, the British economy was indeed revitalized. However, the side-effects were that income disparity widened and the public health and education systems were ruined. The end result was that in the United Kingdom's 1997 general election, the majority of the electorate said "No" to Thatcherism.

What was learned from the 1980s and 1990s was that the closer the market comes to perfection, the higher the risk that market forces will behave violently. What is market violence? It includes the following, in no particular order: widening income disparity, deteriorating public health and education, sharp fluctuations in asset prices, currency crises in developing countries owing to frequent shifting of short-term capital by hedge funds, free competition resulting in a single winner only, and the pollution and degradation of the global environment.

From the mid-1990s, new left-of-centre administrations took power throughout Europe. This was undoubtedly because voters were unable to turn a blind eye to the market violence then occurring and strongly felt a need to "control" the market. Of course, a single government cannot play the role of controlling the market by itself. Local governments and non-governmental organizations (NGOs) must supplement the government's efforts. One should not view society according to the old schema of a "market versus government" dichotomy, but rather as a three-legged stool that is supported by the market, government and civil society. By doing so, one can for the first time conceive of a democratic, market-economy-based society in which the three actors of business, national government and local governments/NGOs stand in a triangular relationship with each other.

Reconsidering the role of government

The simplistic notion that government is wise but the masses are foolish – often referred to as "the presuppositions of Harvey Road" (Harrod, 1951: 192–193)[1] – is the tacit assumption of Keynesian economics. The reason, it was believed, that Keynesian fiscal and monetary policies could be applied effectively in the past was that the "wise government" was able to deceive the "foolish masses". Although that may have been true in the past (say economists of the rational expectations school), the masses are no longer foolish and so the government's predictions tend to be wrong.

In fact, one thing that has been learned from past experience is that government is far from wise. But to then say that "we can't rely on the government, so leaving it up to the market is the best policy" is too simplistic a leap of logic. Before placing all one's trust in the market, one should search for the means to make government wiser.

It is an unmistakable fact that the effectiveness of recent fiscal and monetary policy has declined in comparison with the past. For precisely that reason, a structural policy (that is, a change in the economic structure) is needed to supplement fiscal and monetary policy in order to stabilize the macroeconomy and remove inequities. That is to say, rather than simply connoting fiscal and monetary policy, public policy should be seen as something broader in scope. In that sense, the era of Keynes has indeed come to an end.

Nonetheless, one must question the meaning of the following statement by Keynes in today's context:

> Let us clear from the ground the metaphysical or general principles upon which, from time to time, laissez-faire has been founded . . . We cannot . . . settle on abstract grounds, but must handle on its merits in detail what Burke termed "one of the finest problems in legislation, namely, to determine what the State ought to take upon itself to direct by the public wisdom, and what it ought to leave, with as little interference as possible, to individual exertion". (Keynes, [1926] 2004: 36)

From this point on, the government will probably have to leave "jobs that are too small for government to handle" to local governments and NGOs, and "jobs that are too big for government" to some type of international organization. From the perspective of improving administrative efficiency, it is better to disperse those "jobs that are too small" for the central government downward to lower levels and to consolidate those "jobs that are too big" at a higher level. As a result, the central government will naturally become a "small government".

Controlling the power of the market so that it does not become violent is likely to become a high-priority task for government in the future. Protection of the global environment is one element of that task.

5-5-2 The environment and the economy

A chronology of the global warming issue to date

It was in June 1988, at the World Conference on the Changing Atmosphere hosted by the Canadian government and convened in Toronto, that the shocking results of simulations were announced: if emissions of carbon dioxide (CO_2) continued to rise at the current pace, by the end of the twenty-first century the atmospheric concentration of CO_2 would be more than twice its present level, with the result that the average global temperature would rise by 3°C and sea levels would rise by 60 cm. Just prior to this conference, global environmental issues were taken up for the first time at the G7 Summit, which was also held in Toronto. Concern about the global environment continued to grow and expand with each year, as shown by the fact that in 1989, at the Paris G7 Summit of the Arch, one-third of the "Economic Declaration" was devoted to global environmental issues.

Going back a step in time, *Our Common Future*, the report of the United Nations Brundtland Commission (the informal name of the World Commission on Environment and Development, chaired by former Norwegian Prime Minister Gro Harlem Brundtland), was published in 1987. This report marked the first appearance of the term "sustainable development", which it defined as follows: "Sustainable development is development that meets the needs of the present without compromising the ability of future generations to meet their own needs" (WCED, 1987: 43).

Following these events, the United Nations Conference on Environment and Development was held in 1992 in Rio de Janeiro, where the Framework Convention on Climate Change was adopted. In 1995, the first Conference of the Parties (COP) to the convention was held in Berlin, and at the 1997 COP3, held in Kyoto, the Kyoto Protocol was adopted, committing 40 industrialized nations to cut their average annual emissions of greenhouse gases (GHGs: carbon dioxide, nitrous oxide, methane, two types of fluorocarbons, and sulphur hexafluoride) during the five-year period from 2008 to 2012 by at least 5 per cent over their 1990 emissions.

Thus, from the late 1980s to the 1990s, the protection of the global environment came to be seen as one of the highest-priority global issues of the post–Cold War era. The fact that the Kyoto Protocol committed industrialized nations to cut their GHG emissions can be counted as an example of the United Nations imposing limitations on state sovereignty and of the consolidation of state authority at a higher level (that is, that of an international organization).

A new constraint on the market: "Sustainability"

There is not necessarily a consensus on the meaning of sustainability. The Brundtland Commission nonetheless sounded a warning bell that the twentieth-century model of industrial civilization that made a goal of mass production, mass consumption and mass waste was not sustainable. In other words, the "contradictions" inherent in twentieth-century industrial civilization – the depletion of resources, the pollution of the environment, the worsening of North–South issues and so on – would become tangible within the next few decades, leaving little hope of further economic growth. It was in response to this warning that the 1988 Toronto Summit took up the topic of the global environment.

When debating the connection between the environment and the economy, the constraint of sustainability carries decisive weight. When one says that the twentieth-century model of industrial civilization is unsustainable, one has to clarify what timeframe one is speaking about. Certainly it is not a question of one or two decades. As one example, consider whether or not economic development that accompanies the mass consumption of fossil fuels, an exhaustible resource, is actually sustainable. The reader will probably think that it is "unsustainable", but the following counterargument can be made.

First, if one limits the timeframe to no more than 30 years, it is impossible within that timeframe for the oil supply to be completely exhausted. In that sense, at least for the coming 30 years, economic growth that accompanies the mass consumption of fossil fuels is sustainable. The argument is unquestionably correct on that point – but only if one accepts the premise of a no-more-than-30-year timeframe. If one stretches that timeframe to 50 or 100 years, it becomes a completely different story.

Second, it is said that the reserve-to-production ratio of oil (the proven reserves of crude oil divided by the annual amount produced) is roughly 40 years, but this number is practically meaningless. That is because it does not take into consideration at all the possibility of the exploration and development of new oil fields in the future, and it also ignores the market (price) mechanism's function of adjusting supply and demand. In addition, one must keep in mind that 40 years in the future new energy sources may be developed that one cannot even imagine today. Although being concerned about what will happen 40 years from now and diligently starting to prepare for it may be a case of "well prepared means no worries", one would be ill advised to assume that a proposition about the uncertain, distant future is the unequivocal truth and to act hastily on that assumption. The market is more intelligent than the government, so this argument goes, and it will take a sufficiently long-term perspective as needed. If the oil supply were indeed depleted, the intelligent market would raise the price appropriately. The high price of oil would then provide the incentive for the exploration and development of seabed oil fields, which is an expensive undertaking. That would also increase the amount of oil being produced from existing fields since the cost of drilling would climb along with oil prices. Moreover, the high price of oil would encourage the development of alternative energies, which would also restrain the demand for and supply of oil.

To summarize the above argument: leave it to the market mechanism, and both energy and environmental issues will be successfully resolved. Thus, say the market fundamentalists, the artificial countermeasures taken by governments and international organizations do more harm than good.

A rebuttal of the market fundamentalists' optimistic view of environmental issues

This subsection tries to rebut the above rebuttal by the market fundamentalists and techno-utopians. The market's timeframe is not as long as market fundamentalists claim. Because business managers do not think beyond their own tenure, the time horizon for businesses is short. For individuals as well, it is only rarely that people think about and act on the basis of how they can optimize the entire span of the remaining decades of their lives. In other words, all of the actors who participate in the market think about things from the perspective of perhaps a 10-year timeframe at most. To put this in other terms, when individuals and businesses seek the current value of future gains, they generally tend to set the discount rate rather high.

Looking back at the past, cases where various problems facing humankind have been overcome through the power of science and technology are too numerous to mention. The predictions for technological development related to energy and the environment, however, have for some reason not been met at all. Forecasts for the implementation of the fast-breeder reactor and nuclear fusion are classic examples. The outlook for the implementation of carbon dioxide capture and storage (the process of isolating carbon dioxide in the smoke emitted from the chimneys of thermal power stations and other sources and turning it into a solid form that is then sequestered at the bottom of the sea or in the ground) is also doubtful. The development of space-based solar power is also at a standstill. And it is not clear when home-use fuel cells will come into widespread use by overcoming their excessively high price.

Whether oil supplies are depleted sooner or later, energy- and environment-related technological development is an essential endeavour for "sustainable development". However, the chances are extremely slim of this type of technological development coming about endogenously, without government support, as a result of market competition between private companies. The probability of private companies launching large-scale technological development projects related to energy and the environment is very low indeed because they require enormous research expenditures, because they would take a long time to develop, because the success of development is uncertain, and because the future outlook for energy supply and demand is unclear.

If lifestyles, modes of transportation and other aspects of daily living are changed to more energy-conserving modes, one can expect CO_2 reductions that are no less significant than what such technological developments would bring. However, whether it is the electricity-consuming lifestyle or the automobile-dependent transportation system, there is a very strong element of habitual human behaviour involved. Generally speaking, it is hard to change habits overnight. It takes a long time.

Internalizing externalities

It has been seen that the fundamental theorem of welfare economics – that a perfectly competitive market economy is "efficient" – is dependent on many assumptions and conditions. If these assumptions and conditions are not fulfilled, the result is "market failure" in the sense that the market fails to distribute resources efficiently. The successive diminution of marginal costs, the existence of externalities and the existence of public funds, uncertainties and other factors can all induce "market failure", but what is important in regard to the environment is the existence of externalities.

The players in the market are households (that is, consumers) and firms. The precondition of Adam Smith's thesis that "each actor will pursue his or her own self-interest" is embodied in neoclassical economics as the hypothesis that "households maximize utility and firms maximize profits". It is "assumed" that the utility function is a function only of the amount one consumes, that the profit function is a function only of the volume of investment and production, and that the behaviour of other actors (consumption, production) is irrelevant. If these conditions are not met, then an externality exists. If the behaviour of some actors increases the utility or profits of other actors (that is, if it exerts a positive impact on another actor), then it is called an *external economy*; if it decreases utility or profits (that is, if it exerts a negative impact on another actor), it is called an *external diseconomy*.

The problem with external diseconomies is that one actor inconveniences another actor "for free". It cannot be asserted that inconveniencing another and not paying money (in other words, forcing the injured party simply to put up with the situation) means that the party who created the inconvenience bears no responsibility. Yet, even if the party who created the inconvenience feels compelled by a sense of moral obligation to undertake some kind of compensatory measure, nothing can be done if there is no system in society for that.

It therefore becomes necessary to *internalize* those external diseconomies, in other words to create a market for trading externalities, or to devise some method of compensating externalities. It is the government's role to create a market or devise such measures. If one leaves it all up to the free market, then, as long as each actor aims at the pursuit of his or her own self-interest, the idea that a market for externalities will form on its own is a hopeless fantasy.

If the environment is polluted, what will happen? If the air is polluted with harmful chemical substances and particulate matter, this will cause respiratory ailments. If oceans and waterways are polluted by factory wastewater, this will disrupt the environment and injure human internal organs through the fish that are consumed as well.

If the expense of restoring people's health is not borne by the companies that emit the pollutants, then it becomes necessary for those expenses to be covered by the individuals whose health is affected, by insurance or, in some cases, by the government. The cost of emitting pollutants that accompanies one unit of production and is borne by a party other than the manufacturer is called the *social marginal cost* and is differentiated from the *private marginal cost*, which is the cost borne by a private business for the production of one unit. The marginal cost required to produce one unit should be the sum of the private marginal cost and the social marginal cost but, because companies bear only the former cost, the transaction volume to balance supply and demand (where marginal utility and marginal cost become equal) is relatively excessive. As a result, the environment is polluted and the "bill" for the social cost of repairing it is passed on to the victims and the government.

One measure for reflecting social marginal costs in the market is to impose a tax that is precisely equal to that social marginal cost. By doing so, the marginal cost to the company will equal the sum of the private marginal cost and social marginal cost, and the "market failure" can be corrected.

Regulatory measures and economic measures

Measures to address global warming fall into the following three categories: voluntary curbs, regulatory measures and economic measures.

Voluntary curbs

When companies and consumers proactively make efforts to reduce GHG emissions of their own volition, these are customarily called "voluntary curbs". What is the inducement for voluntary curbs?

Amartya Sen, winner of the 1998 Nobel Prize in Economics, labels the actors who appear in neoclassical economics textbooks – the consumer who thinks only of maximizing utility and the company that thinks only of maximizing profits – as "rational fools", and states that, along with utility and profit maximization, consumer and corporate behavioural norms include commitment and sympathy as well. It is precisely because consumers and companies are not "rational fools" that they carry out "voluntary curbs".

The market economy is said to reflect the sovereignty of the consumer. If environmental awareness rises among those consumers who possess market sovereignty, then environmental awareness within a given company will be incorporated into the quality of that company's products. Suppose, for example, that a consumer looking to buy a passenger vehicle with a budget of ¥2 million compares the products of five companies – A, B, C, D and E – and unhesitatingly chooses the car made by company A even though the features, performance and other aspects of all five companies' cars are just about the same. When asked why she selected company A's car, the buyer responds, "because company A is more concerned about the environment than the others". That is a consumer in a mature consumer society.

Consumers who are committed to environmental protection count the company's concern for the environment as one aspect of a product's quality. When the company's concern is reflected in consumer behaviour, the contradiction between corporate environmental efforts and profit maximization no longer necessarily exists, because spending money on environmental efforts increases sales and, as a result, increases profits.

It is unrealistic, however, to expect all companies and consumers to place priority on commitment and sympathy over profits (utility) and voluntarily to make an effort to fight global warming. If the governments of those countries that are obligated to reduce GHG emissions under the Kyoto Protocol are to fulfil their obligation, they must take measures of some kind as well.

Regulatory measures

Regulatory measures are those that impose some type of constraint or obligation on consumers and companies. For example, requiring convenience stores to close at 11 pm or prohibiting gasoline stations from conducting business on Sundays are regulatory measures.

Economic measures

A typical example of an economic measure is a carbon tax that taxes fossil fuels (coal, oil and natural gas) based on their carbon content. Another example is the greening of the automobile taxation system, which taxes ownership of automobiles in inverse proportion to the car's fuel efficiency (kilometres per litre).

Assuming a free-market economy, then placing priority on economic measures and devising appropriate regulatory measures to make up for any shortfall is probably a reasonable way to counter global warming. It is appropriate for a society with a free-market economy not to prohibit consumers who want to ride in large cars with low fuel efficiency from doing so, but instead to have them pay a high tax. If a carbon tax were introduced, companies and consumers that voluntarily make an effort to reduce GHGs would receive the benefit of a relatively lower tax burden. In other words, the carbon tax system could be expected to have the effect of setting up economic incentives in the market that would limit the consumption of fossil fuels.

Three of the Nordic countries – Finland, Denmark and Sweden – introduced carbon taxation in the early 1990s, and Germany, the United Kingdom, France and Italy have introduced slightly anomalous forms of carbon taxation. All of these countries have adopted the revenue-neutral principle that the increased revenue from carbon taxation will be used to decrease the rate of income taxation or the burden of social insurance fees.

Opposition to carbon taxation

Those who oppose the adoption of the carbon tax system give the following three reasons for their opposition, but, as will be shown below, these arguments can all be rebutted.

First, it is argued, even if a carbon tax system were introduced, the impact on the consumption of fossil fuels (that is, reduction in CO_2 emissions) would be very slight at best. Electricity, gas and gasoline are daily necessities, so the price elasticity of demand is probably very small (in other words, there would be almost no change in demand in response to changes in price).

However, this argument holds true only in the short term; in the medium to long term, demand for the types of secondary energy mentioned above is quite elastic in terms of price. That is because, even if the price of gasoline goes up, the number of people who would immediately shorten the distance they drive their car, thereby reducing their consumption of gasoline, is probably not that large. However, several years later, when they go to buy a new car, they will probably prefer a vehicle that gets the best possible gas mileage. In addition, the automakers will concentrate their efforts on developing fuel-efficient cars. Therefore, when one takes machinery replacement into consideration as well, the price elasticity of mid- to long-term energy demand (3–10 years) is significantly greater than zero.

Second, it is said, carbon taxes slow economic growth. Those who will pay carbon taxes based on carbon content are those selling fossil fuels. If the carbon tax system is introduced, the price of all consumer goods and services will increase. As a result, real personal consumption expenditures will surely decrease. If the government locked up the carbon tax revenues in a safe and left it, then that would be the end of the discussion. However, according to the principle of revenue neutrality, if personal income tax were decreased by an amount corresponding exactly to the carbon tax revenues, then disposable income would increase and personal consumption would therefore rise as well. But, in fact, there is no way of knowing whether the offsetting increases and decreases would result in a net gain or a net loss unless it is actually tried.

In any case, it is probably safe to assume that the absolute value that would result from this offset of increases and decreases would be small. If the government were able to skilfully use carbon tax revenues towards measures to address global warming, then, although personal consumption expenditures might decrease, governmental expenditures would increase, so whether overall domestic demand decreased or increased would depend on the marginal propensity to consume, the investment multiplier, the purpose to which it is applied, and so on.

What would in fact have a negative impact on the macroeconomy would be the use of carbon tax revenues to decrease budget deficits. However, reduction of the budget deficit would lead to a drop in interest rates, which would encourage capital investment by private sector companies and housing investment by individuals. As a result, compared with doing nothing at all, it is possible that the consequence would be an increase in gross national expenditure.

Third, the argument is made that production costs would rise for industries, such as iron and steel, that consume large amounts of fossil fuels in the manufacturing process, thereby weakening the nation's international competitiveness, decreasing exports and having a negative macroeconomic impact. That is absolutely true. In order to avoid this negative impact, there are a number of measures that should be implemented. When exporting items such as iron and steel, the carbon tax can be repaid based on the carbon emissions intensity reported. When the same items are imported, a carbon tax should be charged, again based on the carbon emissions intensity reported by foreign exporters. In other words, iron used domestically would be subject to a carbon tax, but iron to be exported would not be taxed. In this way the indicated negative impact can be avoided. If questions are raised about the time and energy required to report carbon emission units, or about how to handle products such as automobiles that are a composite of iron and other raw materials, then one can do as Sweden did and exempt raw materials industries that consume high levels of energy.

To the extent that proper compensation is made, carbon taxation should be neutral in macroeconomic terms. However, if there are loser industries such as the coal industry that must bear the negative impact of global warming measures, then there are also winner industries such as the environmental industry that manufactures and sells machines and equipment to counter global warming. Even within a single industry, there will be divisions into winner companies and loser companies. Automakers that are in the lead in developing fuel-efficient vehicles are winners, and those that lag behind are the losers.

The introduction of the carbon tax brings to the market new kindling to spark competition. Fierce competition will develop in the area of technological developments such as fuel-efficient automobiles, energy-conserving household appliances, solar panels, fuel cells and micro gas turbines. This is because the success of such developments will decide who the winners and losers are. Accordingly, global environmental issues will become a driving force spurring the economic development of industrialized nations in the twenty-first century.

Emissions trading

At COP3, held in Kyoto in December 1997, then US Vice President Al Gore stated in his speech, "I am instructing our delegation ... to show increased negotiating flexibility if a comprehensive plan can be put in place, one with realistic targets and timetables, market mechanisms, and the meaningful participation of key developing countries" (Gore, 1997). Gore's message actually was: "with the condition that emissions trading and other measures are accepted, the US government is prepared to accept the obligation to make significant cuts." In fact, the Kyoto Protocol did accept the introduction of emissions trading systems.

For example, Japan was obligated to reduce its average GHG emissions by 6 per cent over 1990 levels in the five-year period from 2008 to 2012 (the first commitment period). To rephrase this, Japan was given emissions rights equal to five times 94 per cent of the 1990 emissions total (over the term of the first commitment period). Similarly, Russia, which has a 0 per cent reduction obligation, was given emissions rights equal to five times its 1990 total emissions.

In 2000, Japan's CO_2 emissions were 10.4 per cent higher than in 1990, so cutting down to an average emission level of 6 per cent less than the 1990 level during the first commitment period is considered an extraordinary task. Conversely, Russia's CO_2 emissions in 2000 were roughly 30 per cent lower than those of 1990. If there were a global market for emissions trading, Russia would participate in the market as a seller and Japan would be a purchaser of emissions rights.

First, in the case of Japan, if the market price of emissions rights were set at ¥5,000 per ton of carbon, for example, then domestic opportunities for reducing emissions could be taken advantage of where the marginal cost of reducing CO_2 emissions that would be equivalent to a reduction of 1 ton of carbon would be less than ¥5,000. On the other hand, rather than make use of domestic opportunities to cut emissions where marginal costs would exceed ¥5,000 per ton, it would be a better plan to buy emissions rights on the market. In the case of Russia, if there is an opportunity to cut emissions domestically where marginal abatement costs would be less than ¥5,000 per ton of carbon, then it can earn revenue by increasing the amount supplied to the market.

In this way, emissions trading is an economically rational system. If countries where the marginal abatement costs are relatively high look at the price set on the emissions trading market, they can estimate whether they can reduce those costs by cutting emissions domestically by a certain amount and buying a certain amount of rights on the emissions trading market. In that sense, the adoption of the emissions trading system in the Kyoto Protocol was a response to Gore's demand for the adoption of a comprehensive plan that includes market mechanisms.

Note

1. Harvey Road was the location of the Keynes family residence in Cambridge.

REFERENCES

Gore, A. (1997) "Remarks as Prepared for Delivery for Vice President Al Gore", Kyoto Climate Change Conference, 8 December. Available at <http://clinton3.nara.gov/WH/EOP/OVP/speeches/kyotofin.html> (accessed 3 September 2010).

Gray, J. (2002) *False Dawn: The Delusions of Global Capitalism*, revised edn. London: Granta Books.

Harrod, R. F. (1951) *The Life of John Maynard Keynes*. London: Macmillan.

Keynes, J. M. ([1926] 2004) *The End of Laissez-Faire: The Economic Consequences of the Peace*. Great Minds Series. Amherst, NY: Prometheus Books.

Keynes, J. M. (1936) *The General Theory of Employment, Interest and Money*. London: Macmillan.

Smith, A. ([1776] 1904) *An Inquiry into the Nature and Causes of The Wealth of Nations*, 5th edn, ed. E. Cannan. London: Methuen and Co.

WCED [World Commission on Environment and Development] (1987) *Our Common Future*. Oxford: Oxford University Press.

5-6

Social science and knowledge for sustainability

Jin Sato

5-6-1 Introduction

What does social science have to offer to sustainability? There are already working examples of the application of social scientific tools to sustainability-related problems, such as environmental cost/benefit analysis, emissions trading, conflict resolution techniques, institutional analysis of common property management and re-appreciation of indigenous knowledge. These contributions, ranging from economics to anthropology, are no doubt significant and often contain practical implications; however, they also force us to wonder whether contributions by the social sciences should be limited to providing practical tools and techniques alone.

The social sciences are often viewed as "too soft" and indirect in problem-solving, as well as weak in their predictive powers as compared with the natural sciences. This perception generates "natural science envy" among social scientists, who believe that social science, like natural science, should try to pursue universal laws with rigorous empirical methods, emulating the experimental sciences as far as possible.

However, a strong voice has emerged in the form of the "Perestroika movement" in political science, which argues that the social sciences are different in character and therefore should not attempt to model themselves on the natural sciences. Proponents claim, for example, that the social sciences have an advantage over the natural sciences in addressing questions of values and power – such as what is "desirable" – and informing debates with the aim of achieving better judgement. In this light, attempting to imitate the natural sciences simply downgrades the very nature of social scientific investigation (Flyvbjerg, 2001).

This section argues that there is a gap between the kind of knowledge required for sustainability problem-setting and the knowledge expected of science. The former demands attention to the particular – and the way particulars combine to form a context – whereas the latter tends to see particulars only as parts from which to infer the general, which is a more important qualification for scientific practice. Bridging the gap between the universal and the particular should be a primary aim of sustainability science, and this is the area in which social science can contribute most effectively. This case will be made by extrapolating from the recent debate in social science revolving around "practical wisdom" to explore the future direction of social science, particularly as it pertains to the sustainability of humankind. It seems that the full potential of the social sciences has yet to be exploited in addressing environmental governance in general, and sustainability in particular.

The section will begin with a quote from a well-known activist and researcher of pollution problems in Japan, Dr Ui Jun:

> The perception of pollution is felt with the whole body as a total experience by the victim, whereas the perception by the polluter is partial, capturing only quantitative aspects such as the concentration of pollutants or number of victims. Suppose a person claims to be a "third party" and attempts to listen to both parties in a fair way. He would end up positioning himself on the side of the polluters by standing in the middle between the "total" [but particular] and the "partial" [but general]. (Ui, 2000: 51)

Partial knowledge wins because it is closer to the "universal", whereas total knowledge loses because it is too specific, regardless of how "total" each particular element might be. Ui hints at a kind of hierarchy in our knowledge system where the universal and quantifiable are privileged over (and thus more convincing than) the local and qualitative, regardless of the intimacy of the data as perceived by the victims themselves. This problem is deeper than it appears, since much experiential knowledge goes unexpressed or undocumented even in qualitative terms.

It is naïve to assume that a mere shift in social science can make a significant difference in how humans interact with nature. However, as an examination of the history of nature governance confirms, the impact of the social sciences and the particular vision that they provide is often more influential than it might first appear. To assess the magnitude of this impact, one can ask what particular knowledge is produced and how it is applied in policies and practices. On a more subtle level, one can examine how the promotion of a particular mode of thinking may serve to downplay or nullify the very knowledge needed to address sustainability problems. This latter effect – that is, one type of knowledge dominating over another – is seldom noticed and therefore seldom examined explicitly. It is this second type of problem that is examined in this section.

The problem to be highlighted is the depreciation of experiential or tacit knowledge in relation to other types of knowledge. Experiential knowledge is important because it is everywhere in our lives – the way languages are spoken, the way doctors practise medical care, the way teachers teach in schools and the way farmers grow crops are all informed by the unwritten guiding principles of tacit knowledge.

The appreciation of tacit or experiential knowledge brings us back to the statement by Ui. How does one appreciate and study experiences that are not well accounted for in the analytical framework of scientific debate? In this section, it is argued that the major contribution of social science is not to imitate the natural sciences in aiming for the universal and predictable; rather, social science deals with what is variable and contextual, which aims towards informed judgement.

The outline of the argument in this section can be summarized in three points: (1) sustainability concerns the governance of natural resources and the environment, yet at its very foundation lies the governance of *knowledge*; (2) the development of science has devalued the kind of knowledge (that is, practical wisdom) needed to address environmental problems; and (3) revitalizing this practical wisdom requires a connection between non-contextual scientific knowledge and contextual knowledge.

5-6-2 The formation of the "governable" and the depreciation of experiential knowledge

It is apparent that natural and engineering scientists play a more prominent role in current debates on the environment and sustainability than do social scientists. This trend is not new. The history of "development" contains numerous examples of the preference for scientific knowledge over contextual knowledge (commonly called indigenous or local knowledge). Forest conservation is one example. Because there are many types of trees and soil conditions, it is almost impossible to develop a universal formula that can be applied to all forests at all times. The best knowledge seems to come from the "person on the spot" who has daily interactions with the forest and the ecosystem surrounding it. Furthermore, local people depend on various resources simultaneously. In a village located in Chiang Rai Province close to the Mekong River in Thailand, for example, the author encountered villagers who had access to three primary resources (fish in the river, minor forest products in the forests and rice fields) depending on the season, resource conditions, labour availability and other factors. Local people knew that these resources depend on each other as much as people depend on them – in other words, they accurately perceived this web of resources as an interconnected system.

The history of resource policy, however, demonstrates exactly the opposite trend, which James Scott calls "state simplification" (Scott, 1998). According to Scott, state simplification is the actualization of high modernist ideals commonly observed in the process of state-making:

> Certain forms of knowledge and control require a narrowing of vision. The great advantage of such tunnel vision is that it brings into very sharp focus certain limited aspects of an otherwise far more complex and unwieldy reality. This very simplification, in turn, makes the phenomenon at the center of the field of vision far more legible, and hence, far more susceptible to careful measurement, calculation, and manipulation. (Scott, 1998: 11)

What is called economic development is not facilitated by increased exchange and division of labour at the dispersed level of the economy, but is derived from a top-down process of state simplification – of units, names, locations and occupations – to make them more "legible". It is not difficult to imagine how this process privileges scientific knowledge while downplaying local and experiential knowledge. Statistics also play a significant role in this process, as Agrawal observed when forests became an object of government in India:

> Forests are an example of territorialized entities summarily and reductively represented by specific figures – area of land, number of species, volume of product, etc. Indian forests would not only have been unknowable without statistical representation, but such representations helped constitute the very category of Indian forests. (Agrawal, 2005: 33)

Statistics and numerical representation are not merely simple tools for state simplification that allow central planning, but represent an important means of redefining aspects of nature to be governed and how they should be governed. Development of a general system is also a process of reducing the influence of individuals and the particular knowledge those individuals possess (Agrawal, 2005). It is a process that downplays locally specific and often unrecorded knowledge, however relevant it might be to carrying out daily activities in a unique setting.

5-6-3 The "tragedy of the commons" and the realm of judgement

It is not only anthropologists and rural sociologists who have been made aware of the importance of experiential knowledge, or, to put it more generally, the "realm of judgement". The importance of judgement that is beyond the reach of science and technology was effectively put forward in a famous article by Garrett Hardin (1968) entitled "The Tragedy of the Commons", although this aspect of the article is rarely appreciated. The way most commentators engage themselves with the "tragedy of the commons" is by referring to the metaphor used by Hardin. Herders share a common parcel of land (the commons) on which they are all entitled to let their cows graze. Herders' individual utility calculations will lead them to put as many cattle as possible on the grazing pasture, since this will serve their interests, even though this will degrade the productivity of the pasture. The less appreciated aspect of the article, however, is probably closer to what Hardin had in mind: acknowledging the existence of problems with "no technical solution". Hardin argued, for example, that "the population problem cannot be solved in a technical way, any more than can the problem of winning the game of tick-tack-toe" (Hardin, 1968: 1243).

Hardin introduces the idea that it is not mathematically possible to maximize two (or more) variables at the same time: "Maximizing population does not maximize goods. Bentham's goal is impossible" (Hardin, 1968: 1244). Science does not tell us who and how much should be sacrificed for the sake of the general good; such decisions require deliberation and ethical judgement.

5-6-4 Things "variable" and practical wisdom

Whether the question is about managing natural resources such as water, forests and minerals or about consumption behaviour such as buying environmentally friendly goods, science plays only a partial role in defining humans' behaviour. It is the realm of judgement in particular contexts that determines action and its effects. Hardin himself acknowledged this point, saying that "the morality of an act is a function of the state of the system at the time it is performed" (Hardin, 1968: 1245).

The important types of knowledge required to address environmental issues are not simply "scientific" or "technical". According to Aristotle, there is a third category of knowledge called *phronesis*, or "prudence", which may also be translated as "practical wisdom":

> We may grasp the nature of prudence [*phronesis*] if we consider what sort of people we call prudent. Well, it is thought to be the mark of a prudent man to be able to deliberate rightly about what is good and advantageous ... But nobody deliberates about things that are invariable ... So ... prudence cannot be a science or art; not science because what can be done is a variable and not art because action and production are generically different. For production aims at

> an end other than itself; but this is impossible in the case of action, because the end is merely doing well. (Aristotle, 1976: 1140a24–1140b12)

The above distinction clearly marks the differences in the nature of the knowledge that is pursued in intellectual activities. Aristotle claims that the conflation of inherently distinct types of knowledge – that is, practical wisdom under the pretence of science or vice versa – is a basic mistake that should be avoided. If social science is primarily responsible for the domain of *phronesis*, it should discard its natural science envy and forget about universal laws, focusing instead on the values and power that lie at the heart of human experience (Flyvbjerg, 2001).

In considering whether to accept Flyvbjerg's suggestion wholeheartedly, it is useful to be aware of the three important features of practical knowledge (an important element of experiential knowledge) that distinguish it from the dominant emphasis of scientific discourse. First, practical knowledge is integrative rather than analytical. Second, it is knowledge that one must learn by *doing* instead of by reading textbooks or memorizing rules. Third, it pays attention to the particular instead of heading straight to the universal. In tackling the question of whether "political science can be taught", the eminent political philosopher Isaiah Berlin said:

> Good politicians grasp the unique combination of characteristics that constitute a particular situation – this and no other. What they are said to be able to do is to understand the character of a particular movement, of a particular individual, of a unique state of affairs, of a unique atmosphere, of some political combination of economic, political, personal factors; and we do not readily suppose that this capacity can literally be taught. (Berlin, 1996: 45)

Why did practical wisdom, despite its long-acknowledged importance, lose its place in social scientific tradition? Is it because the efforts to capture tacit or experiential knowledge were not sufficient? This may be one reason; however, the author believes the main cause is to be found elsewhere – in the dogma of efficiency.

5-6-5 Efficiency and technical necessity as barriers

Giving value to experiential knowledge is a matter not just of embracing it as a distinct field of knowledge, but of using it to challenge and provide alternatives to more conventional explanations of social phenomena that have seldom been questioned. Therefore, in order to revitalize the social sciences, one must appreciate not only local/tacit/experiential knowledge but also the forces that tend to downplay this knowledge. This subsection highlights two dominant frames of reference that contribute to the devaluation of tacit/experiential knowledge: efficiency and technical necessity.

It begins by examining the logic of efficiency and how it might obfuscate human experience or render it obsolete. The so-called "Summers Scandal" is a useful case through which to consider the force and limits of arguments based on efficiency. Lawrence Summers, then Chief Economist at the World Bank, signed an internal memo that was leaked to *The Economist* magazine:

> Just between you and me shouldn't the World Bank be encouraging more migration of the dirty industries to the LDCs [less developed countries]? A given amount of health-impairing pollution should be done in the country with the lowest cost, which will be the country with the lowest wages. I think the economic logic behind dumping a load of toxic waste in the lowest-wage country is impeccable and we should face up to that. (*The Economist*, 1992: 66)

The point here is not about a simplistic argument whose assumptions invited criticism from various angles, but about the effects inherent in a logically consistent argument given the assumptions (for example, the plausibility of comparing the worth of lives based on how much people earn). Arguments based on efficiency have the effect of blinding one to the other side of the story. A further strength of the efficiency argument is that it can cover up questions of equity and distribution as long as the resulting enlarged "pie" has the theoretical possibility to compensate the losers. However, it is in the struggle over distribution that human experiences express themselves most fully, and a richer explanation of how one thing leads to another in this context can be provided by social science.

Various technologies that help improve efficiency generate a similar effect of halting debates that might challenge existing power and values, as observed by Winner:

> It is characteristic of societies based on large, complex technological systems, however, that moral reasons other than those of practical necessity appear increasingly obsolete, "idealistic", and irrelevant. Whatever claims one may wish to make on behalf of liberty, justice, or equality can be immediately neutralized when confronted with arguments to the effect, "Fine, but that's no way to run a railroad" (or steel mill, or airline, or communication system, and so on). (Winner, 1986: 36)

The doctrines of efficiency and technological necessity are so dominant that they in effect shut the door on the exploration of alternative possibilities. Of course, in modern society, it is impossible to deny these norms completely. It is also true that they have successfully sustained and directed modernization and economic development. However, sustainability in its radical form must question the trends that placed the sustainability issue on the table. If anyone is to challenge the basic assumptions of development, it must be academics in the social sciences and the humanities.

5-6-6 Social science for connecting science and practical wisdom

An important function of social science is to recover the realm of judgement and practical wisdom as a distinct field of academic contribution. This is not only because this area of knowledge has been neglected and downplayed in the social sciences, but because the problem of sustainability demands a new way of thinking. This new way must place emphasis on integration rather than analysis, experience rather than logic, and attention to the particular rather than blind pursuit of the universal.

Given the overwhelming dominance of scientific discourse based on positivist orientations, however, this is easier said than done. One cannot expect a "mass production" of practical wisdom – it belongs to the category of "tacit knowledge" precisely because of the difficulty of reproducing such knowledge verbally and explicitly. No matter how unorganized practical wisdom may be, it exists in vast quantities, waiting to be tapped. However, it is important to realize that the ultimate concern must be not simply increasing the quantity of knowledge, but making things work towards sustainability; how things work, in the domain of action, must rely on practical wisdom as a guiding force. Therefore, the upgrading of practical wisdom requires the orientation of science towards actual problems, and it is around these problems that a balanced distribution of science, technology and practical wisdom must be mobilized.

When one speaks of orientation towards actual problems, *whose* problems, exactly, is one talking about? Problems related to sustainability are those not of experts alone but of the general public. Their solutions must also come from the public, since, regardless of what science encourages one to do, action must ultimately be taken by each individual "on the spot". Hayek (1945) was one of the first to make this point pertaining to knowledge:

> Today it is almost heresy to suggest that scientific knowledge is not the sum of all knowledge. But a little reflection will show that there is beyond question a body of very important but unorganized knowledge which cannot possibly be called scientific in the sense of knowledge of general rules: knowledge of par-

> ticular circumstances of time and place. It is with respect to this that practically every individual has some advantage over all others in that he possesses unique information of which beneficial use might be made, but of which use can be made only if the decisions depending on it are left to him or are made with his active cooperation. (Hayek, 1945: 521)

The distributional characteristic of experiential knowledge explains why democracy is important, not only intrinsically but also instrumentally, in promoting sustainability. To make use of such tacit knowledge, as Hayek rightly points out, one must bestow on people the right to take actions according to their own will. At the very least, active cooperation is needed from individuals; otherwise, their tacit knowledge will go to waste.

Social science should be aware of the kind of knowledge it produces. Sustainability science requires action (including non-action), and action is contextual. However, science is generally non-contextual, or at least seeks to avoid being contextual. To bridge these two approaches is the true purpose of sustainability science.

REFERENCES

Agrawal, A. (2005) *Environmentality: Technologies of Government and the Making of Subjects*. Durham, NC: Duke University Press.

Aristotle (1976) *The Nicomachean Ethics*, translated by J. A. K. Thomson. Harmondsworth, UK: Penguin.

Berlin, I. (1996) "Political Judgment", in I. Berlin, *The Sense of Reality*. New York: FSG.

Flyvbjerg, B. (2001) *Making Social Science Matter: Why Social Inquiry Fails and How It Can Succeed Again*. Cambridge: Cambridge University Press.

Hardin, G. (1968) "The Tragedy of the Commons", *Science* 162: 1243–1248.

Hayek, F. A. (1945) "The Use of Knowledge in Society", *American Economic Review* 35: 519–530.

Scott, J. C. (1998) *Seeing Like a State: How Certain Schemes to Improve the Human Condition Have Failed*. New Haven, CT: Yale University Press.

The Economist (1992) "Let Them Eat Pollution", 8 February.

Ui Jun (2000) "Utility of Knowledge in Pollution", in *Discourse* [Gensetsu]. Tokyo: University of Tokyo Press (in Japanese).

Winner, L. (1986) *The Whale and the Reactor*. Chicago: University of Chicago Press.

5-7

The human dimension in sustainability science

Makio Takemura

5-7-1 The role of the humanities in the problems of sustainability

It was only about 20 years ago that people first developed an awareness of sustainability as a major issue. But times change rapidly, and people are now widely aware of this as the most pressing problem facing global society. In terms of sustainability-related phenomena, experts point to various global environmental problems such as global warming and contamination of water and soil, but many political and economic problems, such as regional conflicts and the North/South divide, are also involved in sustainability. To address such issues, the cutting-edge achievements of both the natural and the social sciences should naturally be mobilized. But the humanities also have a role to play in examining the thought, cultures and ways of seeing and thinking that underlie these problems, as well as in identifying problems still on the horizon, conceptualizing a desirable state of human existence and global society, and offering a direction towards solutions.

To start with, how should one define the characteristics of the humanities? It goes without saying that the humanities are a branch of learning concerned particularly with human existence. Moreover, whereas the social sciences examine the entirety of human relationships and the organizations, institutions and other phenomena connected to them, the humanities examine the logical structures and the shapes of the sensibilities of the thought or culture that underpin the objects of social scientific enquiries. They also examine people's sense of values. Unlike actual organizations and institutions, the thought and culture under study do not have a visible existence, but rather are expressed through language and are perceived as semantic structures revealed in etiquette, customs and the like. Most of the research on thought and culture is highly suited to literary research, of which philology is a basic component. Although there is some field research, it consists mainly of interviews and the verbalization of non-linguistic expressions. Thus a major characteristic of the humanities is that they involve research centring on things such as meaning and value in a verbalized domain. They also involve research on the human mind that brings forth these things, with the theme of such investigations being the essence of creativity found there, and the very concepts of truth, goodness, beauty and the sacred, among others. All in all, the objects of humanities research are those things at the fountainhead of spiritual activity, and often this research consists of a search for what constitutes the depths or core of human existence.

These objects of research are by nature timeless, perennial questions, which is why, throughout history, people have studied the philosophy of Plato, Augustine, the Buddha, and Laozi and Zhuangzi (Taoism). Sometimes progress itself is the target of critical examination. This research on spiritual values, which continues to fascinate with each passing era, is perhaps an area exclusive to the humanities and not found in other academic disciplines.

The most important role of the humanities is, then, to fundamentally reassess the political and economic principles and existing values and other ideas that permeate and govern present-day society, and to show the direction towards reform. The humanities must, on their own initiative, examine, study and fundamentally criticize the current era. As an example of such criticism, Hisatake Kato (1991), an authority on environmental ethics in Japan, makes the following observations:

> The market economy is inadequate because resource depletion and waste accumulation are external to economic relationships.
>
> Democracy is inadequate because it has no binding authority to protect the interests of future generations or people in other countries.
>
> Fundamental human rights are inadequate because they are too narrow, failing to address the human responsibility to protect non-human life.

These fundamental observations are very interesting because all the institutions regarded as ideal by modern society are criticized as inadequate or flawed in principle. These observations articulate very clearly that there is now an acute need for a totally new view of human beings and of society for the future. Since the beginning of the modern age and up to the present, respect for the independent individual and an atomistic view of the individual have pervaded social principles. Of course, the dignity of the individual, including basic human rights, must always be protected, but the fact that excessive individualism and the principle of unrelenting competition eventually bring about the unravelling of society is clear from the US economic meltdown in the autumn of 2008.

It seems that behind this individualism there is a scientific methodology that tries to break existence down into its most basic elements. That is the standpoint of modern rationalism, which is based on the dualism of subject and object and divides, manipulates and dominates the subject. It is often said that this standpoint made mass production and mass consumption possible and made humans materially affluent, but at the same time, with its attendant mass waste, it has polluted and damaged the global environment. The driving forces that have propelled this environmental damage are the attitudes of humans and of society that have encouraged, as a good thing, the competitive pursuit of economic profit by a small class of people (capitalists and industrialized countries). The problem of the environment and that of disparities – in other words, the crisis of nature and the social chaos of contemporary society – are both rooted in modern rationalism.

Therefore, if one aspires to solve the problems of the global environment and of contemporary society, it is essential to change the paradigm of modern rationalism. A pioneering effort in this regard was the deep ecology movement advocated by the Norwegian philosopher Arne Næss. Deep ecology called for the replacement of objective logic and reductionism with a relational world view, of anthropocentrism with biocentrism, and of an atomistic view of humans with an expansionist and holistic view; it advocated the radical transformation of modern rationalism as a paradigm. Social ecologists levelled blistering criticism at this movement, claiming that deep ecology offered no concrete policies for running society. However, the argument of social ecology – that, until the domination of humans by other humans is eliminated, there will be no solution to the human domination of nature – is not necessarily in the mainstream today. The reasons are that, first, the people involved in various ecology movements have come to respond with unproductive arguments regarding the course of politics, and, second, a measure of progress has been achieved in solving environmental problems through technology, such as the development of energy-saving and pollution-free technologies, so there is now a tendency to forget to examine more fundamental underlying problems. Undoubtedly, many people now are optimistic and do not think much about such things because, even though they may recognize the seriousness of environmental problems, they assume that, in time, science and technology will solve them.

Meanwhile, however, various distortions are being exposed in contemporary society that call once more for a fundamental reassessment of the principles that govern this society. Does the paradigm of our era need to be changed and, if so, how should it be changed? The humanities must keep asking this question, and keep looking for an answer. Even if deep ecology has fallen out of fashion, the fundamental questions it poses are eminently worth full consideration. The importance of the humanities' unique task – to closely examine and study the paradigm that guides the era, and to make recommendations for reform – should be recalled, especially now, when technology is at its height.

5-7-2 A perspective on sustainability science based on the humanities

In 2006, the Integrated Research System for Sustainability Science (IR3S) was officially launched to conduct integrated research with the participation of five universities and four (later six) cooperating institutions. With "sustainability science" as the main subject of research, IR3S aims to reveal the interactions among global, social and human systems, to determine the mechanisms that are bringing about the failure of these three systems and the interactions among them, to restructure each system from the perspective of sustainability, and to present measures and a vision for restoring their interactions. With the idea that all related academic disciplines should be mobilized in this undertaking, IR3S is working towards a "transdisciplinary academic framework" that transcends even interdisciplinary research.[1]

To make such systems the subject of academic enquiry, perhaps a fundamental reappraisal of the conventional reductivist method is needed. Here it is desirable first to clearly position each academic discipline within scholarship as a whole, to systematize it, then to reflect on the methodology and significance of each individual discipline and, having done that, to carry out research with an awareness of its connections with other disciplines.

With that in mind, and after thinking through the methodology of the IR3S approach to sustainability science, it is possible to view this methodology in terms of the solutions associated with different systems as follows:

1. solutions through scientific and technological advances (e.g. the development of energy-saving and pollution-free technologies; global system research);

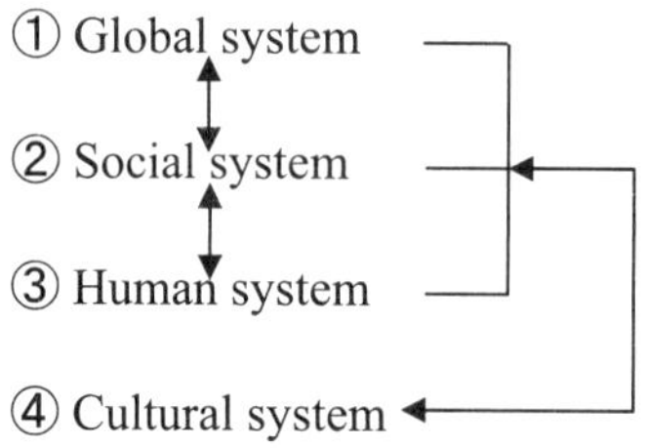

Figure 5.7.1 The methodology of the IR3S approach to sustainability science.

2. solutions through social system transformation (the transition to a cyclical society; social system research)
3. solutions through lifestyle changes (reconsidering human livelihood; human system research)
4. solutions through establishing a new view of human beings and a worldview (awareness of the meaning of life; cultural system research).

Of course, it is the interactions among these systems that must be investigated and studied (see Figure 5.7.1), but awareness of the structure underlying these inter-system interactions must also be deepened.

The humanities are closely connected to (3) and (4) above. Domain (4) in particular is concerned with the aforementioned concepts of truth, goodness, beauty and the sacred, as well as with values, and is therefore a domain that only the humanities can handle. When thinking through the methodology of sustainability science in this manner, it is obvious that (4) determines how (3), (2) and (1) will turn out. Or one could say that applying (4) to (3), (2) and (1) will make their problems evident. Accordingly, perhaps one's view should be that the humanities, the social sciences and the natural sciences are not originally all on the same plane, but rather that the humanities support and orient the social sciences and the natural sciences, thereby forming a multi-tiered structure. Concrete measures for new methods of running society and using science and technology should be conceived on the basis of the fundamental thinking about the future of global society suggested by the humanities.

In actuality, however, there is no such structure; the natural sciences develop their findings independently, and the social sciences have lost their philosophical way, being carried away with a utilitarian attitude and calculations of things such as efficiency. In that sense, the humanities are in practice powerless in the face of the out-of-control rush of modern rationalism. But, precisely because of that, the humanities must restore their awareness of the deep horizon of human existence and make an appeal to other disciplines. That, it seems, could make possible the restructuring of various systems and the restoration of system interrelationships to which sustainability science aspires.

The positioning of the humanities within the whole of learning or within the transdisciplinary framework of sustainability science can, for the time being, be viewed as defined above, but there is one more possible perspective. Philosophy and literature truly arise from the depths of an individual's spirit, and they reveal such things as the meaning of life for the individual and the substance of affluence. But, to go further, the current issues are not a matter of an abstract "mind" or "individual"; rather, they are actually connected to the problem of the "self". Because the humanities are concerned with how one gains awareness of this irreplaceable "self" and its meaning, they evince attributes absent from other academic disciplines.

Especially amidst the global crisis of today's global society, the view of the "self" is being re-examined. Soul-searching over lifestyles is now a major challenge and, in societies on the verge of crisis or in societies where desperate poverty still afflicts much of the population, assumptions about how people should live their lives are fundamentally called into question. Underlying that is the urgent question of just what kind of existence the "self" is. Is the self found in things or in the mind? Is it an atomistic existence or a relational existence? Is it separate from the environment or an integral part of it? In these and other ways, humans' very preconceptions, which are based on the conventional modern view of the self, are being called seriously into question. The central theme of deep ecology is, without a doubt, self-realization by means of enlarging the self. With the realization that the self is not just a mind–body individual, but that all things connected or related are the self, people feel that the sufferings of others, who are also part of the self, are their own suffering, and on that basis they make judgements and act. As each individual appreciates the reality of such a self, it is possible to work towards reforming society. Such was the attempt to transcend the modern, Western view of the self by thought systems such as the philosophy of Spinoza or of Gandhi or of Zen.

In today's society, heavily permeated as it is with information technology, people are constantly bombarded with information provided by the Internet. People are often carried away in a torrent of chaotic information, losing their self and just having fun surfing on this ocean of information. When people lack firm beliefs or values, and totally give themselves over to external sensory and perceptual stimuli, they are not aware of the nihilistic state they are in – one in which their self is manipulated by information, and which can only be described as having fallen into doubly deep nihilism. Each person's inherent life, noble intent and autonomy must be awakened from this condition, and to do that it is essential to give deep thought to the self.

In fact, enquiring into the self entails a fundamental re-examination of the totality of the science that has guided modernization. This is because the methodology of science bases itself on the dualism of subject and object: it asks no questions about the subject, and directs its attention exclusively towards the object as its focus. There, too, science involves itself with the focus by dividing it and then putting it back together again. However, the self is never in the objective world; it is rather the subject itself. Therefore, investigating the focus and the subject at the same time fundamentally changes the paradigm of conventional objective and logical scholarship. Even if one investigates the focus within a system without dividing it, the all-important self is lost as long as one always assumes the viewpoint of placing the self outside the system, and perceives and studies the world system as one's focus. If the expansion of the self in deep ecology also goes no further than an objective understanding achieved by objectively connecting the self and the world and then expanding the self, then it is certainly not perfect. If one is involved only objectively, then perspectives on and awareness of the self itself, the subject itself and life itself are lost, and people remain unaware of the things that should be the most important. When that is the case, it would seem that ultimately it is not possible to solve the fundamental flaws in the worldview that governs the modern era. By taking a fresh look at the self, however, one can fundamentally and critically examine modern rationalism, which has got along without asking about the self. In doing so, one may also gain the prospect of recombining the knowledge of all learning. This is an undertaking that, for the time being, can be carried out only by the humanities.

If the hope is to envision a new social order at this point, one must dispassionately reconsider what kind of existence human beings were originally supposed to have, and this is something that should come into view in the process of asking about the self. Now is the time to ask in-depth questions about the meaning of this singular self that lives in a global society.

Thus the humanities face the challenge of fundamentally rediscovering and restructuring the true aspect and meaning of this singular existence called the self. It is no longer self-evident that this self exists independently, as was once assumed, and indeed it has already been amply asserted that the self exists within relationships with other existences. Accordingly, one can anticipate that investigating the self will take place as part of an investigation into the whole of the relationships between the self and other existences, which in fact should also be an investigation of the world. If that line of enquiry is extended, one will no doubt perceive the relational structure between the humanities and other sciences. In other words, even if the humanities focus on the self, they will never be entirely closed. Because the humanities thoroughly investigate the entirety of what is called the self, they must break out into other fields. In this way the humanities will naturally find pathways connecting them to the social sciences and the natural sciences.

5-7-3 The possibilities of Buddhist thought in sustainability science

Regardless of time and place, the self has always been investigated in a variety of ways in the humanities, and especially in philosophy and literature. However, enquiring into the self is enquiring into a subject that cannot be objectified; to say the least, pursuit of merely objective recognition of the self is not sufficient. In that sense, knowledge about the self is complete precisely because it is different from ordinary knowledge. Hence philosophy, which is based on self-awareness and intuition, and especially the wisdom and faith of religion produce a deeper awareness of the self. Among religions, Buddhism in particular has always been the way of "self-clarification" rather than the way of belief in transcendent entities. This seems to be a good time to take a fresh look at the essential self from the perspective of Buddhism and other religions and bodies of traditional knowledge.

Among religions, Buddhism is an especially remarkable thought system. Where most Indian philosophy emphasized the *atman* (self), Buddhism expounded the non-self. It enquired into the self, and arrived at the awareness that the ego does not exist. What is this non-self thought, then? Is it something whose truth can be advocated, even today? This subsection will describe the view of the self in Buddhism, a traditional Oriental thought system, and offer a suggestion for the future direction of the humanities.

The non-self thought of Buddhism was first explained in terms of the "five aggregates and non-self".[2] The five aggregates are the five physical or mental elements known as matter, perception, mental conceptions, volition and consciousness. The mind itself is nothing but a complex of these different elements. "Five aggregates and non-self" means that, even if these elements form a temporary harmonious unity, there is no constant and unchanging autonomous existence (that is, an eternal, invariable, self-mastering self). This teaching is said to liberate people from attachment to an objective, eternal self and consequent suffering.

This self-liberation from attachment to an objective, eternal self is thought to open the avenue to awareness of the self as subject. But

Hinayana Buddhism attempted to find the solution to deliverance from transmigration by means of breaking attachment to the self and thereby entering the stillness of nirvana.

Subsequently it was discerned that even the elements of the five aggregates do not have eternal substance, but rather are mere phenomena that arise from causation. The five aggregates were analysed in more detail and given many elements – for example, the five groups of the 75 *dharma*s (according to Vasubandhu's *Abhidharma-kośa-bhāsya* ["Abhidharma Storehouse Treatise"]) and the five groups of the 100 *dharma*s (according to the doctrine of consciousness-only); *dharma* here means a constituent element of the world.[3] However, those elements too are perceived as having no nature of their own. Here is where the *Prajñāpāramitā Sutra* ("The Perfection of Wisdom Sutra") and other Mahayana Buddhist texts set forth the idea of the emptiness of all phenomena. This teaching attempts liberation not only from attachment to the self but also from attachment to things. Liberation from attachment to the self and things gives rise to profound wisdom accompanied by awareness that transcends objective knowledge. One is then not enthralled or controlled by the self or by things, thereby enabling free activity and the attainment of self-realization in the true sense of the word. What is achieved here is activity based on perfect awareness of the subject, always and naturally functioning as a subject that benefits itself and others. This is one of the basic stances running through Mahayana Buddhism.

Later, the doctrine of consciousness-only was established,[4] and it explained that each individual is made up of eight consciousnesses. In addition to sight, hearing, smell, taste, touch and thought, there are the *manas* consciousness and the *alaya* consciousness. The self is revealed as an existence not only consciously but also subconsciously. The *manas* consciousness clings to the idea of self-existence but, when one is well-trained and this consciousness changes to wisdom, it becomes what is called cognition of intrinsic equality, which discerns the true nature that self and others are equal. The *alaya* of *alaya* consciousness means storehouse, and is the consciousness that stores all past experiences. Generally these consciousnesses themselves have the subjects of sensation, perception and other senses, and are not just transparent subjects. Sight consciousness generates matter within itself and sees it, whereas thought consciousness produces all perceivable objects within itself and knows them. The objective aspect within a consciousness is called 相分 (*xiang fen*, "images of objects in the consciousness"), and the subjective aspect is 見分 (*jian fen*, "function of the consciousness perceiving *xiang fen*"). Each of the eight consciousnesses has a *xiang fen* and a *jian fen*.

What kind of *xiang fen* does the subconscious *alaya* consciousness have? Apparently that will always be in the world of the unknowable, but it has been determined that the *xiang fen* has a "body with sense organs", a "physical world" and "seeds" (see "Discourse on the Stages of Concentration Practice" in *Taishō Shinshū Daizōkyō*, 1927a: 580a). "Body with sense organs" refers to the body with its five sense organs, in other words, the physical body. "Physical world" refers to the material world. "Seeds" are what one might call information, which is stored in the *alaya* consciousness. Seeds are the causes that bring about actual sensation and perception as in seeing, hearing, thinking and otherwise perceiving. Therefore, at the root of the self, each person has an *alaya* consciousness, which maintains that person's bodies, environments and causes of sensation/perception.

This way of thinking is not easy to understand right away. But when one thinks about the life of the self, one can expect it to be found in the midst of the interchanges and interrelations between the body and the world via eating, excretion, breathing and other bodily functions. Such being the case, the self should be regarded as the whole of the body and its environment. In the final analysis, the self is the whole of the interrelations and interchanges between the subject and the environment, with the focus on the body. The self is certainly not the existence of the body and mind alone.

Generally the source of this consciousness-only thought is said to be the statement "the three worlds are only in the mind" in the *Avatamsaka Sutra* ("The Flower Adornment Sutra") (*Taishō Shinshū Daizōkyō*, 1925: 558c). It was China's Huayan school that organized and systematized the thought of the *Avatamsaka Sutra* and established Huayan thought. This sutra sets forth the Huayan idea of causation that one is all, all is one, and a single thing contains everything, while everything contains a single thing. This observation means that all things exist within limitless relationships, the causation of things constantly influencing one another.[5] This awareness reveals the worldview (known as the *dharma*-world) of the unhindered blending of phenomena, in which all things are infinitely related to one another and in that sense exist as themselves while permeating and fusing with one another.

"Things" here are not objective existences, but some kind of subject–object correlation, in other words, none other than the self on each occasion. Thus, the unhindered blending of phenomena means that all people, the self and others, exist within a limitless relationality. In other words, the self has always included all related others; others are actually the self.

The inclusion of all others in the self is clearly set forth in the esoteric Buddhism of Kūkai (a Japanese monk also known as Kobo Daishi). Kūkai's "Principle of Attaining Buddhahood with the Present Body" (*Sokushin-jobutsu-gi*) states: "The hanging jewels reflecting one another are called attaining Buddhahood." This is explained as follows. "This metaphor shows that the great activities in body, speech and mind of all Buddhas and honored ones are mutually interpenetrated. Myself, the Buddhas and all living beings are mutually interpenetrated, so those selves are this self and this self is those selves. Buddhas are human beings and human beings are Buddhas. They are different and the same" (Kūkai, 1978: 507). A poem that extols the tenth stage of the development of mind ("Mysteriously Arranged Mind"), in Kūkai's "Precious Key to the Secret Treasury" (*Hizo hoyaku*), says: "The vast Buddhas are in my mind and the vast honored ones are in my body" (Kūkai, 1978: 465–466). The commentary on the introduction to the tenth stage, in Kūkai's "Ten Abiding Stages on the Secret Mandalas" (*Himitsu mandara jujushinron*), says: "When one attains the tenth stage he will realize the deepest ground of his mind and understand the quantity of himself as real, because he will find in his mind the Garbhadhatu mandala and the Vajradhatu mandala" (Kūkai, 1978: 397). This means that the Buddhas and honoured ones are in our minds; in other words, all others are the self. In the two mandalas, the Mahavairocana Buddha is placed at the centre, with many Buddhas and honoured ones arranged geometrically around Mahavairocana. All of this is in fact within one's own mind.

In a mandala drawing one sees only what the faces and bodies of the Buddhas and honoured ones look like, but the self is the whole of the mind, body and environment. That the self and all others have a limitless relationality means that a single "whole of the mind, body and environment" (of a self) and all other "wholes of the mind, body and environment" mutually permeate and fuse with one another, while at the same time each is an irreplaceable subject. This internal reality is pictured in a mandala as a symphony of Buddhas and honoured ones and others in conjunction with such an environment. What is more, that whole is the self. The substance of the self reaches its pinnacle here.

This has been a simplified discussion of what Buddhism says about the self. What can be said about the relationship with sustainability science from this view of the self?

If one objectively relates to the ego and to things and changes the way one becomes attached to things – that is, if one fundamentally transforms one's objective logic (in which the denial of self-knowledge is used as an intermediary) – one will discern the relational state of the world and its completely equal true nature (*sunyata*) and become aware of the reality of the self.

As such, the self has always been the whole of others. From this follows the ability to think from the viewpoint that others are the self. When people become aware of the true self, including subjects themselves, from a relations theory perspective they naturally function as a subject that benefits itself and others. That leads to a reassessment of self-centredness and grants access to a viewpoint by which people think seriously not only about contemporary others but also about others of future generations.

Each individual is also a whole, comprising mind, body and environment. Accordingly, the whole of others is the whole of all minds and bodies and the environments in which they are placed. This leads one to think from a viewpoint in which each individual's environment is also the self. That can be a driving force in the attempt to coexist with the environment, protect the rights of nature to the maximum, and work towards setting the natural environment itself to rights.

The viewpoint described above holds that life is relational and whole. From this viewpoint, people will perhaps consider the relationships of wholeness among all organisms (that is, the ecosystem) from a perspective that does not divide things. That will lead to a reconsideration of reductionism based on objective logic.

The true nature of the whole of all minds, bodies and environments is the wisdom of the Buddha. This brings people to think from a viewpoint that reveres, respects and cherishes all selves, others and their environments. The aim of people will be to endeavour to manifest the substance of the Buddha's wisdom (virtue). And that can lead to a shift from a viewpoint valuing quantity to one valuing quality, and a reconsideration of the vain desire to dominate and encroach upon the environment and others.

5-7-4 Modern society and the significance of the humanities

As has been seen, enquiring into the self is actually enquiring into the world itself. One of the roles of the humanities is to examine such fundamental questions in order to illuminate the state and problematic nature of contemporary mainstream knowledge itself, point out the inadequacy of the objective logic that dominates today's scholarship, and urge a shift in the paradigm of knowledge concerning the self and the world. By so doing, the humanities can illuminate a desirable way of living for the individual and a desirable image for society, and on that basis point out the problems of real-world society and show the way towards improvement. It goes without saying that one should look at cultures and thought systems in all times and places for ideas on desirable ways of living and desirable societal images. The humanities must once again unearth the forgotten truths to be discovered there and use them to advantage in the modern era. This task is a vital role for the humanities.

Here a view of human beings and the world based mainly on Buddhism has been explored from the author's stance as a specialist in Buddhist studies, but also more generally with regard to traditional culture and thought. The aim has been to convey a view that goes beyond the modern, atomistic viewpoint of the individual, opposes competition, oppression and discrimination, does not necessarily take the position of human superiority, and sometimes even takes the perspective of the ordinary, secular world and relativizes values from a transcendent viewpoint. Especially in the world of religion, the individual is often not considered complete as an individual but becomes complete by means of some existence that transcends the individual, and that which transcends the individual is held in reverence. Moreover, others likewise become complete within that same transcendent existence, and there is an awareness that the self and others essentially share a community-type existence. This does not apply to human beings alone, for one can see the same arrangement in all other living things. Therefore deep sympathy arises between the self and others, so that they tremble together in fear of the afflictions they have in common and naturally come to consider the plight of the weak among them. In time, this leads people to action in protest against oppression and injustice. The harmonious coexistence of many kinds of people in society is impossible without such thinking and action.

In many cases, this kind of religious thought, as well as other traditional views of human beings and social thought, expresses principles and a worldview that are fundamentally different from the elements that constitute the system of "contemporary society". The elements of contemporary society, which include individualism and the competition principle, are rooted in a stance that emphasizes reason, choosing one of two sides in a conflict, divide-and-rule and objective logic. Richly expressed here is a desirable state of the world that is no longer visible owing to an overemphasis on reason. It seems that this desirable state is a relationistic worldview, or a biocentric, organismic worldview, or a discernment of phenomena-as-reality, life-as-death, or the like, and is therefore an expression of life rooted deeply in human existence. For this reason, henceforth one should, from a global viewpoint, intensify dialogue and mutual understanding among thought systems based on religions and traditional cultures, subject them to philosophical and logical scrutiny and develop them into a universal knowledge that has currency in contemporary society. The new knowledge thus obtained will surely generate important clues for ideas on how to redesign the social system with a more human orientation and how to support the sustainability of global society from the deepest horizon. Especially now, with the collapse of a social system that is based on individualism and the competition principle and that has operated according to the supremacy of the market economy, the formation of a social order based on new values is a matter of urgency. At this time, the role of the humanities, which should set the orientation for this new order, is of crucial importance. Society is at a point where people should thoroughly investigate the essence of their own lives, confirm fundamental values and, having done that, create a blueprint for a new society.

At this stage one should proceed to a sketch for designing the social system, and here there must be dialogue and integration with the social sciences. Only when this is achieved will it be possible to bring a concrete transformation of the very state of society into the realm of attainability, to guide the orientation of science and technology, and to make a substantial contribution to realizing sustainability. However, because contemporary conventional wisdom thinks that expansion of the individual is a matter of course and assumes that the individual consists only of one's own mind and body, it will not be easy to introduce a view of the self based on the Buddhist ideas of non-self and the indivisibility of self and others, and to bond that with today's social sciences. Although this is an extremely difficult challenge, no effort should be spared in pursuing it.

The significance of the humanities should perhaps be sought in their potentially profound influence on the thinking and ways of living of each individual (self) rather than in their direct involvement in the whole of society. Nowadays the lifestyles of individuals are called into question from a global standpoint, but exactly what kind of lifestyle is expected of us? For example, there is often mention of the 3Rs – Reduce, Reuse, Recycle – in connection with environmental problems and sustainability. However, Kyoji Okamoto (2008) argues that one more R is needed: Refuse, meaning that people should refuse to buy the products of a company that does not act in line with social justice. This will be briefly discussed here. The idea now is that, just because an entity is a business, it should not engage exclusively in the pursuit of profit, but rather its activities should uphold social justice and help realize society's common benefit. For example, if a company makes children and other people in poor countries work under abysmal conditions, and markets the products made by them at an unreasonably low cost, it may be said that the company is not discharging its social responsibility. That is to say, social responsibility is based on maintaining and bringing about social justice globally and, further, on endeavouring to help create a fulfilled society. In this vein, questions will arise not only about what kind of products today's businesses should make and how they should produce and offer them, but even about what kinds of lifestyles they propose. This is called "corporate social responsibility".

It would appear that the term "corporate responsibility" is used nowadays with "social" dropped because mere social responsibility might exclude consideration of environmental problems involving natural systems, despite the fact that it is necessary to consider and implement measures to, for example, combat global warming and protect ecosystems and biodiversity. Not only that, but the term "sustainable development" is now frequently used as an extension of this trend. Although it may seem strange to use the word "development" in this context, sustainable development here should be taken to indicate that in one's economic activities one is resolved to exclude that which is not mindful of global sustainability. Sustainability here does not mean simply environmental conservation. Sustainable development now takes into consideration the sound maintenance and development of global society as a whole, aims to eliminate poverty in society, reconsiders the nature of consumption in developed countries, and includes among its goals the protection of the ecosystem and biodiversity. Sustainable development is the slogan for a viewpoint that endeavours to bring about social justice.

Apparently "corporate" as used above means not only corporations but also smaller firms, municipalities, national governments and other organizations. In fact, because the word "corporate" includes meanings such as organizational, group or collective, organizations such as universities can also be included.

In this day and age, even something as simple as the refusal of individuals to buy products made by unscrupulous companies could well lead to the exclusion of bad companies from society and aid in the transformation of society into a form that embraces social justice. As represented by the act of refusal, changes in the consciousness and behaviour of many individuals can induce changes in businesses and other organizations, which in turn can bring changes in society. Surely this is a crucial and optimistic viewpoint. The question, then, is how individuals can be helped to awaken their autonomous thinking and their own sense of values. There is no doubt that the humanities' investigations of the self and human existence will be a driving force behind that endeavour. Therein lies a major role for the humanities in contemporary society.

5-7-5 Conclusion

The humanities can reveal the root structure of human existence. In doing so they can fundamentally criticize the state of real-world society and at the same time offer a vision of what society should be like. Furthermore, they can show what kind of behaviour is required of human beings and guide them to practical action.

The social role of the humanities today can be found in fundamental criticism of the inhumane state of contemporary society, as exemplified by excessive individualism and the competition principle, institutions that positively evaluate only efficiency and achievements, the unrelenting pursuit of self-interest and the physical and mental domination of many losers by a few winners. At the same time, from a stance that advocates a deeper and broader view of human beings and the world, and with academic composure, the humanities should also be able to make proposals about the desirable state of society and how individuals should live. Even if the humanities cannot be directly involved in running contemporary society, they can do such things as articulate ideals for institutional design from a critical perspective. Or they can trace the meaning of the individual from a perspective that is as broad as imaginable. Additionally, they can offer their unique point of view, which is nothing less than an enquiry into the irreplaceable self itself. This is where the significance and role of the humanities are to be found.

However, if the humanities stop there, their influence on the whole of society can only invite pessimism. To make use of the humanities' findings on the scale of society, they must be integrated with the social sciences. This will perhaps guide the natural sciences, too, in a new direction. In that sense, dialogue and integration with the social sciences are of critical importance. It is the author's intention to persevere in that effort.

Notes

1. See "About Sustainability Science" on the IR3S website: <http://en.ir3s.u-tokyo.ac.jp/about_sus> (accessed 17 June 2010).
2. It is said that the Buddha expounded on the five aggregates and non-self thought in his first discourse.
3. The *Abhidharma-kośa-bhāsya* summarizes the thought of Sarvastivada, one of the early Buddhist schools. *Dharma* is a term of many meanings, including truth, teachings, providence and law, but here it is defined as "true immutable nature, a model which lets people understand itself", meaning that which maintains the self itself, that is, the elements of the world.
4. It is said that the doctrine of consciousness-only was brought to completion by Maitreya, Asanga and Vasubandhu. According to Hirakawa (1979), Asanga's dates are 395–470 CE and Vasubandhu's dates are 400–480 CE. Asanga wrote the *Māhāyana-samgraha*, and Vasubandhu's works include the *Vimśatikāvijñaptimātratāsiddhi* ("Twenty Verses on Consciousness-Only") and the *Triṃśikā-vijñaptimātratā* ("Thirty Verses on Consciousness-Only"). They explained the world only with the eight consciousnesses of sight, hearing, smell, taste, touch, thought, the *manas* consciousness and the *alaya* consciousness. At the same time, they expounded the five groups of the 100 *dharma*s and became representative of *abhidharma* (analysis of the world's constituent elements) in Mahayana Buddhism. The consciousness-only thought system brought to China from India by Xuanzang formed the Dharma-character school, which was also brought to Japan.
5. See "Essay on the Five Teachings" in *Taishō Shinshū Daizōkyō*, 1927b: 503a. The chapter on 義理分斉 (*yi li fen qi*, "the realm of the Huayan school doctrine") explains the logic in 十玄縁起無礙法門義 (*shi xuan yuan qi wu ai fa men yi*, "the teaching doctrine of ten profound aspects of unobstructed dependent origination") and 六相円融義 (*liu xiang yuan rong yi*, "the doctrine of six characteristics completely interpenetrated").

REFERENCES

Hirakawa, A. (1979) *A History of Indian Buddhism*, vol. 2. Tokyo: Shunjusha.

Kato, H. (1991) "Introduction", in *A New Recommendation for Environmental Ethics* [*Kankyo rinrigaku no susume*]. Tokyo: Maruzen Library (in Japanese).

Kūkai (1978) *Complete Works of Kōbō Daishi* [*Kōbō daishi zenshō*], vol. 1. Koyasan (Wakayama Prefecture): Esoteric Religious Culture Research Institute (in Japanese).

Okamoto, K. (2008) *Evolving CSR: A Perspective on "Reform" that Transcends the Doctrine of Corporate Responsibility*. Tokyo: JIPM Solutions (in Japanese).

Taishō Shinshū Daizōkyō (1925) vol. 9, ed. Professor Dr J. Takakusu and Professor Dr K. Watanabe. Tokyo: Taisho Shinshu Daizokyo Kanko Kai (Society for the Publication of the Taisho Tripitaka).

Taishō Shinshū Daizōkyō (1927a) vol. 30, ed. Professor Dr J. Takakusu and Professor Dr K. Watanabe. Tokyo: Taisho Shinshu Daizokyo Kanko Kai (Society for the Publication of the Taisho Tripitaka).

Taishō Shinshū Daizōkyō (1927b) vol. 45, ed. Professor Dr J. Takakusu and Professor Dr K. Watanabe. Tokyo: Taisho Shinshu Daizokyo Kanko Kai (Society for the Publication of the Taisho Tripitaka).

5-8

The integration of existing academic disciplines for sustainability science

Kazuhiko Takeuchi

This chapter has examined how various academic disciplines might contribute to the formation of sustainability science: the physical sciences as represented by climate system science; agriculture-related science and technology; economics and the other social sciences; and the humanities, most notably philosophy. Here, in summation, how these diverse existing disciplines might be integrated into the framework of sustainability science will be considered.

Today natural scientists find themselves increasingly compelled to expand their focus to areas previously considered the domain of the social sciences. A case in point is the study of global warming by climate system scientists, who can no longer analyse the dynamics of climate change strictly in terms of natural phenomena when the human impact on the atmosphere grows ever more conspicuous. Conversely, social scientists cannot ignore the issue of global warming in their discussions of present-day politics or economics, and must therefore include the natural sciences in their purview. These developments have created unprecedented conditions demanding the integration of academic disciplines.

Efforts are currently under way to respond to these circumstances by building a platform upon which natural scientists, social scientists and technologists can converse with one another. So far the most successful example of such a platform is the Intergovernmental Panel on Climate Change (IPCC). The process of compiling the IPCC's assessment reports exemplifies two attributes that are precisely those of sustainability science: the synthesis of findings based on a process of abductive reasoning that transcends conventional analysis, as Yoshikawa has described in Section 5-2, and the maintenance of a neutral stance in the presentation of those findings to society. In the case of the IPCC reports, the differing views of scientists are compared on the basis of the latest scientific data, and conclusions reflecting a final consensus are conveyed in the form of a neutral message from the scientists to international society.

IPCC uses models as a framework for integrating the work of researchers in the natural and social sciences and technology. These models indicate, for example, what subsequent advances will be needed in renewable energy or energy conservation technology to prevent global temperatures from rising by more than 2°C, and they allow the costs of such measures to mitigate global warming to be estimated. The success of the IPCC approach has influenced efforts to deal with other global environmental problems; in the area of biodiversity, for example, plans are under way to establish an Intergovernmental Platform on Biodiversity and Ecosystem Services under the auspices of the United Nations Environment Programme.

Integration of knowledge in this manner is easy enough for the purpose of devising mitigative measures against global warming. Integrating the knowledge needed to devise adaptive measures, on the other hand, demands a different strategy. That is because the means chosen to adapt to global warming depend heavily on the natural and social conditions of the affected region in question. These vary markedly from region to region, as do levels of scientific, technological and socioeconomic advancement. Hence it is difficult to devise adaptive measures with any substantial degree of universal applicability. The most effective way to deal with this problem is to organize multidisciplinary groups of researchers to conduct empirical solution-oriented research geared to each locality. This is precisely what Sato means when he cites the need for "practical wisdom" in Section 5-6. And, in fact, this need dovetails with another salient characteristic of sustainability science: a problem-solving orientation that respects the diversity of traditional forms of knowledge and rejects the oversimplification of socioeconomic systems.

One of the implications of such a stance is a concern with ethics, morals or norms, a concern that is also an important aspect of sustainability science. Hence a role can be seen here for the humanities, which deal in the ultimate questions of human existence. IR3S has sought the participation of philosophers because it believes that their perspectives are indispensable to sustainability science. In Section 5-7, Takemura argues that a vital role for the humanities today is to elucidate the fundamental structure of human existence and use this as a basis from which to criticize the norms of contemporary society and point the way towards a better social paradigm. However, a methodology for applying this perspective to the integration of existing areas of knowledge has yet to be developed.

Expectations are higher than ever for the integration of separately evolved disciplines that is crucial to the development of a science of sustainability. Establishment of a transdisciplinary discipline of the sort that sustainability science aspires to be should also contribute to ameliorating the conflicts among disciplines that have arisen through their ongoing fragmentation. Offsetting the negative effects of the relentless specialization that has been the fate of modern science is one of the most valuable functions that an integrated discipline such as sustainability science can serve. If it succeeds in this task, sustainability science should trigger a significant mid-course correction not only in scientific research but in education as well.

6

Education

6-1

Overview of sustainability education

Mitsuhiro Nakagawa, Michinori Uwasu and Noriyuki Tanaka

6-1-1 Issues in the twenty-first century and the quest for a new educational paradigm

The twentieth century was an era of rapid advancement in science and technology and of the expansion of regional economies into a worldwide network (in other words, a global economy). Production, transportation and consumption grew drastically throughout the world, and the global population rose from 1.7 billion at the beginning of the twentieth century to 6.0 billion in 2000. Yet, hidden behind this dramatic economic growth, various issues remain unsolved.

Poverty and starvation are still serious problems in developing countries. Developed countries still face economic issues such as income inequality and unemployment during cyclical recessions. Wars, ethnic conflicts and human rights abuses such as racial discrimination are observed even now in many parts of the world. Other urgent issues range from global environmental problems such as global warming and resource and energy depletion, to mental health problems in advanced countries such as school absenteeism, bullying and suicides among the middle-aged and elderly, all of which have emerged especially in the second half of the twentieth century. Though these problems may be superficially separated, they all seem to reflect two common characteristics of the present era.

One characteristic is the "fallacy of composition". This occurs when the pursuit of benefits by each individual leads to unexpected problems in the aggregate. At the individual level, people have expanded their consumption in a quest for a more comfortable life. As a result, scarce resources are being consumed to a point exceeding the environmental carrying capacity of the Earth. The depletion of natural resources and energy, global environmental problems, the coexistence of food abundance and hunger, and the North/South divide can all be considered a result of this "fallacy of composition". In the past era of small-scale regional economies, quasi-harmonious economic growth could be achieved through the division of labour and trade. The current economy, however, has become so expansive that such "quasi-harmonious" adjustments can no longer be expected through market mechanisms. It is now necessary to introduce a global system to ensure that human activities do not exceed the Earth's environmental capacity and to distinguish, from the viewpoint of comprehensive wisdom, what should be promoted from what should be restricted.

The other characteristic may be described as "fragmentation of the holistic workings of organic life". This manifests itself in symptoms of fragmentation or disconnectedness caused by the division of labour and narrowed specialization in human activities, which result from the never-ending pursuit of higher productivity and efficiency in the process of modernization. Examples include the threat to food safety from the spread of bovine spongiform encephalopathy (BSE) and avian or swine influenza, the degradation of ecosystems and the loss of biodiversity. Moreover, the feelings of alienation and isolation, various kinds of psychological mal-adaptation and the loss of a sense of satisfaction or purpose in life observed in advanced countries can be attributed to people's disconnection from the lives that surround them and to the fragmentation of holistic human activities in societies with a high level of division of labour.

As long as the current industrial society pursues segmentation and meticulous management with the objective of higher efficiency, the emergence of specialized fields in the educational system that prepares students for life in this society cannot be avoided. The idea of training highly skilled specialists advocated by universities today reflects their attempt to increase not only domestic productivity but also international competitiveness in the global economy by helping students develop specific skills in certain fields.

However, sustainable development cannot be achieved unless people overcome the problems of the fallacy of composition and the fragmentation of the holistic workings of organic life that lurk behind the problems faced in the twenty-first century. Therefore, increasing attention has been paid to new educational programmes that aim to train people who can lead the way towards sustainable development.

A crucial step in correcting fallacies of composition or the fragmentation problem is to recognize that all things are basically connected with people's existence. A paradigm shift from traditional atomism and reductionism to relationalism and a holistic view of the world seems to be necessary.

6-1-2 International trends in sustainability education

Just as variations exist in the definition of sustainability, so education relevant to sustainability is diverse in its principles and content. In general, education from the standpoint of sustainable development is both an objective and a means. It supports the balancing of environmental conservation and development by offering educational opportunities to all and focusing on such issues as poverty and the gender gap (United Nations, 2006). Environmental education in elementary schools can be considered a form of education about sustainability even if it does not specifically mention the word "sustainability", because global environmental problems are closely connected with poverty, development and gaps between social groups. University-level higher education also emphasizes the environmental aspects of various academic fields, and students can usually study sustainability issues in environmental studies courses as well as in courses in existing disciplines with titles that contain the word "environmental". Throughout the present chapter, "sustainability education" generally refers to education relevant to the idea of sustainability unless it is defined otherwise.

The need for a paradigm shift in education has been identified in the field of international education by the United Nations and has been advocated as part of "Education for Sustainable Development" (ESD). At the World Summit on Sustainable Development (WSSD) held in Johannesburg in 2002, Japan proposed the Decade of Education for Sustainable Development (DESD). This became a turning point that added momentum to a movement to reorganize UN-led educational programmes dealing with environmental issues, development, peace and human rights from the standpoint of global sustainability and to integrate them under the ESD umbrella. In 2005, the UN General Assembly adopted the DESD International Implementation Scheme and international promotion of ESD began.

As UNESCO (2007: 6) explains, "ESD prepares people to cope with and find solutions to problems that threaten the sustainability of the planet". Abe (2006) points out that the characteristics of ESD are its interdisciplinary nature, its comprehensiveness, its value basis, its critical thinking and problem resolution, its diverse methods, its participative decision-making and its regional ties.

Responding to this international trend, the Japan Council on the UN Decade of Education for Sustainable Development (ESD-J) was established in 2003. ESD-J is Japan's first interdisciplinary non-governmental organization and has among its members many ESD-related organizations engaged in issues involving the environment, development, human rights and youth education. Their participation in such an organization reflects the fact that ESD comprises educational efforts that form bridges between individual citizens and society, human beings and nature, local regions and the world, and the present and the future.

Looking at trends in sustainability education programmes in higher education worldwide, there are many academic programmes dealing with sustainability issues, mainly in Europe and the United States, but there is a lack of official statistics or organized information on them. These programmes can be categorized into (1) interdisciplinary programmes incorporating engineering, forestry and other fields of study (that is, environmental studies) and (2) new programmes dealing with sustainable development and sustainability science. The former include the School of Forestry and Environmental Studies at Yale University, the School of Natural Resources and Environment at the University of Michigan and the Earth Institute of Columbia University, and the latter include programmes in sustainability science under the Integrated Research System for Sustainability Science (IR3S), the programme in environmental studies and sustainable science at Lund University, Sweden, and the School of Sustainability at Arizona State University. The number of higher education programmes in sustainable science has grown in recent years with the emergence of global environmental issues, financial crises and other problems that threaten the existence of global society.

Banas (2007) has conducted a survey of programmes in sustainability in different parts of the world that illuminates the characteristics of sustainability education in various countries and regions. According to the survey, sustainability education is integrated into many programmes in the social sciences, such as business administration, public policy and economics, in Europe and the United States. This is an interesting finding because it contrasts with the fact that, in Japan, sustainability education is mainly incorporated into engineering programmes. The reason is probably that sustainability education in Europe and the United States is based on sustainable development as advocated by the United Nations, whereas in Japan it is based on environmental problems associated with natural resources, energy and industrial waste as well as the concept of the "three Rs" (reduce, reuse, recycle). In this context, the idea of a "harmonious society", advocated by China in recent years, aims at development balancing a harmonious relationship between the environment and society; hence sustainability education in Chinese higher education may be mainly organized in relation to the social sciences. Thus the understanding of and goals associated with sustainability depend on the current situation and the way problems are approached in each country, which are reflected in the characteristics of its educational programmes.

6-1-3 Sustainability education at IR3S

As part of this international trend in sustainability education, IR3S has aimed at establishing academic centres for sustainability education as well as for research in sustainability science. The IR3S member universities have launched sustainability education programmes to train the next generation of specialists who will play an important role in the sustainable development of human society. Here, the six characteristics of these IR3S-related sustainability education programmes are summarized: (1) a holistic view, (2) understanding the global, social and human systems, (3) "T-shaped" training, (4) diverse educational methods, (5) cultivating global citizenship and (6) spiritual education. Details of the educational programmes at each university are found in later sections of this chapter.

A holistic view

Twentieth-century science and technology are characterized by a reductionist approach that attempts to solve problems of the whole by decomposing them into simple parts based on unilateral criteria, then combining optimal solutions for each part. Behind the successes achieved through this positivist, reductionist approach, however, is a growing accumulation of insoluble problems.

Sustainability education is oriented towards emphasizing the relationship between the whole and its individual components, and towards training students to understand themselves and the world from a holistic viewpoint. It uses as its model the attributes of a living organism, as opposed to a machine, and focuses on the process of change observed in the multi-layered relationship between humans and the world. Also, similarly to the "knowledge science" (Sugiyama et al., 2008) and "open systems science" (Tokoro, 2009) proposed in recent years, it emphasizes development of the ability to synthesize learned knowledge through recognition of multidimensional values and advancement of the sharing, reorganization and integration of knowledge.

Understanding the global, social and human systems

Since its inception, IR3S has sponsored projects oriented towards comprehensive research and education on the global, social and human systems (Komiyama and Takeuchi, 2009). This approach is based on a recognition that each of these systems has problems potentially leading to its unsustainability, and that additional problems arise from the disordered relationships between the three systems. IR3S believes that the key to creating a sustainable human society is to deepen understanding of the interactions between these systems and to restore healthy relationships between them.

Sustainability education places importance on training a new type of professional who can take the lead in restoring the connection between human beings and nature, citizens and society, local regions and the world, and the ego and the self.

Training "T-shaped" professionals

"T-shaped" educational training is so named because it aims at both in-depth specialization in a particular field and a broad familiarity with the knowledge surrounding that field. These attributes are represented by the vertical and horizontal parts of the letter T, respectively. If students on a given graduate programme can also participate in a sustainability education programme as a complementary programme at their graduate school, this enables them to study their field of specialization simultaneously with other encompassing subjects.

Students trained in this manner come to understand the social context surrounding their specialized field from a macroscopic viewpoint and realize their responsibilities as specialists from a historical perspective. They are encouraged to open themselves to dialogue with colleagues in different fields in addition to mastering their own specialized area of knowledge.

Diverse educational methods

Miller (1996) categorized modes of instruction in school education into three types of interaction between the curriculum or teachers and students – transmission, transaction and transformation – and argued the importance of combining these three types organically in holistic education. Sustainability education incorporates this idea and offers three types of curriculum: lectures focusing mainly on the transmission of knowledge, seminars involving active interaction between the instructor and students, and fieldwork in which the instructor and students achieve a deep understanding of issues and grow as specialists by going into the field, gaining direct experience with local residents and thinking, feeling and taking action together. Through these various types of curriculum, students are expected to gain an understanding of the essence of sustainability issues with their bodies as well as their minds, and to acquire the awareness and will needed as specialists to work towards the attainment of a sustainable society.

Cultivating global citizenship

Global citizenship has been promoted in basic education, the idea being to foster the development of global citizens who can appreciate both cultural diversity and the universal values in each culture. Yoshida (1999) argues that cultural centralism, cultural relativism and anti-cultural cosmopolitanism must be overcome through global citizenship education. Respecting diverse cultures around the world requires not only recognition of the differences and relative values of each culture but also an effort to unearth, through dialogue, the universal values concealed by cultural differences. This effort necessitates the cultivation of an outlook capable of understanding differences and relative values. A first step towards acquiring such an outlook is to develop a familiarity with and in-depth understanding of one's own culture. In other words, a cosmopolitan perspective must be deeply rooted in people's relationship to their own culture. The inculcation of such a perspective through global citizenship education can lead to the "internal globalization" of people.

Spiritual education

Mahatma Gandhi famously stated that "Earth provides enough to satisfy every man's need, but not every man's greed". Thus, sustainability education that aims at the realization of a sustainable society must also stress spiritual education.

Currently, university-level education emphasizes scientific thinking, reason and rationality. Priority is also given to strengthening students' social egos so that they can adapt themselves to a competitive society. Irrational things such as the physical senses, emotions, affects and instincts are suppressed, and the ego and the natural self are separated from one another. For this reason, some young people are unable to feel joy or vitality in their lives, lose their sense of themselves and succumb to feelings of fear and aggression towards others. Sustainability education must aim at training students not only in logical thinking but also in all-round human attributes including the physical senses, emotions, intuitions and spiritual impulses that have traditionally been treated lightly in school education.

Sustainability education should promote the restoration of connectivity between nature, society and the individual by fostering trust, pride and the joy of living in young people so as to instil in them a sense of wonder towards nature, compassion for others, and self-awareness.

Sustainability education with these general characteristics has already begun at the IR3S member schools. Graduates of these programmes receive a joint graduation certificate and begin their career as specialists with a basic understanding of sustainability science. It is hoped that educational programmes of this sort will proliferate and that the trend towards a new form of university education integrated from the standpoint of global sustainability will grow stronger.

REFERENCES

Abe, O. (2006) "What Is Education for Sustainable Development?", in Japan Holistic Education Society (ed.), *Promoting Sustainability Education*. Osaka: Seseragi Publishing (in Japanese).

Banas, S. (2007) *A Survey of University-Based Sustainability Science Programs*. Supplement for Forum for Sustainability Science Programs Roundtable, AAAS 2007 Annual Meeting, 17 February, San Francisco. Available at <http://sustsci.aaas.org/files/University%20Survey%20V2.pdf> (accessed 17 June 2010).

Komiyama, H. and K. Takeuchi (2009) "IR3S, the Emergence of Sustainability Science and the Formation of Global Media Networks". *Energy and Resources* 30(2): 24–28 (in Japanese).

Miller, J. (1996) *The Holistic Curriculum*, 2nd edn. Toronto: OISE Press.

Sugiyama, K., A. Nagata, A. Shimojima, K. Umemoto and T. Hashimoto (2008) *Knowledge Science: 81 Keywords for Knowledge Creation*. Tokyo: Kindai Kagakusha (in Japanese).

Tokoro, M. (2009) *Open Systems Science*. Tokyo: NTT Publishing (in Japanese).

UNESCO (2007) *The UN Decade of Education for Sustainable Development (DESD 2005–2014): The First Two Years*. Paris: UNESCO. Available at <http://unesdoc.unesco.org/images/0015/001540/154093e.pdf> (accessed 17 June 2010).

United Nations (2006) *Millennium Development Goals Report 2006*. New York: United Nations. Available at <http://mdgs.un.org/unsd/mdg/Resources/Static/Products/Progress2006/MDGReport2006.pdf> (accessed 17 June 2010).

Yoshida, A. (1999) *Holistic Education: Japan's Experience and New Ideas*. Tokyo: Nippon Hyoronsha (in Japanese).

6-2

Core competencies

Makoto Tamura and Takahide Uegaki

6-2-1 Introduction

The previous section discussed the historical background and various issues associated with sustainability science in the twenty-first century, and confirmed that new education programmes are needed to develop the human resources capable of building a sustainable society. Therefore, alongside Education for Sustainable Development (ESD), which has grown in popularity internationally in recent years (see, for example, UNESCO, 2005, 2007), there is a need for education that can provide a comprehensive understanding of the global, social and human systems and their interrelationships. It is also desirable that the search continue for other diverse teaching methods as well.

In order to identify what is desired of these new forms of education, the final destination must first be defined, that is, the goal that education should be striving to reach. In other words, sustainability education must make clear what sort of human resources development it aims for. By articulating the aims in terms of the human resources to be developed, one can identify and share objectives and ultimate goals with regard to the desired state of sustainability education.

For this reason, when defining the human resources desired, the core competencies required of these resources must be clarified. Asking what capabilities should be acquired is a question essential to human resource development, one that enables a specific vision of the human resources desired to be shared, and also provides clues for the implementation of this vision in practical education.

This section will look closely at the core competencies required of sustainability education and present a model of these competencies for discussion in later sections. This section mainly focuses on the core competencies in the context of sustainability education in Japan. For a discussion about other countries, see, for example, Segalàs et al. (2009).

6-2-2 Goals for human resource development in sustainability education

First, the type of human resources that sustainability education aims to develop will be identified by examining the differences between sustainability education and conventional education.

Difference between the goals of conventional education and of sustainability education

Conventional education in Japan has largely focused on the accumulation of knowledge. This tendency is particularly prominent in higher education (see, for example, Central Council for Education, 2003). Undergraduates and graduates dedicate themselves to the acquisition of segmented expertise by taking specialized and segmentalized courses from the time they enter universities and graduate schools. Conventional undergraduate and graduate education has adopted this approach because the main goal of this education is to impart expertise. The root of this tendency is, no doubt, a presumption that higher education best serves society by producing human resources who are specialists who can apply the latest specialized knowledge in specific fields.

However, the fundamental issue addressed by sustainability science is how to reconstruct the global, social and human systems into appropriate forms in order to establish a sustainable society, based on a recognition that there are mutual contradictions among these three systems. Therefore, the human resources trained by sustainability education must have in their minds a long-term vision for building a sustainable society and must acquire competencies that can contribute to solving the problems that are actually occurring in various areas of society.

Goals of human resources concerning a sustainable society

The background provided above reveals some conspicuous features of the human resources to be cultivated through sustainability education.

First, the orientation towards action by people trained in sustainability education is motivated not so much by academic goals as by the chance to solve real and present problems. This orientation derives from the original impetus for sustainability science itself, that is, the awareness that a variety of problems actually exist and solutions are needed for them. Indeed, this is often mentioned as the primary distinction between discipline-oriented and problem-oriented education (see, for example, Gibbons et al., 1994).

Second, the fields in which these human resources will be active are not limited to specific areas of expertise as in the case of conventional education, but rather cover a wide range of fields, both domestic and international. These individuals are expected to be active in a variety of roles in organizations such as public organizations, regional governments, corporations, non-governmental organizations and non-profit organizations, and to uphold the functions of these various organizations. This is because sustainability science assumes that the current problems the world faces may originate in the breakdown of interactions among the global, social and human systems, and it endeavours to approach a variety of fields simultaneously in a broadly defined, problem-oriented framework.

Third, sustainability education is targeted to students who belong to almost any field of academic endeavour. This is because the human resources to be cultivated through sustainability education are not necessarily "experts in sustainability science". As mentioned above, human resources are needed who will take an active role in a wide variety of fields. It is important that people with diverse forms of expertise undertake different approaches according to their field while at the same time maintaining the larger perspective of working towards a sustainable society. The above discussion is summarized in Table 6.2.1.

6-2-3 Core competencies required for human resources

Given the requirements for sustainability-educated human resources described above, which competencies are required for these human resources? What is important here is that, first, the required "competency" does not always take the form of knowledge and, even when it does, it is not always perceived as a specific conventional discipline. In addition to expertise, sustainability education requires comprehensive competencies that combine holistic knowledge, skills and mind. This subsection will describe these competencies.

Table 6.2.1 Differences in the human resources required between conventional education and sustainability education

	Conventional education	Sustainability education
Orientation for action	Discipline oriented	Problem oriented
Field	Field in which students can make use of their expertise; specific job descriptions	Local to global fields; various job descriptions
Target	Students who belong to specific faculties, or candidates seeking expertise in specific academic fields	Students who belong to all fields and candidates seeking expertise in various areas of knowledge who also aspire to help build a sustainable society

Holistic knowledge

The required competencies include knowledge as a matter of course. However, this refers not to a single genre of knowledge that fits into the fixed framework of a specific academic discipline, but to knowledge that encompasses a broad spectrum of fields ranging from the natural sciences to the social sciences. In this sense, "holistic knowledge" is indispensable for sustainability education. To be sure, one human being cannot cover all academic fields, and the academic position of an expert will become ambiguous if he/she attempts to address too many areas of knowledge in piecemeal fashion. Therefore the role of holistic knowledge required here must be clarified.

Holistic knowledge mainly helps provide human resources with the competency to detach and position themselves. In other words, this competency enables the students (1) to place their own accumulated experiences and knowledge in a larger context, thereby defining their positions, and (2) to identify the issues to be dealt with. For example, conveying one's own specialized expertise to someone outside that field requires the ability to explain it by combining one's own expertise with surrounding fields. This also includes the ability to carefully consider the interactions between different fields so as to avoid raising unnecessary problems in the pursuit of specialized knowledge.

In recent educational debate, the significance of holistic knowledge has been cited in the context of criticism of the conventional segmented "foxhole" model. In this debate, the "T-type" model is sometimes proposed as an educational model that combines both expertise and holistic knowledge. What should be emphasized here is not simply that having the required competencies means having T-type knowledge, but that these competencies need to function with an appropriate degree of objectivity.

Skills

The required competencies also include skills. Higher educational expertise naturally includes various skills, and students completing a major course of study acquire special skills in some form. However, the human resources it is sought to develop should also have skills specific to sustainability education.

The first and foremost of these skills are communication skills, which include the ability to understand the emotions of others, to see perspectives other than one's own and to build relationships with others. Whether taking a local or a global perspective, it is necessary to co-operate with different people with different viewpoints in order to actually resolve problems. Communication skills are crucial to building these cooperative relationships and maintaining them in a sound and appropriate manner.

The second area of required skills is collaboration skills. These are skills in building relationships using communication skills and linking these relationships with various other related factors. With the complex and all-encompassing challenges involved in attaining a sustainable society, the ability to link individual problems to other problems and to recognize patterns is critical. Collaboration skills also include the ability to recognize the relationships between different activities occurring at the same time and to bring together the various stakeholders involved.

The third skill set is problem-solving skills. Unlike the ability to link factors or build relationships, which are germane to dealing with individually occurring problems, problem-solving skills include the ability to single out problems and find clues for their resolution. Because efforts to achieve sustainability include a variety of actors, conflict can arise even when care is taken to avoid it. Whereas communication skills can help to prevent conflict before it occurs, problem-solving skills enable us to manage unavoidable conflicts in the best manner possible.

Mind (heart)

Along with holistic knowledge and specific skills, mind can be considered an essential competency for human resources. Here, the meaning of "mind" is closer to will, heart or attitude than to self-consciousness or reason. Mind as envisioned here includes the following elements:

(1) Motivation. This refers to the mental strength and endurance to bring to completion actions that are initiated.
(2) One's own beliefs. A new activity often runs into failure or barriers of some sort. Maintaining one's motivation without giving up requires a strong belief in or rationale for one's own practice.
(3) Synchronic consciousness, or internationality. This is the competency of not being restricted to narrow personal relationships, but rather being able to imagine people around the world in situations different from one's own and striving to communicate with them. This also refers to the ability to expand one's sphere of activity even when in unfamiliar areas or foreign countries.
(4) Diachronic consciousness. Behind the ability to motivate people, raise awareness of a problem into a belief and refine the sensibilities of others sharing the same epoch, is an omnipresent love expressed towards the existence of other individuals and other generations.

Expertise

Considered in this way, one might mistakenly assume that the aim of sustainability education is to develop "experts in sustainability science" who have acquired holistic knowledge, skills and mind. The human resources envisioned in this section are not experts who acquire only these competencies, but those who combine their own core expertise with these competencies. Human resources who possess the above competencies but lack expertise in a field in which they can make a contribution will not be able to perform at a level sufficient to solve the problems that actually need to be solved.

Figure 6.2.1 summarizes the desired competencies. As mentioned above, these should be combined with the expertise that is emphasized in conventional education.

The human resources it is sought to develop will above all acquire their own core expertise and additionally acquire the specific competencies required for sustainability science. In other words, by learning holistic knowledge and expertise in addition to mind and skills, these human resources will be prepared to discover, identify and take steps to resolve sustainability issues on their own.

6-2-4 Conclusion

This section has examined differences between the human resources developed through conventional education and through sustainability education, and has discussed the core competencies required of the latter.

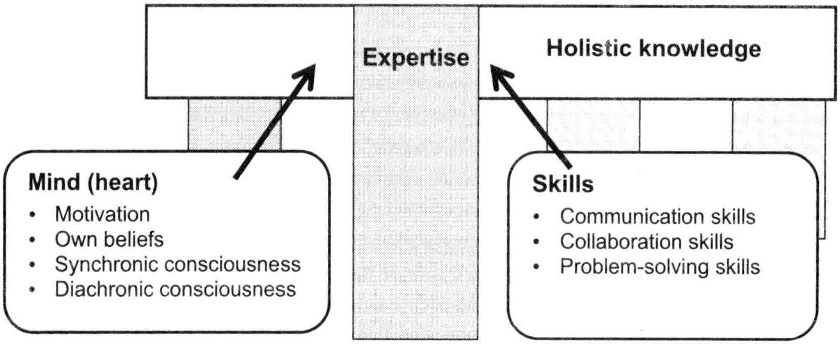

Figure 6.2.1 Core competencies for human resources fostered by sustainability education.

First, the following characteristics of human resources fostered by sustainability education were identified: (1) their activities are problem oriented rather than discipline oriented; (2) their field of action is envisioned as a broad field embracing various occupations as well as various activities appropriate to diverse social perspectives but sharing the goal of building a sustainable society; and (3) students belonging to any discipline may be candidates for sustainability education.

Next, regarding the core competencies required, it was argued that there should be expert knowledge at the core but that equal importance must be placed on (1) holistic knowledge, (2) skills (including communication skills, collaboration skills and problem-solving skills), and (3) mind (including motivation, one's own beliefs and synchronic/diachronic consciousness).

In summation, the aim of sustainability education is to develop human resources with different specialties who, from various social backgrounds and various perspectives corresponding to their individual situations, can share a long-term and comprehensive perspective on the construction of a sustainable society. Through sustainability education, these human resources will be expected to work towards solving actual problems while building collaborative relationships with a broad diversity of actors.

REFERENCES

Central Council for Education (2003) *Modality of the Fundamental Law of Education and Basic Promotional Plan for Education Befitting to the New Times.* Report of the Ministry of Education, Culture, Sports, Science and Technology, Japan, March (in Japanese).

Gibbons, M., C. Limoges, H. Nowotny, S. Schwartzman, P. Scott and M. Trow (1994) *The New Production of Knowledge: The Dynamics of Science and Research in Contemporary Societies*. London: Sage Publications.

Segalàs, J., D. Ferrer-Balas, M. Svanström, U. Lundqvist and K. F. Mulder (2009) "What Has to Be Learnt for Sustainability? A Comparison of Bachelor Engineering Education Competences at Three European Universities", *Sustainability Science* 4(1): 17–27.

UNESCO (2005) *United Nations Decade of Education for Sustainable Development (2005–2014): International Implementation Scheme*, Paris: UNESCO.

UNESCO (2007) *The UN Decade of Education for Sustainable Development (DESD 2005–2014): The First Two Years*. Paris: UNESCO. Available at <http://unesdoc.unesco.org/images/0015/001540/154093e.pdf> (accessed 17 June 2010).

6-3

Pedagogies of sustainability education

Hisashi Otsuji and Harumoto Gunji

6-3-1 Introduction

What sort of real-world situations are likely to be faced by students who have completed a course in sustainability science? The answer to this question will show the direction to be taken in the pedagogy of sustainability education. Pedagogy can be described simply as a "strategy of instruction". However, teaching, instruction and facilitation are sometimes discussed as distinct activities. In this section, the position is adopted that all three are necessary in sustainability education, and a broad view is taken of pedagogy as design that considers all the elements relevant to the praxis of teaching and learning. As far as space permits, not just the ideas and content of sustainability education will be examined but also its methodologies, learners, teachers/facilitators, basic attitudes towards science and nature and the learning environment, while considering situations that the student is likely to encounter after completing a course in sustainability science (see Box 6.3.1).

6-3-2 Desirable abilities

Panoramic knowledge

In the previous section of this chapter, panoramic or holistic knowledge was seen as central to the ability to exercise detachment and place oneself in different positions. This ability demands constant monitoring of

Box 6.3.1 A possible scenario of a situation likely to be encountered by a student after completing a course in sustainability science

An industrial cluster is invited to a large village that has low income but is nearly self-sufficient. Although this is a major project with the expected economic benefits of corporate tax revenues and job creation, there is strong disagreement in the village between supporters and opponents. A considerable amount of money is spent on infrastructure and maintenance for the proposed site – land usage, industrial water, waste water, road construction and so forth – and the village eventually takes on a modern appearance.

However, problems soon develop. The constant stream of heavy trucks pollutes the air with exhaust gases, traffic accidents increase, groundwater is polluted, and the waste incineration facility and landfill site for the ash produced cannot keep up with the growing population. In the new residential areas, the ground has begun to subside in places where the ground level had been built up.

Residents lose interest in their farm fields and the village's elderly residents barely manage to dredge and maintain the agricultural water supply that had saved the villagers from drought over hundreds of years. Even the rice planting is left to the elderly.

A plan is proposed to dam the river a long way upstream in order to secure industrial water and to meet the increased demand for electricity from the growing population in the surrounding area. However, this would require excavation in a residential district with compulsory resettlement. Some people also voice their fond memories of the river in the days when they could swim in it.

With the arrival of cheap imports from developing countries, followed by the global economic recession, factories cannot sell their products and some begin to close down. The population starts to decrease, as does the village's income. The newly opened elementary school remains, but after a few years the older elementary schools in the mountain villages close. The local government officials born in the baby-boomer generation give priority to securing their retirement funds and, hiding behind the excuse of financial difficulties while keeping government coffers untouched, they do nothing to support the new green tourism proposal intended to improve the village's prospects. The village festival had once flourished with the participation of local businesses but is now barely kept going by the elderly residents. The large supermarket, which had dealt a blow to the small village stores and shopping district, pulls out, forcing villagers to go to the neighbouring town for some of their everyday goods.

Box 6.3.1 (cont.)

> Agricultural production becomes unstable, perhaps because of global warming. Whereas smaller typhoons used to bring welcome rain, now a few powerful typhoons cause landslides and damage crops.
>
> Imagine that you have landed a job as a public official in the planning department of this village. Your colleagues sitting beside you have come through a conventional university education. You have received a degree in Sustainability Education. How does your education differ from that of your colleagues? How do you differ from your colleagues in terms of the abilities you possess and the activities that will be expected of you?

one's own thinking so as to maintain a metacognitive, third-person perspective (Flavell, 1976). Practical problem-solving exercises with students from other disciplines are a good way of implementing transformative learning (Mezirow, 1991), which leads students to think reflectively.

This subsection will look at a few areas relevant to this. First, consider the success of science studies in the twentieth century. Science studies taught that science is not neutral and that scientific knowledge is knowledge verified through certain procedures within a particular scientific community. This knowledge does not necessarily state the truth; it is provisional and may be rewritten in the future. Science has also tended towards dualism and reductionism.

Sustainability education will ideally foster such an awareness in graduates, so that when presenting their own specialist knowledge they are possessed of the humility to recognize its presuppositions and limitations without rigidly adhering to it, and also the open-mindedness to accept and impartially compare opinions from other perspectives. In addition, they should come to readily understand that people have different sociocultural values, and that sometimes the agreed-upon and adopted method is not necessarily the one that is the most effective from a scientific or technological perspective.

Students will also have learned that attitudes towards nature are influenced by culture. For example, owing to the influence of the creation story in the Bible's Book of Genesis, which describes the dominion over nature granted by God to man, the West has tended to view nature as a resource. Activities for a sustainable society include reflection on such views of nature, but ultimately regard humankind as an intrinsic part of nature and aspire to a society of harmonious coexistence. This sort of viewpoint can be learned by students together irrespective of their specialties, and may even be presented in the traditional form of knowledge transmission through lectures.

Skills

The previous section of this chapter cited communication skills, collaboration skills and problem-solving skills. Assigning students problems and fostering in them the ability to solve those problems has long been a central aspect of all school education. Understanding a problem is regarded as the process of recognizing relationships that could not previously be grasped.

For example, when trying to understand the wishes of local residents, one must seek to grasp the problem in a structured way by first listening attentively to the voices of people not always accustomed to expressing their ideas in a methodical manner. Local people will not necessarily voice their true feelings in interviews in a one-off visit, and it may require time and the building of trust before they will talk freely. This is often the case in Asian cultures in particular. A commitment to participant observation and dialogue is, therefore, highly valued in the pedagogy of sustainability education.

The practice of learners entering a community has already been established under such terms as "service learning" (Jacoby, 1996)[1] and place-based education (Sobel, 2004). This practice differs from simply learning about the community. Rather than viewing the community objectively as a detached observer, the learner actually enters and lives in that community, becoming one of its members if only for a short time.

Mind (heart)

There is no single absolute method for fostering motivation, one's own beliefs, internationalism and synchronic/diachronic consciousness. However, rather than being transmitted like conventional knowledge, these are qualities that resonate between people. This resonance requires an encounter between the humanity of the lecturer/facilitator and that of the learner, or at the very least an opportunity for the lecturer/facilitator to exhibit his or her own beliefs. An introduction in lecture format to Saussure's structural linguistics can also be useful, since the synchronic and diachronic ways of seeing that emerge have great potential to broaden the mind. There is, however, more to it than just displaying enthusiasm. A quiet walk in the forest giving learners the opportunity to make discoveries on their own could also produce a sympathetic meeting of minds between learner and facilitator.

Specialist knowledge

Conventional knowledge transmission, or the "banking concept" (Freire, 1970), and exam-based formats with their attendant fear of failure are still often effective for acquiring specialist knowledge as well as the panoramic knowledge described above. However, it is good to aim for interactive lectures by creating a space for teacher–learner and learner–learner interaction, rather than providing only a one-sided flow of information. A lecture that involves students from different fields sitting together and at times exchanging opinions is an exciting prospect. Such activities foster the open-mindedness, curiosity, acceptance of the opinions of others and teaching skills that are sought in teachers of sustainability education.

Integrative abilities: Considering groups and organizations

The skills described above are those an individual should possess, but of course the goal of sustainability education is not simply to turn out a succession of super-individuals. Problem-solving must be tackled in society at the organizational level. If colleagues in an organization are highly knowledgeable, a better solution is likely to emerge by pooling those resources to produce results as a group. It is also important to share with colleagues the knowledge and skills one has cultivated by oneself. Every organization is a dynamic entity (Tuckman, 1965).

One method of effectively spreading knowledge among people with different responsibilities at the same level in an organization is the "jigsaw classroom" (Aronson, 1978), which has become popular in school education in the West.[2] This method can also be used for in-service training. Consider an example of knowledge being passed down through different age groups. At the North Vancouver Outdoor School in the Canadian province of British Columbia, high school students stay overnight with elementary school pupils for one week and act as a link between the children and instructors.[3] A long-term cycle has already arisen in which those elementary school pupils go on to become Outdoor School counsellors when they reach high school, and then, as adults, send their own children to the Outdoor School. In addition to normal outdoor education, pupils also learn about the wisdom of First Nation people.

Basic organization theory and learning theory can also be a part of sustainability education. For example, organization theory includes the concept of legitimate peripheral participation (Lave and Wenger, 1991). In terms of this concept, the individual is aware of him/herself as a member of the organization, sometimes as a novice, sometimes as an expert. In this way, the team itself grows.

In the reality of a sustainable society, those who provide knowledge and those who receive knowledge are by no means fixed players. As the learners change, so too can the providers, and, through these mutual changes, the situation on the ground will change dynamically.

Tacit knowledge

As problem-solving experience is built up, a certain degree of intuition begins to develop, together with the ability to see the path to achieving a goal.

In sustainability education, training methods such as the dilemma story (Settelmaier, 2009), which requires decision-making in hypothetical situations, can be used to provide practical experience with such scenarios.[4] By working through case studies or hypothetical situations, students acquire certain skills that come from knowledge of the relevant circumstances. To put it another way, a transformation occurs, a certain stance is formed towards problems likely to be encountered in the future, and with it the student comes to possess the ability to anticipate the unseen (what one might call "second sight"). Even if it cannot be clearly put into words, an intuition operates that steers them in the right direction and their involvement is proleptic in nature. If students can become aware of the double-edged nature of things as having both positives and negatives – that is, the Chinese logic of yin and yang – it is possible to make judgements with foresight and understanding.

6-3-3 Examples of participatory techniques

Some typical participatory techniques will now be introduced and their main features described. As already mentioned, there is a wide range of teaching methods in sustainability education. The traditional lecture is often effective, depending on the content being presented and the motivation of the students. However, since this method will be familiar to most readers, only participatory methods will be discussed, which until now have not been widely used in higher education. The participatory format itself can be categorized in various ways: real problems or mock problems, activities in the classroom or in the real world, activities using linguistic communication or physical experience, and so forth. Games, simulation, role play and planning are used relatively often, and their characteristics are discussed below. Of course, there are many other techniques (for example, Ishikawa, 2008; Kakuta and ERIC, 1999: 63–66), and the techniques themselves are sometimes used in conjunction with each other to make up an actual teaching activity.

Games

Games allow students to learn while having fun. There are a variety of forms, including both physical and thought-based activities. What they all share is the element of fun. Learners acquire a comprehension of complex rules and a mastery of sophisticated tactics when enthusiastically engaged in a game. Therefore, by incorporating the content of sustainability education into games, students can be made to think about combinations of complex rules or conditions while enjoying themselves. The learning content can be incorporated by modelling phenomena (simulation) or having students take roles in the game (role play). Although fun is the key characteristic, it is also important to reflect adequately on what has been learned.

Simulation

By experiencing a mock situation through modelled phenomena, students can learn with a real sense of the processes involved in a given phenomenon. The mock experience is effective when the real phenomenon cannot easily be experienced because, for example, it does not regularly occur, is too far away, would require many years to collect data on, or involves too many elements. In such cases, the question is how far the modelling should go. If the modelling is inadequate, it will probably be difficult to create a mock experience. If the modelling is excessive, the knowledge provided will be sufficient and will obviate the need for the mock experience. Simulation may also be incorporated into a game, and planning may also be made the subject of mock experience.

Role play

Performing a role different from their usual one allows learners to understand and think about problems from the perspective of that role. Role play could be thought of as a form of simulation in the sense that learners are having the mock experience of another person. The expression of opinions in that role is also a feature shared by debate. Debate normally involves dividing a group into proponents and opponents, but in role play a variety of positions can be set up. Role play would seem to be particularly suitable for sustainability education if it enables students to learn through the mock experience of how various interested parties think and how agreement could be reached.

Planning

Planning involves learning through the process of creating a plan for coping with a particular problem. Before a solution can be considered, the problem must first be correctly understood and analysed. Learners must overcome challenges such as coming up with a concrete proposal, assessing its feasibility and determining whether the interested parties can agree to it. This process leads to learning, whether by dealing with a mock situation or with a real one. If the latter, it may involve techniques of learning through participation in the community, such as action research or service learning.

Whatever technique is used, the important thing is that the learners actively participate. The participatory techniques introduced here are all group activities that require communication with other people. As pointed out in the subsection on games, it is important not only to engage positively in the activity itself but also to reflect on what has been learned. In other words, students should be able to participate confidently and actively, communicate appropriately and enhance their learning through adequate reflection. It is the facilitator's job to ensure that this happens. Through sustainability education of a participatory nature, students can expect to acquire an attitude of positive participation, group communication skills and the knowledge content of the activities, while at the same time encountering and picking up the skills of the facilitator in promoting group activity learning.

6-3-4 The continuity of sustainability education in local communities

Finally, it is important to point out the continuity that exists before, during and after university education. The discussion in this subsection concerns higher education because it appears at the high-level specialization stage of the T-type education model. However, before receiving a higher education, learners pass through primary and secondary education. Moreover, in today's knowledge-based society, learning is seen as necessary in all sorts of situations after the completion of undergraduate or postgraduate studies.

A movement related to sustainability education is Education for Sustainable Development (ESD). This type of education is relatively new and hence shares similar terminology (UNESCO, 2005) with sustainability education. Both types seek to educate people to build a sustainable society, and, in terms of desirable abilities and teaching/learning methods, ESD shares many of the attributes discussed in this section. Higher education organizations are naturally involved, and sometimes refer to ESD as HESD.

The decade from 2005 to 2014 has been declared the United Nations Decade of Education for Sustainable Development. The initiative is presided over by UNESCO and is being promoted in every UN member state. United Nations University also approves regional centres of expertise (RCE), which provide ESD to regional communities, and each RCE involves various local stakeholders. These include schools, universities, local governments, businesses, non-profit organizations, non-governmental organizations and community education facilities.

Through this regional involvement, children, too, are beginning to learn through encounters with real problems, sometimes being taught by specialists, sometimes presenting their own opinions in the community. The usefulness of conventional learning will also be rediscovered as a way not merely to advance to a higher-level school but also to provide knowledge, or the necessary foundation for such knowledge, for overcoming real problems. Students wishing to be involved as specialists in creating a sustainable society can then enrol in sustainability education programmes at university or graduate school.

At this stage, students engaged in sustainability education will also learn through participation in the local community, an aspect of pedagogy mentioned earlier in this section. In the role of aspiring experts, they are likely at times to be involved in the education of others within the community, and will thus feel a sense of responsibility and fulfilment as they learn.

After graduation or completion of their studies, these new experts endowed with the abilities demanded by sustainability education can expect to participate in the education of people from a range of sectors in the local community. At times, they may also be involved in the education of the next generation of undergraduate or graduate students, and they will also naturally participate in learning as members of the community.

Universities that provide sustainability education need not only to improve their existing knowledge-transmission-based teaching but also to participate actively in this sort of community development. Outside the university, the spaces where practical learning takes place in the local community are spaces not only where undergraduates and postgraduates learn but also where people learn before becoming students and where people simultaneously work and learn after completing their studies. Viewed in terms of the continuum of learning, the learner acquires expertise in some subject at university or graduate school as part of the entire process of learning in the community, and thereafter continues to participate in the community as a learning member.

If communities grow in this way, the simplistic model of a talented elite being educated in the leader-training institutions of higher education and going on to lead and reform society will gradually become obsolete. For example, in Japan, where 74 per cent of high school graduates go on to higher education (53.8 per cent to university or junior college, both statistics as of 2007), a model is needed that accounts for the fact that most of the people making up a community have received some form of higher education. Since higher education is in reality positioned as the final stage of a citizen's education, there is the potential for adopting more direct and organized strategies.

Sustainability science is a new integrated domain of knowledge. In the pedagogy of sustainability education, the various theories and applications developed in its sub-domains should also be integrated, and a diversity of effective teaching methods that meet various goals should be recognized.

Notes

1. Information on service learning can be found at: <http://www.servicelearning.org/> (accessed 18 June 2010).
2. Information on the jigsaw classroom can be found at: <http://www.jigsaw.org/> (accessed 18 June 2010).
3. Information on the North Vancouver Outdoor School can be found at: <http://www.nvsd44.bc.ca/programs/outdoorschool.aspx> (accessed 18 June 2010).
4. Information on dilemma stories can be found at: <http://www.dilemmas.net.au/> (accessed 18 June 2010).

REFERENCES

Aronson, E. (1978) *The Jigsaw Classroom*. Beverly Hills, CA: Sage Publications.

Flavell, J. H. (1976) "Metacognitive Aspects of Problem Solving", in L. B. Resnick (ed.), *The Nature of Intelligence*. Hillsdale, NJ: Lawrence Erlbaum, pp. 231–236.

Freire, P. (1970) *Pedagogy of the Oppressed*. New York: Continuum Publishing.

Ishikawa, K. (2008) "Contents, Methods and Curriculum", in H. Tanaka (ed.), *Development Education: for a Sustainable World*. Tokyo: Gakubunsha, pp. 19–33 (in Japanese).

Jacoby, B. (1996) *Service-Learning in Higher Education: Concepts and Practices*. San Francisco: Jossey-Bass.

Kakuta, N. and ERIC (1999) *Training Manual for Facilitators in Environmental Education: To Promote, Perform, Create and Participate*. Tokyo: International Education Resource & Innovation Center, appendix (in Japanese).

Lave, J. and E. Wenger (1991) *Situated Learning: Legitimate Peripheral Participation*. Cambridge: Cambridge University Press.

Mezirow, J. (1991) *Transformative Dimensions of Adult Learning*. San Francisco: Jossey-Bass.

Settelmaier, E. (2009) *"Adding Zest" to Science Education: Transforming the Culture of Science Classrooms through Ethical Dilemma Story Pedagogy*. Saarbrücken, Germany: VDM Verlag.

Sobel, D. (2004) *Place-Based Education: Connecting Classrooms & Communities*. Great Barrington, MA: The Orion Society.

Tuckman, B. W. (1965) "Developmental Sequence in Small Groups". *Psychological Bulletin* 63(6): 384–399.

UNESCO (2005) *Teaching and Learning for a Sustainable Future: A Multimedia Teacher Education Programme*, Version 4.0. Paris: UNESCO. Available at <http://www.unesco.org/education/tlsf/> (accessed 18 June 2010).

6-4

Key concepts for sustainability education

Motoharu Onuki and Takashi Mino

Section 6-2 of this chapter discussed the nature of the competencies desirable for sustainability education, which include the ability to understand the diversity of sustainability-related factors and academic disciplines and the complexity of their interactions, the ability to view sustainability problems as a whole system, the ability to think in a transdisciplinary way (that is, thinking that transcends the boundaries of one's own discipline and culture), the ability to communicate, interpret, present, facilitate and form consensus across different disciplines, cultures and languages, and, last but not least, various relevant social skills. Also discussed was why sustainability education should be approached in a problem-oriented, project-oriented, case-oriented manner addressing specific questions or topics that bear relevance to society. This section will review the questions "What is a sustainable society?" and "What is needed for building a sustainable society?" while introducing four key concepts for sustainability education: dilemmas, detachment, dynamics and diversity.

6-4-1 Noticing dilemmas

Problems related to sustainability are typically characterized by a complex yet dynamic interplay of interests, incentives and causalities among multiple stakeholders in various geographical, historical and cultural contexts. For this reason, it is frequently the case that a strategy focusing on just one part of a problem may give rise to another problem elsewhere.

Put simply, by overemphasizing single aspects of a problem, one risks losing sight of other less immediately discernible aspects. When confronted with different perspectives (how one frames a problem), one may find oneself facing dilemmas in one's approaches and solutions. It is important to bear in mind that dilemmas often arise between development and the environment, between global and regional perspectives, and between solutions advanced by the natural sciences and the social sciences.

To solve problems related to sustainability it is therefore critical to develop an integrative understanding of a whole system, rather than attempting to grasp multiple discrete factors that are complex and dynamic in nature. By dealing with a whole system in an integrative way, one can gain insights into questions such as "When one of these factors changes, how will it affect other factors?" and "What might be the extent of such an effect?" To this end, it is worthwhile to promote systemic thinking, which involves exhaustively identifying any relevant factors, describing their interactions and resulting effects, and observing the dynamics of the system as a whole. By comprehensively surveying all relevant factors and characterizing their interrelations, it is possible to view a problem from an advantageous, global perspective. A capacity for systemic thinking from such a global perspective is crucial to gaining an understanding of sustainability-related problems.

6-4-2 Achieving detachment

In dealing with dilemmas and approaching problems by systemic thinking from a holistic perspective, a key prerequisite is the ability to perceive that there exist multiple propositions, interpretations and methodologies, even among actors and knowledge domains seeking to address the same problem or provide solutions. Conventional academic disciplines have evolved by prescribing and predefining their own specific methodologies and targets, and no discipline has existed before that deals with sustainability-related problems as a whole. It is therefore vital to integrate findings from discrete conventional academic disciplines through collaborative input by different specialists. The importance of multidisciplinary work, interdisciplinary work (which emphasizes the overlapping boundaries of knowledge domains) and transdisciplinary work (which aims at achieving synergy and integration across conventional disciplines) has been widely recognized. In attempting to build a transdisciplinary platform that enables integration across essential disciplines through a transition from multidisciplinary cooperation, scientists need to demonstrate the ability of detachment. In other words, although they need to retain a foothold in their own line of thinking (which, in many cases, is dependent on a specific area of expertise and cultural context), they also need to raise their vision to a certain height in order to view the whole system. At the same time, they need to have a thorough understanding of the methods and solutions typically adopted in conventional disciplines, even as they adopt a multi-angled perspective. By listening to the ideas of people from other fields of specialization or different cultural backgrounds and by transforming one's viewpoint, one can promote transdisciplinary interactions and achieve a broader understanding through systemic thinking while maintaining a global perspective on potential dilemmas.

It is often the case that, when one focuses on something with a set pattern of thinking, one loses sight of other ways of perceiving it. In working with specialists from other fields of expertise, it is important to bear in mind that the process of understanding, discussing and presenting solutions to a problem may not always be straightforward, and to exhibit an introspective flexibility that allows one to check one's own narrow views in light of the views of others. When attempting to understand a complex and diverse mix of stakeholders, factors and causalities through systemic thinking, and to integrate inputs from multiple disciplines, there is a particular risk that certain significant phenomena may be overlooked as trivial. Thus, it is necessary to be constantly aware of the perspective each stakeholder brings to a given problem.

6-4-3 Visualizing dynamics

Achieving an understanding of the qualities essential to understanding the diversity, complexity and dynamics of an interconnected body of subjects, factors and causalities is by no means easy. It is, therefore, necessary to exploit different investigative approaches and to mobilize all relevant disciplines into collaboration. Does this mean, however, that once one achieves a grasp of this diversity, complexity and dynamics, and even succeeds in overcoming them, one will be able to build a sustainable society?

On the contrary, it is unthinkable that human societies can survive over a long period as stabilized or static societies. The reason is clear. A society and its human activities are dynamic in nature, with transformational events occurring from time to time. The forces that transform societies are endogenous within the societies themselves. The term "sustainable society" is used here to refer to a society that contains a system capable of continuously and autonomously regulating its own dynamics.

In the modern age, economic incentives (that is, the pursuit of interests by mass production and mass consumption at the expense of the exploitation of resources) are the current driving forces that transform and drive our society. However, a system operating on mass consumption and mass production will one day run out of steam. At this juncture, it is particularly worthwhile to seek an alternative to economic incentives as the transforming force for society.

6-4-4 Diversity supports sustainability

What is meant by alternative driving forces for a society other than economic incentives? Kates et al. (2001: 641) have defined the essence of sustainable development as "meeting fundamental human needs while preserving the life-support systems of planet Earth". If the goals of a society are to be recast from the pursuit of economic prosperity to the quest for environmentally sustainable human development and the improvement of quality of life, what would be the driving forces for such a society?

There is no easy answer to this question. Nevertheless, once again, the clues may be found in diversity and complexity. It is said that globalization has resulted in diminished diversity (not only biodiversity in nature but also cultural diversity in human society). In modern societies in which economic growth through resource consumption is a driving force, diversity and complexity are seen as an implacable enemy of efficiency. However, in societies where economic growth has failed to become a driving force, some residual form of diversity may survive. This residual diversity may have the potential to become a driving force for environmentally sustainable human development. Opportunities to be exposed to the different temperaments and philosophical outlooks of other people and to find new resources in other regions generate a new kind of dynamics, one that is required for a sustainable society. In short, it should not be part of society's grand design to understand diversity, complexity and dynamics in order to reduce them; rather, the goal should be to conserve diversity and complexity by integrating them so as to generate new forces of transformation and bring about change. It is hypothesized that this process may be the very key to the social dynamics that will drive society and promote sustainability.

In educational practice, it is productive to promote training that deals directly with the questions "Is it sound strategy to enhance diversity further?" and "Will there be incentives at work if diversity is further promoted?" These questions are among the basic standards of evaluation. It is reasonable to assume that driving forces towards dynamic change will emerge if people can be produced who are adept at understanding and making discoveries about diversity in new ways.

6-4-5 Summary

This section has discussed the significance of dilemmas, detachment, dynamics and diversity as key concepts in sustainability education. These "4Ds" may be conceptualized as part of a philosophy of sustainability education. As such, they form the roots and stems of a fundamental attitude that "You do not need to accept what you do not like, but you have to respect it." In the future, it is anticipated that the incorporation and integration of these ideas into the contents and pedagogy of day-to-day educational practice will prove a timely and necessary step forwards towards sustainability.

REFERENCES

Kates, R. W., W. C. Clark, R. Corell, J. M. Hall, C. C. Jaeger, I. Lowe, J. J. McCarthy, H. J. Schellnhuber, B. Bolin, N. M. Dickson, S. Faucheux, G. C. Gallopin, A. Grubler, B. Huntley, J. Jäger, N. S. Jodha, R. E. Kasperson, A. Mabogunje, P. Matson, H. Mooney, B. Moore III, T. O'Riordan and U. Svedin (2001) "Environment and Development: Sustainability Science", *Science* 292(5517): 641–642.

6-5

Economics, development and governance in sustainability education

Akihisa Mori

6-5-1 Introduction

This section focuses on education and research involving the economics and governance of sustainable development and discusses unique features that contrast with environmental economics. It then considers what these unique features require of education in the economics of sustainable development and examines ways to address these requirements.

6-5-2 Environmental economics and the economics of sustainable development

The first textbook on environmental economics written by Japanese scholars was *Environmental Economics* by Kazuhiro Ueta and colleagues, which was published in 1991 (Ueta et al., 1991). In the two decades since its publication, various other textbooks on environmental economics by Japanese scholars have appeared, and foreign textbooks have also been translated into Japanese. Universities have introduced lectures on environmental economics and hired full-time faculty members in the field. In addition, there is an increasing number of academic associations that investigate and discuss environmental economics, law, policy, business administration and sociology, and the number of researchers and students participating in such associations has been rising.

At the same time, the research topics and fields that environmental economics covers have also expanded. First of all, greater attention is being paid to intergenerational and cross-spatial issues. In the past, environmental economics dealt with the contemporaneous effects of environmental degradation on residents in a given neighbourhood. However, as seen in transboundary pollution and climate change, an environmental problem occurring at a particular point in time potentially has intertemporal and cross-border implications; in other words, it can affect future generations as well as the entire planet. In proposing projects or policies with potential environmental impacts, it is necessary to consider the effects on and interests of not only a small set of contemporary parties, but also a broader group of stakeholders. In addition, it is necessary to take into account global effects as well as sustainability and intergenerational equity when deciding environmental targets and policy measures.

Secondly, environmental economics is now concerned not only with policy measures but also with governance and social ramifications. Many researchers have recognized that command and control, the traditional model for major environmental policy measures, has become less effective and efficient. Moreover, judicial solutions are known to be too costly. Consequently, more attention is being paid to economic instruments such as taxes and emissions trading, and voluntary approaches such as eco-labels, environmental management systems and information-based programmes. Because these instruments give wider discretion to the private sector, many have started to advocate shared responsibility. In other words, responsibility should be borne not only by the Ministry of the Environment, the central government agency mandated to deal with environmental issues, but also by various actors including other central government agencies, local governments, private companies and civil society. This has led to calls for consideration of issues relevant to environmental governance, such as participation, transparency and accountability, and to social aspects of sustainable development, such as social capital and empowerment. These two kinds of expansion in the field of environmental economics imply a need for the field to evolve into what is referred to as the "economics of sustainable development".

However, even if environmental economics becomes the economics of sustainable development, the fundamental purpose of the field will not have changed since the publication of *Environmental Economics* in 1991. That purpose is (a) to clarify the economic and institutional mechanisms that lead to environmental degradation or prevent sustainable development, (b) to explain economic mechanisms and conditions for realizing both development and environmental conservation, and (c) to design institutions and policies to assist in policy-making that enables various actors to take action. This aim stems from the fact that both environmental economics and the economics of sustainable development are problem-solving oriented. For educational purposes, however, differences in view, goal and logic should be kept in mind.

6-5-3 Differences in view, goal and logic of problem-solving

One influential environmental discourse is green neo-liberalism, which the World Bank has employed in extending sector adjustment loans. It advocates market-based policy instruments and governance under the existing market mechanism. It ascribes environmental problems to the misallocation of natural resources and thus to the undervaluation of land, forests, mineral resources and water, as well as to open access to communities' land and resources and the provision of services at prices below cost. It recognizes that the governments of developing countries have supported the lives of low-income families and have gained a large rent by intentional undervaluation and under-pricing of natural resources, and that the land and resources managed by communities have been virtually made open for access, leading to excessive use and increased illegal usage. Based on such observations, this logic calls for private property rights to natural capital, appropriate valuation and an increase in the price of services as policy instruments to curb excessive use. It also calls for enhancement of monitoring capabilities by granting communities basic rights to use environmental resources as well as the creation or reorganization of government environmental protection agencies, the establishment of national research centres for environmental policies, and the training of groups of environmental specialists as means for efficiently implementing market-based environmental management.

However, measured value depends on the allocation of rights and institutions. For example, automobiles generate negative externalities such as accidents, air pollution, pavement damage and traffic congestion. When pedestrians and bicycles have priority in using the roads, the negative externalities for which one automobile must compensate are several dozen to several hundred times greater than when automobiles have priority. This is because, when pedestrians and bicycles have priority, investment must be made to allow for automobile traffic without violating their rights, whereas such investment is not necessary when automobiles have priority (Uzawa, 1974).

Another influential environmental discourse is ecological modernization. This discourse sees environmental degradation as a structural problem that can be dealt with only by attending to how the economy is organized, but not in a way that requires an altogether different kind of political-economic system (Hajer, 1995: 25). It recognizes that market failure arising from negative externalities causes environmental problems and sees the solution as internalizing externalities within a market mechanism. Negative externalities undermine the function of the market, preventing it from achieving efficient resource allocation. The Piguvian tax is seen as a remedy for this type of market failure. Environmental capacity development is also required that integrates environmental and developmental concerns at all levels, aims to strengthen institutional pluralism, belongs to and is driven by the community in which it is based, and involves a variety of management techniques, analytical tools, incentives and organizational structures in order to achieve a given policy objective. But environmental taxes impose higher political, economic and social costs, at least in the short term, and may arouse fierce opposition. Packaging with well-functioning environmental governance is required to apply pressure on firms vertically through local residents, non-governmental organizations, consumers, stockholders and international organizations, as well as the national government, and horizontally through competitors.

However, the internalization of negative externalities does not necessarily guarantee environmental sustainability. Internalization of negative externalities leads to environmental conservation or emissions reduction up to the level where marginal cost is equal to marginal benefit, given existing technology and knowledge.

The economics of sustainable development, on the other hand, sets a policy goal of ensuring sustainability. The concept of sustainability can be classified into strong and weak sustainability. Strong sustainability calls for the preservation of the physical stock of specific forms of natural capital that are regarded as non-substitutable, that is, critical natural capital. It requires controlling human activities within the limits of environmental capacities, leaving a safety margin, and taking into account uncertainties in and ignorance of environmental impacts. This view of sustainability requires the precautionary principle and preventive measures before there are definite scientific results "proving" that protection of the environment is necessary, or the shift of burden of proof to would-be environmental disrupters to demonstrate that their actions will not result in unacceptable ecological damage.

In contrast, weak sustainability refers to a non-decreasing production base for coming generations that is composed of institutions plus an aggregate of physical capital, human capital and natural capital, or the sum of these three types of capital measured in terms of their shadow prices, that is, inclusive wealth (Dasgupta, 2007). Weak sustainability differs from strong sustainability in its assumption of infinite substitutability between natural and physical capital. Weak sustainability, or the maintenance of the level of consumption for each generation, can be achieved as long as economic rents derived from the exploitation of exhaustible natural resources are invested in other forms of capital capable of yielding an equivalent stream of income in the future. It can also be achieved after attaining a certain level of income even if the environment is damaged by excessive use of natural capital at the initial stage of economic growth.

This view leads to the logic of ecological modernization. Ecological modernization assumes the rationality of capitalism and the market as driving forces for environmental conservation. However, it differs from green neo-liberalism in that it considers firms to be the main cause of environmental damage and supports economic instruments such as environmental taxes and fees as policy measures to advance super-modernization through technological innovation and social structural transformation. Taking into account the experiences of Western Europe, ecological modernization also advocates an optimal policy mix consisting of regulations, economic instruments and voluntary approaches to give firms wider discretion, as well as the creation of an integrated, predictable and comprehensive framework for environmental regulation and management. As a way to convince firms to comply with these policies, this logic calls for higher environmental awareness in civil society and the participation of diverse actors for efficient environmental governance, as well as the creation of ecological lead markets.

6-5-4 Different logics for poverty and environmental degradation

The debate on poverty and the environment provides a good example for students to learn about the above differences in the prevailing discourses. Many people living on land that is infertile, dry, unsuitable for cultivation owing to steep slopes, ecologically vulnerable or prone to floods or other natural disasters are forced to live in severe poverty. In regions with a large population living on ecologically vulnerable land, people tend to overuse such land, rendering it ecologically unrecoverable in the future. This makes people poorer and further accelerates environmental degradation. This is referred to as the "poverty–environment trap".

Traditional views have assumed that the poverty–environment trap is caused by the livelihood of the poor. In other words, poor people in rural areas live in an ecologically vulnerable region, depend heavily on natural resources and do not have alternative means to support their lives. They often engage in low-productivity agricultural practices such as shifting cultivation and slash-and-burn farming. Also, their attempts to compensate for high child mortality and short life expectancy cause relatively high birth rates and population growth. This in turn increases the number of poor people while access to productive land remains limited. People have no choice but to overuse natural resources in order to support their lives and to escape from poverty. A short-term, myopic perspective leads them to abuse these natural resources without making proper investments in them. This results in deforestation, soil degradation, destruction of watershed and vegetation and other environmental damage, which in turn lead to the loss of livelihood because of a rise in physical damage and human disasters caused by floods and droughts, a fall in agricultural productivity and a decrease in income from forest products. To sustain their livelihood, poor people depend further on natural resources, accelerating environmental degradation. Or they may migrate to cities and form urban slums in ecologically dangerous areas, such as the neighbourhoods around factories, further degrading the sanitation of cities.

Based on the above assessment, this view calls for the control of population increases and of short-sighted practices as the means of eliminating the poverty–environment trap and regards economic growth and the assignment of private property rights to land as the most effective policy instruments to this end. The logic behind green neo-liberalism is derived from this view.

In contrast, a more recent view argues that, even if the poverty–environment trap has in fact been growing worse and is caused by the poor, the responsibility rests not only with them but also with institutions and policies. Not uncommonly, and often through their past experiences and traditional local ceremonies, customs and folklore, the poor understand the negative impacts of environmental degradation on their health and livelihood as well as the significant positive effects of access to natural resources and the quality of the environment on their ability to maintain their livelihood. These people have an incentive to conserve the environment. However, the economic rent obtained by exploiting natural resources is mainly distributed to the rich and is used for further exploitation of those resources; it is rarely used to accumulate assets for and reduce the vulnerabilities of the poor. This uneven distribution of wealth, together with the voicelessness and powerlessness of poor people, drives them to the intensive use of natural resources and consequently into situations in which they have to destroy their own assets. In countries where the government does not legally recognize the community's traditional entitlement to common-pool resources, the poor lose the means to mitigate vulnerabilities such as bad weather and natural disasters. In addition, countries in need of funds for new development or to overcome foreign debt have been forced to accept and implement policy reform packages based on the logic of green neo-liberalism advocated by the World Bank and other multinational development agencies in exchange for financial assistance.

The above view suggests entirely different policy implications. The most effective policy is not to engage poor people in activities leading to economic growth, but to increase their assets and reduce their vulnerabilities. This logic calls for the empowerment of local communities and the restructuring and enhancing of traditional regional networks. At the same time, it requires policies and institutions that make the accumulation of assets easier for the poor, including granting and protecting clear and enforceable property or usage rights of local communities and user groups to land and common-pool resources, offering social services and goods that the private sector cannot provide, and improving the transparency of decision-making and accountability. Furthermore, it requires debt reduction and the redesign of rules for international trade to recover the self-decision capacity that governments have been deprived of in the process of debt repayment, structural adjustment and globalization.

6-5-5 An implication for pedagogy

Even if environmental economics and the economics of sustainable development are oriented towards solving problems, they will fall into the category of mere knowledge rather than guidance if students are taught by means of lectures. Students need to learn through an actual decision-making process, but it is rare for students to come across situations where they have to make decisions in real society, even if they undertake an internship.

The case method of instruction offers students the opportunity for simulated experience. It was originally developed as a pedagogy for law school and master's programmes in business administration to identify optimum decisions in a specific context. Usually, a case is described before or during a lecture, along with the backgrounds, strategies and positions of important stakeholders. Through the analysis of context, causes, risks and stakeholders, and through group study and discussion, students are required to propose alternatives or to evaluate decisions. Projects, programmes and policies on the environment and sustainable development can serve as cases for instruction, although their contexts, stakeholders and performances are much more complex and obscure than those of business administration or court cases.

Cases are often taken from decision-making in the past. They contain a variety of views, logic and options that students could use in a specific context, as well as the consequences of that decision. More often than not, however, instructors face difficulties in finding cases that fit their instruction purposes in existing textbooks. They have to seek out cases. Finding new cases necessarily entails evaluation, which does not exist independently of the views and logic behind the decision-making. In environmental economics, evaluation often is concerned with efficiency, employing cost/benefit analysis of a project or policy and valuation of the environment. In the economics of sustainable development, however, evaluation includes not only efficiency but also relevance and effectiveness in terms of the degree of achievement, impact and sustainability, as proposed by the Organisation for Economic Co-operation and Development in relation to development assistance. Recent evaluation emphasizes legitimacy in terms of participation, transparency and opportunities for presenting opinions, as well as processes of stakeholder empowerment and trust-building, vision-sharing and usefulness for policy learning (Crabbé and Leroy, 2008).

The case method of instruction can also help students prepare for field studies. Recently, many universities have included field studies and internships as part of their curriculums. In reality, however, students may easily become fed up with the challenges of the field, lose sight of the focus of their study and end up engaged in aimless surveys. Case studies and the case method of instruction will train students in the methods they can employ to understand and analyse specific fields and cases. However, few teaching materials for case studies and the case method of instruction have been developed so far regarding the environment and sustainable development. Even fewer evaluate cases that consider the three pillars of sustainability: environmental, economic and social sustainability, which in reality can be inconsistent. It is imperative that teaching materials are developed that directly focus on proposals and evaluations that address these three pillars in the case method of instruction.

6-5-6 Conclusion

This section has focused mostly on the economics of sustainable development. The economics of sustainable development is oriented towards solving problems and aims to support policy-making, but it currently involves different logics, including green neo-liberalism and ecological modernization. This is because no universal logic has been established for achieving environmental sustainability while simultaneously enhancing human development and social sustainability. Although no universal logic yet exists to deal with these various problems, case studies and the case method of instruction can be effective teaching methods to prepare students for in-depth fieldwork by instructing them in analytical methodology and allowing them to undergo simulated experiences. The development of teaching materials for the case method of instruction relevant to sustainability remains a challenge.

REFERENCES

Crabbé, A. and P. Leroy (2008) *The Handbook of Environmental Policy Evaluation*. London: Earthscan.

Dasgupta, P. (2007) *Economics: A Very Short Introduction*. Oxford: Oxford University Press. (Japanese translation by K. Ueda, R. Yamaguchi and Y. Nakamura. Tokyo: Iwanami Shoten.)

Hajer, M. A. (1995) *The Politics of Environmental Discourse: Ecological Modernization and the Policy Process*. Oxford: Oxford University Press.

Ueta, K., H. Ochiai, Y. Kitabatake and S. Teranishi (1991) *Kankyo Keizaigaku* [*Environmental Economics*]. Tokyo: Yuhikaku Books (in Japanese).

Uzawa, H. (1974) *The Social Cost of Automobiles*. Tokyo: Iwanami Shoten (in Japanese).

6-6

Practices and barriers in sustainability education: A case study of Osaka University

Michinori Uwasu, Michinori Kimura, Keishiro Hara, Helmut Yabar and Yoshiyuki Shimoda

6-6-1 Introduction

Behind the recent recognition of sustainability as an educational theme are growing concerns about climate change and the increasing environmental impacts of human activities. Sustainability education is essential for building sustainable societies that can overcome global environmental problems in relation not just to the environment but also to existing social and economic systems. However, conventional education at universities has not yet fully addressed the implementation measures required for developing the human resources that can contribute to resolving these problems. Conventional specialized instruction in the modern natural sciences and social sciences is representative of Descartes' reductionism in that its objective is to acquire the necessary skills from the knowledge systems developed through this education, but in doing so no fundamental attempt is made to answer questions beyond the scope of such systems. Hence something is needed that goes beyond the boundaries of conventional professional education in order to provide an education that offers comprehensive solutions and visions pertinent to the issues addressed by sustainability.

Although the rationale behind the development of sustainability science as an academic field has been established, it is necessary to identify the elements of sustainability education that distinguish it from conventional professional education and to understand how these elements can be implemented in reality. This section introduces the barriers encountered when implementing sustainability education at Osaka University, as well as ideas for overcoming these barriers, and discusses the issues and significance of sustainability education in that context.

6-6-2 Barriers to sustainability education programmes

At the Integrated Research System for Sustainability Science (IR3S), the concepts and implementation of sustainability education are being steadily advanced in tandem with the development of sustainability science. However, no matter how worthy the educational goals are, major barriers arise when promoting innovations in education, notably the aversion of systems to change (that is, university organization) and habits (that is, faculty/student awareness). IR3S is developing and operating sustainability education programmes appropriate for universities, but the objectives, concepts and systemic conditions of the programmes have a significant effect on the actual form taken in their educational implementation.

This subsection discusses the barriers encountered in implementing sustainability education at Osaka University in Japan. An educational system called the "advanced associate programme system", comprising a set of programmes, has been established in recent years at Osaka University as a way to meet society's need for diverse human resources with broad knowledge and flexible thinking. This effort has included developing sustainability-related holistic competencies reflective of IR3S educational concepts/goals and relevant to training students specializing in sustainability and the environment, as well as providing for interdisciplinarity and transdisciplinarity among academic disciplines. Several such education programmes have been established, including one entitled "Sustainability Science".[1]

However, there are many barriers to this kind of interdisciplinary/transdisciplinary education in graduate school study, which is traditionally intended to deepen, not broaden, expertise. At Osaka University, where the emphasis in sustainability education is on integration of the humanities and the sciences, the following issues were seen to arise in the promotion of this course of study.

The first issue is the problem of perceptions in Japan of the separation of the humanities and the natural sciences. More than 80 per cent of high schools that have a high proportion of students advancing to university education require their students to choose between a humanities or a science course of study for their high school curriculum (Mainichi Shimbun Science and Environment Department, 2007). Therefore, student awareness of the difference between the humanities and the sciences is extremely high in Japan compared with other countries. Moreover, in the

1990s, many national universities abolished cultural studies in their undergraduate education programmes, and this lack of liberal arts programmes may have served to further intensify perceptions of the difference between the humanities and the sciences.

A second issue is the fact that many sustainability, environmental and engineering courses are offered at specialized graduate schools in Western countries within social science disciplines (for example, business management and political science). Because the method of education in these cases is mainly practical education and most of the students are working members of society, this method of education would not be possible if barriers were erected separating the humanities and the sciences. Furthermore, other kinds of professional graduate schools (clinical psychology, social work, etc.) also appear to achieve a strong fusion between the humanities and the sciences. The presence or absence of this kind of cultural backdrop in higher education appears to significantly affect how well integrated education is promoted beyond conventional education.

Third, there are marked differences in the meaning of education between humanities and science programmes in graduate schools, especially master's programmes. In the sciences, graduate master's programmes are on a continuum with undergraduate programmes. Many students advance to graduate programmes before entering society as engineers in the corporate world, for example. Advancement to a doctoral programme is necessary for a career as a full-fledged researcher. In the humanities, however, most students enter the world of work after finishing a four-year undergraduate programme, and it is generally accepted that humanities students who advance to graduate programmes are preparing for careers as specialists. In the Osaka University School of Engineering, for example, there are on average 923 first-year undergraduate students and 843 first-year graduate students (91.3 per cent of the undergraduate average). In the School of Economics, on the other hand, there are on average 261 first-year undergraduate students, compared with 86 first-year graduate students (33.0 per cent of the undergraduate average).[2]

From the above, it may be surmised that humanities students tend not to show an interest in interdisciplinary education outside their own area of expertise because they already perceive themselves as specialized researchers. This trend was confirmed in interviews with faculty members at Osaka University teaching integrated humanities/science education courses. It seems that graduate students regard learning about advanced specialized fields and learning about interdisciplinary fields such as sustainability science to be mutually contradictory activities. However, what is important for society is not so much the state of education at universities or student knowledge, but rather student flexibility and the ability to work across different fields.

First-year students at Osaka University usually show an interest in sustainability and environmental issues because the university offers them introductory lectures and seminars on those topics. However, as the students work through their undergraduate major coursework, pre-graduation research and graduate research, and become specialists in a specific field, their perspective narrows and they show less interest in interdisciplinary/integrated education as they consider their career paths. Work must be done to ensure that students recognize the needs of society as described above. This is an important problem from the perspective of ensuring student diversity and maintaining student motivation, not just at Osaka University or Hokkaido University, both of which spread their sustainability education programmes across their entire curriculums, but also in full major programmes in sustainability science at schools such as The University of Tokyo.

6-6-3 Ideas to overcome barriers

From the above it is evident that ideas for enhancing the effectiveness of sustainability education for students are needed in order to conduct sustainability education under current conditions and move towards achieving its major objectives.

In the sustainability education programme at Osaka University, even lecture classes include time for discussions among students on diverse topics to raise students' awareness of the limitations of their own disciplines and to improve their ability to communicate in different fields. The education programme also encourages students to acquire basic knowledge in different disciplines through group work on themes demanding interdisciplinary competency. Additionally, students participate in field trips where sustainability science is put into practice and the importance of participation in society at large is emphasized. Efforts are also made to show the students the importance of being able to flexibly integrate knowledge from different fields in order to solve problems effectively. It is important to interest students in the demands and expectations of society with regard to sustainability science as well as in what they will learn and what they will be able to contribute to society through such learning. To this end, the authors, who provide sustainability education training at Osaka University, have interviewed new students about the specific details of their interest in the programme and what they hope to learn, so as to gain a better understanding of new students' responses to the programme.[3]

First, the primary motivations of the students can be summarized as falling into three response types: "interested in the environment", "want to broaden horizons" and "want to think about what sustainability is".

Although most students wanted to learn about how sustainability is specifically related to their own research, some were interested in such topics as network-building and involvement in international activities (such as international short education programmes and English-language lectures). When asked, from an integrated educational perspective, about why they wanted to learn about something outside their academic field, most students had no particular awareness of the integration between the humanities and the sciences but simply responded that they "didn't really know what students study in the humanities (or the sciences), and wanted to try studying that".

From specific responses one can see the need for science students to understand the knowledge and perspective of fields such as economics or law (for conducting cost/benefit analysis or emissions credit trading, etc.), and for humanities students to learn about environmental engineering (that is, technological development and dissemination). This highlights the difference between the humanities and science perspectives. For example, some students were interested in learning, from a humanities perspective, about what competing solar cell technologies existed and which technologies were not adopted. However, many members of the science faculty researching solar cells did not broach the subject of comparing technologies, even though they were comfortable talking about the specific technologies from the standpoint of electricity generation mechanisms. The gap in expertise that has arisen between the humanities and the sciences regarding popular themes presents a significant issue for the quality of sustainability education.

Because sustainability education incorporates broad themes and multiple perspectives, lectures are frequently presented in omnibus style. However, the survey showed that this variety often causes confusion for the students. Although it is desirable for faculty members to have diverse perspectives with regard to sustainability, there should be at least some shared core of sustainability science. This means that faculty development is an important educational concern. The gap between student and faculty perspectives may be caused by a failure on the part of faculty members to examine their own areas of expertise in a macro/objective fashion. However, many faculty members are researchers with specialized teaching experience and have not had this kind of training (or, even if they have, it is mostly on an individual or private level). The detachment and holistic competencies that are valued in sustainability education are excellent examples of this perspective at work. Thus it is important that not only students but also faculty members endeavour to improve their holistic competencies.

Finally, some students use sustainability education as a tool for improving their practical competencies. In addition to studying English or another foreign language, developing presentation skills and acquiring an environment-related education, another incentive for students may be the fact that more corporations in recent years have introduced sustainability report publications. This suggests that some students are studying sustainability as a job hunting strategy to make themselves more attractive to employers. To answer this type of demand, the education programme of the Research Institute for Sustainability Science includes lectures, discussions and seminars in English. Lecturers from universities both inside and outside Japan, including the IR3S member universities, as well as from corporations, international organizations and non-profit organizations, are actively invited to lecture or to hold seminars. Workshops are also held on the topic of sustainability education itself, presenting an opportunity to discuss trends in sustainability education overseas or the status and role of international organizations with regard to global environmental issues. These efforts are thought not only to meet the needs of students but also to contribute to spreading awareness about sustainability science among campus faculty members.

6-6-4 Improving sustainability literacy through programme education

The shared core goal of IR3S is to improve holistic competencies and knowledge. This shared component goes beyond the realm of conventional education. What viewpoints and knowledge, then, are specifically needed? At Osaka University, the aim is to define the improvement of holistic competencies and knowledge as sustainability literacy and to improve literacy not only for all university students but among faculty members as well. As is discussed in detail in other sections, sustainability literacy comprises many elements, all of which are important. In terms of its status as a minor programme at Osaka University, these elements are treated as problems and issues that involve especially ill-defined and complex systems. Therefore it is important to emphasize the development of competencies that will facilitate the improvement of students' skills and knowledge in sustainability issues.

Specifically, through methods such as exercises and discussion, students are encouraged to deepen their understanding of the cause-and-effect relationship among problems such as trade-offs, approaches to backcasting (scenario based) and the role played by policies and technologies, as well as to improve their communication skills within other fields (Uwasu et al., 2009). For instance, through group work problems, students build visions and scenarios characteristic of sustainability science in the programme's core subjects.[4] Vision-building is not something done in conventional professional education, but understanding the significance of building visions is one way to experience the benefits of sustainability education. A specific example is a group discussion on future visions and scenarios for themes such as building low-carbon-emission cities.

According to a survey questionnaire administered after one such exercise, most students responded that they felt a holistic perspective, a lifecycle perspective on products, a long-term perspective and dialogue/communication skills were important competencies in understanding sustainability. The survey results showed that students thought the exercise was a good opportunity to expand their holistic perspective and communication abilities. Exercises such as these are thought to contribute to an understanding that it is insufficient to build a social vision based only on environmental sustainability, that a single problem has many different viewpoints, that there are opportunities to contribute to the formation of this vision no matter what one's major field of study, and that reaching consensus is important.

Excursions can deliver similar educational advantages. To date, Osaka University has commissioned field trips to Hyogo Eco Town (autumn 2008) and a home appliance recycling centre (summer 2009). The important point of these excursions is that "a picture is worth a thousand words". Simply seeing manufacturing processes and experiencing advanced levels of engineering first-hand has an impact on students from any field of study. However, the experience of interacting with on-site engineers and other programme participants holds even greater significance to programme students. For instance, even if they already understand that recycling is promoted for the benefit of resources and the environment, talking with site personnel helps them further understand that there are many issues involved other than the environment, such as employee working conditions or the relationship between the factory construction process and the local district. Even if they understand the benefits to the environment from a macro perspective, they can learn that a variety of socioeconomic issues exist on the micro level.

The on-site visit and group work activities offered by the sustainability education programme at Osaka University are undertaken primarily according to the problem-based learning (PBL) method. This is a method by which activities such as group projects are first begun without providing any detailed instruction; only after the task has proceeded some way, is the significance of that task then explained in relation to the theory of sustainability science. Through PBL, students gradually learn the link between the task activity and its objective (for example, the significance of visions and scenario-building, the concept of sustainability science) (Martens, 2007). PBL is an effective teaching method to equip students with core competencies in sustainability science.

6-6-5 Discussion

Sustainability education is an interdisciplinary subject that has a strong capacity for providing supplemental education to students specializing in other fields. In addition to the breadth of the subject matter and the complexity of the problems considered in sustainability education, other important elements of this capacity are interdisciplinarity, theory and practice, internationality, local/global perspectives, norms and values (subjective perspectives) and verification (objective perspectives). However, sustainability education is above all expected to contribute to solving the problems that present barriers to building a sustainable society. For example, research on the problem of global warming is being conducted across a broad range of natural science and social science disciplines, and, in order to arrive at solutions, it will be necessary to interpret all of this research comprehensively, then use it to help form a social consensus. Therefore, sustainability education is extremely relevant to the future handling of issues that go beyond existing academic systems, in that these will likely take on a stronger flavour of interdisciplinary or transdisciplinary education (Lattuca, 2001).[5]

As a way of promoting this transdisciplinary approach, the sustainability science effort at IR3S is gradually developing distinctive theories and frameworks of thought such as knowledge-structuring and backcasting. Currently, the speed at which academic fields are fragmenting and specializing is alarming, and no one person understands the total body of knowledge in each field. Obtaining the holistic capabilities that are one of the core competencies promoted by IR3S is a significant form of competency development for actively utilizing existing expertise and knowledge, and this competency is thought to contribute to structuring knowledge for effective use in building a sustainable society.

However, in order to implement transdisciplinary professional education, it is important that there be a shift in awareness on the part of faculty members as well as in the academic environment overall. For instance, in addition to the importance of holistic capabilities, faculty members must understand teaching methods such as PBL that are not frequently used in professional education. Faculty members who teach sustainability also have dual roles teaching in other disciplines and are frequently appointed for limited terms. This presents significant problems such as policy and organizational barriers. Nonetheless, from a long-term perspective, it is clear that faculty development is extremely important for promoting sustainability literacy. At IR3S, involvement in multiple activities (not just teaching but also participating in research and symposiums) could be considered a form of faculty development. In actuality, faculty members deeply involved in IR3S sustainability programmes often have difficulty communicating closely with dual-role faculty members, so an organization such as IR3S plays an important role as a communication platform for sustainability educators.

Finally, the importance of networks as an element of sustainability education that goes beyond conventional professional education is considered. IR3S is fulfilling an important mission as a network by establishing a collaborative academic programme, and is especially significant for its role in promoting such a programme. While maintaining programme diversity, five IR3S member universities have set up courses as part of this collaborative sustainability science programme with shared concepts and shared subjects, and it is important that this movement is growing both inside and outside Japan. In Europe, the Copernicus Network already facilitates collaboration among universities for sustainable development. Currently over 320 universities in 38 European countries participate, and practical teaching and research collaboration is ongoing within the network to encourage sustainable development (Copernicus Campus Sustainability Center, 2006). It will be difficult to build such a network in a culturally and linguistically diverse region such as Asia but, when considering how to promote sustainability in Asian countries that do not participate in Western academic networks, the IR3S network in Japan has particular significance given Japan's key role in Asia. The authors look forward to the expansion of sustainability education and broadening collaboration throughout Asia.

Notes

1. Its official name is the Osaka University Graduate School Advanced Associate Program System. In April 2008, 10 interdisciplinary programmes were launched, including "Sustainability Science"; 20 programmes had been launched as of 2009.
2. As of May 2008.
3. To date, 21 students were interviewed in 2008 and 14 students in 2009. More than half of the students had an engineering background, but the backgrounds of the other students were diverse, with students from physics, economics, business management, public health and the human sciences.
4. By demonstrating the certainty and actual likelihood of events based on visions that have a scientific foundation, a sense of hope and confidence can be imparted to companies and households. This sharing of hope and confidence could serve as a driving force to change society.
5. Akashi (1997) similarly classified interdisciplinary research. According to Akashi's classification, sustainability science is thought to fit into the group "sharing new knowledge through joint work in multiple academic systems".

REFERENCES

Akashi, H. (1997) *Introduction to Interdisciplinary Research: Keywords for the Ultrainformation Age.* Tokyo: Cosmo Two One (in Japanese).

Copernicus Campus Sustainability Center (2006) *Copernicus Guidelines for Sustainable Development in the European Higher Education Area: How to Incorporate the Principles of Sustainable Development into the Bologna Process.* Oldenburg, Germany: Copernicus Campus.

Lattuca, L. R. (2001) *Creating Interdisciplinarity: Interdisciplinary Research and Teaching among College and University Faculty.* Nashville: Vanderbilt University Press.

Mainichi Shimbun Science and Environment Department (2007) *Life in the "Sciences": Sciences White Paper 2.* Tokyo: Kodansha Bunko, pp. 13–59.

Martens, P. (2007) "Problem-based Learning", Maastricht University (unpublished).

Uwasu, M., H. Yabar, K. Hara, Y. Shimoda and T. Saijo (2009) "Educational Initiative of Osaka University in Sustainability Science: Mobilizing Science and Technology towards Sustainability", *Sustainability Science* 4: 45–53.

6-7

Field study in sustainability education: A case from Furano City, Hokkaido, Japan

Nobuyuki Tsuji, Yasuhiko Kudo and Noriyuki Tanaka

6-7-1 Introduction

Understanding problems from a global perspective is essential to the examination of sustainability associated with such issues as global warming, deforestation, self-sufficiency in energy and food, and population problems such as an ageing population. However, sustainability is an extremely broad concept, and considering issues only from a global perspective does not necessarily lead to improvements in sustainability for specific local regions. Therefore, in the midst of accelerating globalization, it has become crucial to discuss how local regions, which are affected by globalization in no small way, should make efforts towards achieving sustainability.

Even in regions enjoying a relatively ideal process of development, various changes have unavoidably resulted in the increasing diversification and segmentation of regional actors. The broad range of the sustainability concept includes a wide variety of actors, and thus even within a given locality it is difficult to acquire a sufficient grasp of the relevant actors and their vested interests. This creates a tendency towards inconsistency in policy design and implementation that often hinders the realization of sustainable regional development. A discussion of issues concerning regional sustainability would thus contribute significantly to the advancement of sustainability science and would also be immensely important in terms of sustainability education.

Conducting field studies effectively and continually in a particular region is, however, difficult without a sound, mutually beneficial relationship between the region and a university. In the example discussed in this section, the Sustainability Governance Project (SGP) of Hokkaido University established a friendly relationship with Furano City in Hokkaido (see the Appendix for an overview of the city) as part of a research project, and was able to collaborate in training people who would contribute to the revitalization of a regional city by offering basic data useful for improving the sustainability of the city as well as the results of a survey of residents' opinions. The case presented here offers know-how for the development and implementation of a sustainability education programme with a local partnership that can be applied to other countries and regions.

6-7-2 Overview of the implemented programme

SGP offered a practical course in sustainability science in which students were required to write concrete policy proposals on regional sustainability for Furano City by conducting interviews and group discussions themselves. This was one of the elective courses for the educational programme "General Survey of Sustainability Science" offered by the Center for Sustainability Science, Hokkaido University. Using the case of Furano City, the course was designed to train students to propose policies, based on local information, that would improve the sustainability of the city. The students were divided into three groups to analyse agriculture, tourism and commerce, respectively, and they were asked to consider sustainability from the perspective of each sector. The course thus centred on problem-based learning. There were 12 students (7 male and 5 female), all graduate students in a master's or doctoral programme; 9 were Japanese and 3 were non-Japanese, so both the Japanese and English languages were used in the course. Most of the students were majoring in engineering or agriculture, with the others in arts or sciences. At the time of taking the course, the students had completed two basic courses on sustainability in natural science and social science given by the Center.

The course began with an intensive two-day study session (with an overnight stay) at the Furano satellite campus of Hokkaido University. The session was conducted by three professors (including one outside professor) and by assistants with varying areas of expertise who supported class instruction and group work – one assistant professor (in ecology) and three doctoral research fellows (in agricultural economics, ethics and agricultural systems). In addition, five officials from Furano City Hall (from the agriculture, commerce and tourism, and planning sections) participated in lectures and group discussions. The study session consisted of a series of lectures on various topics by the professors, followed by an introduction to a variety of information on Furano City (geography, economic activities, historical background and general conditions of the city), advice from the doctoral research fellows, complementary explanations from the city officials and group discussions. The students were divided into groups, each well balanced in terms of nationality, field, gender and age. Comments and information provided by the city officials in the group work represented the views of practitioners in the municipal government and were deemed to have contributed to making the early-stage discussions focus on real, concrete problems.

After the study session in Furano City, the students gave several interim presentations at the university's main campus and received appropriate advice. During the same period the student groups voluntarily visited Furano to conduct field studies and interviews in relevant sectors and to collect supplementary information. Each group then proposed policies at a final presentation attended by officials from Furano City Hall and staff from the Furano Tourism Association. The students received comments from those attending as well as from their instructors and incorporated these into their policy proposals. Their final reports were sent to the City Hall and the Tourism Association. The Furano City officials and Tourism Association staff who cooperated with the programme commented that the discussions with the students were tremendously beneficial and led to new perspectives on Furano. The education programme was deemed a success, bringing benefits to both the university and local collaborators. This experience confirms the notion that a necessary condition for successful implementation of an education programme with sufficient local cooperation is to include both educational opportunities and research activities that can be expected to benefit the local region. Using similar approaches, education programmes with local partnerships can be implemented effectively in other regions.

6-7-3 Establishing local personal networks through research activities and incorporating them into education programmes

One of the characteristics of the Furano education programme was the cooperation obtained from city officials as advisers. Their cooperation was possible because SGP had already been conducting research on sustainability in local governments with a focus on Furano City. To build a material and energy flow model for agriculture, interviews were conducted at the City Hall, at farms and at the Furano branch of the

Agriculture Extension Centre. In this process, a cooperative relationship began to emerge between Furano City residents and the individuals affiliated with SGP at Hokkaido University.

A comprehensive investigation of multiple sectors was subsequently conducted, focusing on how relevant actors perceive the structure of problems that they face regarding sustainability. Research on the issues to be resolved and the feasibility of the necessary consensus-building for Furano City continued for two years. The researchers conducted individual interviews with about 20 local stakeholders in the agriculture, tourism, business, waste and recycling, social welfare, healthcare and media sectors, and clarified problems identified by different actors as well as the structure of the problems. Based on this clarification, the study listed issues that Furano City should consider and also analysed stakeholders' mutual expectations (Motoda et al., 2009). This showed the importance of developing relationships and human resources to strengthen cooperation between different sectors, government revenues, the nurturing and utilization of local culture, and interactions among local actors, in addition to activities in the agriculture and tourism sectors. A portion of the research results was submitted to the mayor of Furano City as a proposal. In addition, the research results were provided to Furano residents through public lectures and research report seminars. The research showed the critical role of cooperative relationships between the agriculture, tourism and commerce sectors – the three key sectors in this region – in driving sustainability in Furano, which led to the planning of the education programme for students. Because of the success in establishing favourable relationships in this manner, full cooperation was offered when the students conducted interviews as part of their education programme at the City Hall, Tourism Association and local Chamber of Commerce.

6-7-4 Students' final reports

The following are summaries of the final reports submitted by the student groups analysing the agriculture, tourism and commerce sectors, respectively, of Furano City.

Agriculture group

The current problems faced by the agriculture sector include a declining agricultural population, ensuring a supply of safe and trusted agricultural produce, pressure from cheap imported produce, and ensuring long-term stable management of farms. The establishment of an agricultural corporation would contribute to encouraging new practitioners to engage in agriculture, training successors, promoting special agricultural foods produced with the reduced use of pesticides and chemical fertilizers, ensuring food traceability and establishing a "Furano brand" that represents food safety and reliability, which cannot be guaranteed with imported foods. The agricultural corporation would run on a system where consumers prepay for agricultural products before they are planted. This system would provide economic stability to organic farmers whose revenues tend to fluctuate and would enable consumers to buy safe products with trust, which would in turn lead to the securing of long-term, regular customers for the corporation.

Tourism group

The problems in the tourism sector include a lack of interaction and communication between the tourism, commerce and agriculture sectors and insufficient information for foreign visitors. Furano City is characterized by its high-quality agricultural products, its nationally recognized name, an image as an eco-friendly city and abundant flowers in non-winter seasons. The students propose the following measures to realize sustainable tourism: short- or long-term farm stays, programmes for experiencing and learning agriculture, changing the attitudes of farmers and bridging between tourism and other sectors. It is suggested that an entity funded by Furano residents be established to promote the above. In addition, through the association of residents with the new entity, it is hoped that their willingness to participate in various tourism-promoting efforts would increase.

Commerce group

Currently recognized problems in the commerce sector are a decreasing population, Furano residents' low levels of satisfaction with regard to the city and the fact that tourists, whose numbers are said to reach 2 million annually, do not visit the central district of the city. The creation of jobs and the revitalization of the central district are suggested as solutions. Proposed plans include transmission of information through websites and free magazines, especially about shops in the central district; an expansion of the Challenge Shop programme through which people opening new stores are supported by the city; and an attempt to attract customers by promoting the central district as a place for art. These proposals are based on the idea of emphasizing small rather than big businesses. A statement in the report by this group – "sustainability means to keep changing" – has many interesting implications.

A common theme among these three groups is information transmission via the Internet. The agriculture group attempts to solve problems by establishing an agricultural corporation with investments from consumers and stabilizing its management through multifaceted operations involving production, sales, processing and tourism. The tourism group regards agriculture as a tourism resource and identifies residents, farmers, the city administration and the commerce sector, as well as of course tourists and tour companies, as stakeholders. Starting with a viewpoint from a particular sector, the students in the end proposed solutions from a flexible perspective, indicating that they have begun to develop a sustainability-conscious mindset.

Appendix: Overview of Furano City

Furano City celebrated its centennial in 2003. The city has relationships with The University of Tokyo and Hokkaido University: the former's University Forest and the latter's Eighth Farm have been located in the city. Also, the name of some districts in the city includes the word *Goryo* (Imperial Estate), which shows that the city has strong connections with Japan's Imperial Family. The city is located at the centre of Furano Basin, which is approximately in the centre of Hokkaido and is rectangularly shaped with east–west and north–south dimensions of 32.8 km and 27.3 km respectively. With an area of 600.97 km^2, Furano ranks thirteenth in area among the 35 cities in Hokkaido. The eastern and western parts of the city are mountainous, and the University Forest of The University of Tokyo is located in the southern part. The Sorachi River runs through the centre of the city, providing a rich natural environment.

The weather is mainly continental, with a large temperature difference between night and day as well as between summer and winter, with hot and humid summers and winters with large snow accumulations. The highest and lowest temperatures are around 35°C and –30°C respectively, with an average annual temperature of about 6°C. The city has annual rainfall of approximately 1,000 mm.

Public transportation is provided by a railway line, with trains taking two hours to Sapporo and one hour to Asahikawa. National roads connect the city with the central and eastern parts of Hokkaido, and the city is a key traffic junction. Access to air transportation is provided by Asahikawa Airport, 50 minutes away from Furano. As of March 2008, the city had a population of 24,560. This is 30 per cent less than its peak of 37,000 in the early 1960s, but the fall in population has levelled off since 1975. As the population has declined, it has also aged.

The city's major industries are agriculture and tourism, the latter being greatly influenced by skiers and television dramas featuring the region. The city has been actively engaged in reducing waste and in recycling resources in recent years. Since 2001, 14 different types of waste have been collected separately and recycled based on the principle of reduced incineration and landfill disposal, and the recycling rate has reached more than 90 per cent. The city is also known as "the navel of Hokkaido", "the city of skiing", "the city of wine" and "the city of the drama *From the North Country*".

Acknowledgements

The authors gratefully acknowledge the support and cooperation received from Furano City for education and research. The education programme was conducted as a part of SGP, Hokkaido University, and received financial support from the Special Coordination Funds for Promoting Science and Technology. Special Research Funds from the Office of the President of Hokkaido University supported part of the implementation cost of the programme.

REFERENCE

Motoda, Y., Y. Kudo, H. Shiroyama, H. Kato and N. Tsuji (2009) "Regional Sustainability and the Structure of Issues Recognized by Relevant Actors: A Case of Furano City, Hokkaido", *Shakai Gijutsu Ronbunshu* 6: 124–146 (in Japanese).

6-8

Sustainability education by IR3S universities

Takashi Mino and Yoshiyuki Shimoda

6-8-1 Introduction

The Integrated Research System for Sustainability Science (IR3S) defines sustainability science as "an academic discipline that seeks to understand the interactions between global, social and human systems, and proposes comprehensive solutions and ideas for sustainability" (Komiyama and Takeuchi, 2006: 3). Based on this definition, IR3S has been engaged in research, education and industry collaboration in its mission to build this new academic discipline. In the field of education, IR3S has also made efforts to develop a curriculum for graduate-level courses in sustainability science and to implement an education programme using that curriculum.

At the time of its inauguration, IR3S envisaged creating education programmes at its five participating universities targeting different audiences. Because of the large number of varied situations and circumstances within society in which the concept of sustainability is important in decision-making, it was considered desirable to introduce sustainability education in different fields of specialization in order to produce graduates from a range of disciplines who understood sustainability science and could apply their knowledge in those diverse situations. The IR3S sustainability education programmes that emerged offer diverse learning experiences that utilize the unique features of each university, ranging from minor programmes open to all graduate students to specialized master's or doctoral programmes leading to a master's or doctoral degree.

The sustainability education programmes at the IR3S member universities operate in close cooperation and share an educational core maintained through meetings between programme leaders. A survey of this core is presented in subsection 6-8-7. At the same time, each university's programme differs in format, objectives, teaching staff and target students. For example, The University of Tokyo and Ibaraki University have set up new graduate programmes offering a master's or doctoral degree in sustainability science, while Kyoto University, Osaka University, Hokkaido University and Ibaraki University have built their own frameworks incorporating elements of sustainability science either as a part of existing educational courses or as a minor programme. In this way, the IR3S group as a whole shares ideas and joint programmes, but each university also provides its own unique programme, defining its own target students and goals. Table 6.8.1 presents a summary of the education programmes provided by the five IR3S universities.

6-8-2 The University of Tokyo

The Graduate Program in Sustainability Science (GPSS) at The University of Tokyo was launched in October 2007 as an interdepartmental programme of the five departments in the Division of Environmental Studies (DES), Graduate School of Frontier Sciences (GSFS), with the collaboration of the Transdisciplinary Initiative for Global Sustainability (TIGS) and IR3S.[1] Because GPSS was started as a graduate programme that awards a master's degree (Master of Sustainability Science), the objective of GPSS is to educate students to be professionals and researchers who can take an active role in efforts to achieve sustainability, rather than to serve as a minor programme imparting sustainability skills to students who are majoring in other disciplines (engineering, economics, sociology and so on). The core competencies required for professionals in sustainability science are substantially the same as the add-on sustainability skills required in existing disciplines; however, GPSS focuses especially on communication, facilitation and mediation skills for connecting different disciplines and fostering transdisciplinarity.

To achieve this objective, all GPSS lectures and courses are held in English and GPSS conducts its own entrance examination in English, thereby enabling international students to complete the programme without having to acquire Japanese language skills. (Studying the Japanese language is recommended for understanding Japanese culture, but is not

Table 6.8.1 Summary of education programmes at the IR3S universities

University	Name of programme/course		Certificate/ degree
The University of Tokyo	Graduate Program in Sustainability Science (GPSS)	An independent master's and doctoral programme associated with the Graduate School of Frontier Sciences. Students are selected through the programme's own admission scheme.	Master/Doctor of Sustainability Science
Kyoto University	Sustainability Science Course	A course in the Department of Environmental Management, School of Global Environmental Studies. Open to students from other departments in Kyoto University.	Certificate (10 credits)
Osaka University	Sustainability Science Education Program	A part of Osaka University's advanced associate programme system. Open to all master's students.	Certificate (8 credits)
Hokkaido University	Hokkaido University Inter-department Graduate Study in Sustainability (HUIGS)	An inter-graduate course of Hokkaido University. Open to all graduate students.	Certificate of Completion of the HUIGS programme (8 credits)
Ibaraki University	Sustainability Science Course	A regular master's course of the Urban System Planning Course of the Graduate School of Science and Engineering. An independent entrance examination is held.	Certificate of Sustainability Science Course (30 credits)
	Sustainability Science Program	Composed of minor programmes offered by all graduate schools of Ibaraki University. Open to all graduate students.	Certificate of Sustainability Science Program (10 credits)

a requirement for completion of GPSS.) As a result, GPSS has been accepting about 20 students every year with diverse academic backgrounds from all over the world. In fact, two-thirds of GPSS students are international students and GPSS is therefore considered an international graduate programme and a frontrunner in the internationalization of the Kashiwa Campus of The University of Tokyo, where GPSS is located. The original fields of specialization of the students range from sociology, economics, international studies, development studies, psychology and language to engineering, bio-science, agricultural science, environmental science and so on. Through study and serious discussion of specific sustainability issues with these diverse colleagues, students in the programme are expected to acquire the necessary competencies, including communication, facilitation and mediation skills and a transdisciplinary viewpoint. For this purpose, GPSS emphasizes hands-on experience, fieldwork and practical exercises as well as lecture-based courses.

The GPSS curriculum consists of three parts: courses oriented to knowledge and concepts, practical courses oriented to experiential learning and skills, and thesis work. The knowledge and concept courses include core courses that provide a holistic view of sustainability and cover relevant knowledge and disciplines associated with sustainability issues, as well as a variety of elective courses selected from a wide range of academic fields, spanning the humanities and sciences, which have been part of DES. The core courses consist of original courses designed for GPSS through coordination with TIGS and IR3S, courses selected from the Environmental Management Program of DES, and courses specifically offered to GPSS by GPSS-supporting departments in DES. A significant portion of the knowledge and concept courses are based on collaboration with DES, since DES has already been working to establish environmental studies as a transdisciplinary field.

The practical courses include exercises intended to foster basic attitudes of acceptance of diversity and respect for minorities, as well as practical skills for action in the real world. These are participatory in nature. Through exposure to diverse student groups and ideas in group discussions and dialogues, students become acquainted with a variety of perspectives among their fellow students and learn the importance of diversity and minorities while acquiring various sustainability-related skills, rather than simply gaining knowledge of the subject matter. The coursework includes: training in the holistic thinking needed to assess sustainability-related issues from a holistic point of view; acquisition of the facilitation and negotiation skills necessary for building consensus; exercises to foster the understanding of cultural diversity that is essential to cross-cultural communication; and a wide range of case studies dealing with various examples of global, international and regional problems.

In their thesis work, students are encouraged to take a transdisciplinary approach to complex sustainability problems by integrating the diverse approaches, ways of thinking and key principles of different disciplines. In some cases, professors from different departments may collaboratively supervise one student as a team. For those who wish to pursue a higher degree in relevant disciplines, the GPSS master's thesis work thus provides a unique experience.

The GPSS management committee, which consists of representatives of the relevant departments in DES, meets every month to discuss how to manage and improve the curriculum. In this way, GPSS is educating students to be professionals and researchers in sustainability science. In October 2009, GPSS started a doctoral programme. Those who complete this programme will be awarded a PhD in Sustainability Science. The doctoral programme tries to contribute to the establishment of "Sustainability Science" through doctoral research.

6-8-3 Kyoto University

In order to systematically tackle complex and wide-ranging environmental problems from the global to the local level, Kyoto University established an independent graduate school in 2002 consisting of the Hall of Global Environmental Research (research unit), the Graduate School of Global Environmental Studies (GSGES, education unit) and the Grove of Universal Learning (research and education support unit). One of the features of this organization is a system known as *kyoudou bunya* (collaborative labs). This system makes it possible for faculty members of other departments to collaborate on education and research without cumbersome procedures. Through this collaboration system, over 100 faculty members from other departments at Kyoto University are involved in research and education at the Hall and the School.

To promote the creation of a sustainability science course, the Kyoto Sustainability Initiative (KSI) was established in 2006 by the GSGES and seven research institutes (Institute of Economic Research, Humanities Research Institute, Center for Southeast Asian Studies, Institute for Chemical Research, Institute of Advanced Energy, Research Institute for Sustainable Humanosphere and Disaster Prevention Research Institute). KSI then created an educational course through collaboration between these institutes. Since Kyoto University did not, at that time, have a framework or system for implementing university-wide graduate school lectures, KSI decided to establish a Sustainability Science Course in the GSGES. Since 2007, subjects in the social sciences, humanities and natural sciences taught by staff from KSI-related departments (KSI-provided subjects) and subjects related to sustainability science taught by staff from other departments (KSI collaborative subjects) have been taught on the Sustainability Science Course as part of the Course in Environmental Management of the GSGES. Since 2008 KSI has also targeted students from other departments within Kyoto University, and has developed an educational programme that goes beyond the departmental framework.

The objective of this programme is to develop human resources capable of playing an international role in creating a sustainable society and, more specifically, to develop professionals who are able to fully understand the diverse, international and interdisciplinary nature of the concept of sustainability and thereby conceive and implement solutions to problems with wide-ranging perspectives that are not limited to existing fields. In terms of content, the GSGES provides lectures on "Global Environmental Law and Policy", "Global Environmental Economics", "Ecosystem Management" and "Environmental Education and Ethics" as compulsory subjects, with optional subjects provided by other research institutes. Students who have taken at least three subjects provided by KSI and a total of at least five subjects provided by KSI and KSI collaborators are able to obtain a certificate. Moreover, to promote the integration of the humanities and the sciences, students majoring in science courses are encouraged to take humanities courses and vice versa.

A feature of Kyoto University's Sustainability Science Course is that faculty members from various fields provide a wide range of subjects, as shown in Table 6.8.2. Another feature is that there is very little gap in terms of content between the research and education fields. This is because sustainability science has not yet been systematized as an academic field and has no standard textbooks, meaning that staff members develop lectures by reorganizing their field of research from the perspective of sustainability. The relationship between KSI's priority research fields and sustainability science education at Kyoto University is shown in Table 6.8.3. The priority research fields are as follows:

- A measures aimed at creating an optimal *recycling-based society*
- B environmental countermeasures based also on economic and technical analysis of *climate change* countermeasures
- C maintaining *intergenerational and intragenerational equity* in order to achieve sustainable development
- D environmental governance in order to improve the effectiveness of environmental countermeasures and *environmental risk management*

As can be seen in Table 6.8.3, many courses are related to priority field A, but not many courses are necessarily related to priority field C. This reflects the difficulties involved in education and research in the latter field.

Table 6.8.2 The 17 subjects provided by Kyoto University's Sustainability Science Course, AY2007 to AY2009

1	C	Global environmental law and policy
2	C	Global environmental economics
3	C	Management of global resources and ecosystems
4	C	Environmental ethics and environmental education
5	O	Introduction to Chinese history of the natural environment
6	O	Nature–human interaction in the Southern Himalayas
7	O	Environment and society in Southeast Asia
8	O	Evaluation methodology on advanced energy systems
9	O	Sustainable energy systems
10	O	Science for diagnostics and control of humanosphere
11	O	Science for creative research and development of humanosphere
12	O	Environmental chemistry and biochemistry
13	O	Catchment processes and sustainable management
14	O	Sustainable catchment management for better protection of lakes, reservoirs and surrounding ocean areas
15	O	Economic analysis of disaster risk management
16	O	Integrated disaster and environmental risks for the development of sustainable society
17	O	Frontier of sustainability science (IR3S joint subject)

Notes: C: compulsory; O: optional.

Through KSI, a distance learning system has been developed. Domestically, this system has been used to provide lectures within Kyoto University (between the Yoshida and Uji campuses) and between universities participating in IR3S. Internationally, it has been tested with good results at conferences with the Institut Teknologi Bandung, the University of Indonesia, Vietnam National University, and the Bangkok Liaison Office of Kyoto University's Center for Southeast Asian Studies.

As a result, the Vietnamese Academy of Science and Technology and Vietnam National University in Hanoi were added to the list of collaborative universities in 2007, the operation of a distance learning system between Kyoto, Thailand, Hanoi and Jakarta was tested, and a lecture was given from Kyoto to Hanoi. In China, demonstration lectures on the economics of sustainable development at Renmin University of China and Fudan University, both locally and remotely, proved the feasibility of the distance learning system as well as raised the awareness of staff and students at these universities.

In 2008, a lecture entitled "Environment and Society in Southeast Asia" was distributed to the Institut Teknologi Bandung via the distance learning system, which raised awareness of sustainability science in Indonesia. Between 2008 and 2009 a series of lectures on "Economics of Sustainable Development" was given at Fudan University as a special subject

Table 6.8.3 The relationship between KSI's priority research fields and the Sustainability Science Course subjects

	A Recycling-based society	B Climate change	C Intergenerational and intragenerational equity	D Environmental risk management
1	Strong	Strong	Strong	Strong
2	Strong	Strong	Strong	Strong
3	Strong	Strong	Weak	Somewhat weak
4	Strong	Strong	Somewhat strong	Somewhat weak
5	Strong	Somewhat strong	Weak	Somewhat weak
6	Strong	Somewhat weak	Somewhat weak	Somewhat weak
7	Strong	Somewhat strong	Strong	Strong
8	Strong	Strong	Weak	Somewhat strong
9	Somewhat weak	Strong	Weak	Somewhat weak
10	Strong	Somewhat strong	Weak	Somewhat weak
11	Strong	Somewhat weak	Weak	Somewhat weak
12	Somewhat weak	Somewhat weak	Weak	Somewhat weak
13	Somewhat weak	Somewhat strong	Weak	Strong
14	Somewhat strong	Somewhat strong	Weak	Strong
15	Somewhat strong	Somewhat weak	Somewhat weak	Somewhat strong
16	Somewhat strong	Somewhat weak	Somewhat weak	Somewhat strong
17	Strong	Strong	Strong	Strong

Note: Relationship with priority research field:

(white)	Weak
(light shading)	Somewhat weak
(medium shading)	Somewhat strong
(dark shading)	Strong

by an invited lecturer, raising awareness of sustainability science in China. With the distribution of KSI's education content to the Asia region in mind, all lectures have been video-recorded.

Cases can now be seen where the knowledge acquired through the Sustainability Science Course is being applied through internships. The Environmental Management course of the GSGES requires an internship of 3–5 months for a master's degree and 10–12 months for a doctorate. The internship programme aims to provide practical experience at actual sites both in Japan and overseas, and continues to be expanded. Students acquire knowledge from lectures in classrooms before starting their internships. In many cases, however, the students are overwhelmed by the problems they encounter on-site and feel that the knowledge they acquired in class is of no use in solving these problems. As a result, the students, with help from their tutors and other staff at the internship, find themselves engaged in an intense effort to determine how to solve these problems and thereby complete their thesis. It is hoped that this process will identify problems that need to be solved and, in turn, create new knowledge that enables their solution, thereby opening up new interdisciplinary fields. At the School of Global Environmental Studies, students who visit Southeast Asia on internships, for example, typically attend lectures on subjects relevant to the internship, such as "Environment and Society in Southeast Asia". In this way, experts in a variety of fields give lectures on the Sustainability Science Course that provide students with opportunities to encounter a wide range of fields and contribute to improving the quality of the internship system.

In terms of results, two students completed the KSI Sustainability Science Course and obtained their certificates in 2009. One of the remaining challenges is how to increase the number of students taking this course, not only from the GSGES but also from other research departments. It is hoped that students who complete this course will contribute to the creation of a sustainable society in a variety of fields.

6-8-4 Osaka University

In April 2008, the Research Institute for Sustainability Science (RISS) at Osaka University launched the Sustainability Science Education Program (also known as the RISS Program), a new graduate education programme offered by Osaka University as part of the IR3S education programme. RISS offers this as a minor programme in Sustainability Science, and any students enrolling in Osaka University's master's programme are eligible to enrol in it. The mission of the RISS Program is to provide students from different academic backgrounds with opportunities to deliberate sustainability issues from a variety of perspectives. The Program also attempts to maintain a diversity of instructors from many academic fields in the curriculum. This aids the dissemination of the concept of sustainability science among Osaka University's faculty members as well as students.

The RISS Program was established as part of Osaka University's advanced associate programme system, a unique system in graduate school education launched by the university in April 2008 that reflects the university's awareness of the need for interdisciplinary education. As of academic year 2009, the RISS Program consisted of 26 courses, comprising 4 Sustainability Science core courses (2 required) and 22 Sustainability Science associate courses (elective). These 26 courses are offered through

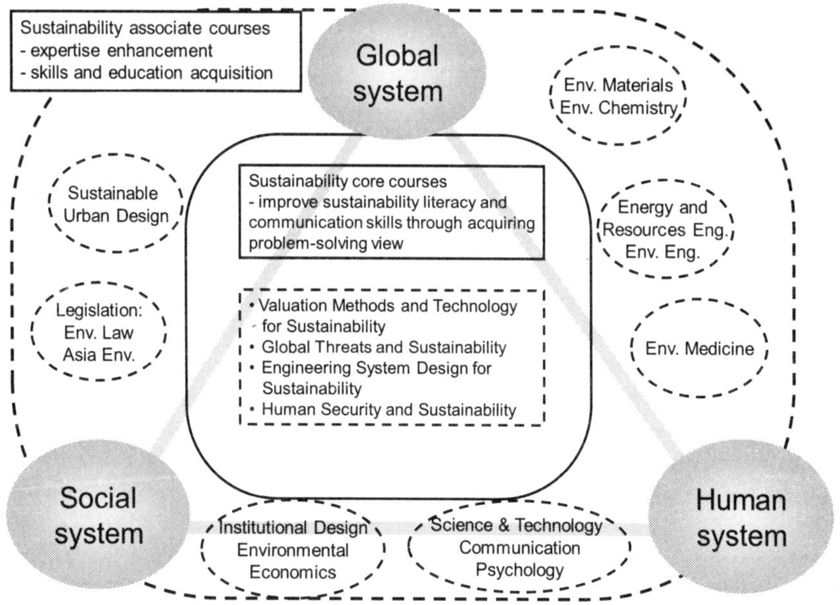

Figure 6.8.1 Overview of the RISS Program, Osaka University.

eight graduate schools at Osaka University, covering a wide range of academic disciplines and approaches (see Figure 6.8.1). The requirements for Program completion are as follows: students should first obtain 8 credits (4 courses) in total from the curriculum courses, of which 4 credits (2 courses) should be from the core courses; students then must meet their master's programme requirements.

The four core courses in Sustainability Science are intended to provide students with opportunities to learn skills and different perspectives essential to understanding the interactive mechanisms within and among the global, social and human systems. Through specific examples, students acquire a holistic knowledge of sustainability issues such as global warming, energy, food and water. Students also learn the use of tools such as lifecycle assessment and the importance of trade-offs between different dimensions, as well as the role of uncertainty and dynamics. The core courses comprise lectures that are primarily provided by faculty members of Osaka University and the other IR3S universities, as well as group discussions, projects and field trips.

The associate courses deal with topics related to sustainability. The current associate courses already existed in the ordinary master's curriculums before the RISS Program started. After investigating the content of most courses in the master's programme at Osaka University, suitable courses for the associate courses were selected.

As of the spring semester of 2009, 39 students were enrolled in the RISS Program. These students were from five different schools: Engineering, Engineering Science, Economics, Human Sciences and Medicine. Of the 39 enrolled students, 32 were from the School of Engineering but belonged to different departments such as Sustainable Energy and Environment, Civil Engineering, Mechanical Engineering, Material Sciences and Business Engineering.

6-8-5 Hokkaido University

The basic concept of the sustainability education programme offered by Hokkaido University was established by one of the working groups of the Sustainability Governance Project (SGP) at Hokkaido University in 2006. The idea is to foster holistic capabilities among graduate students at the disciplinarily separated graduate schools. Based on this notion, the Hokkaido University Inter-department Graduate Study in Sustainability (HUIGS) programme is designed to enhance the ability of students to think "deeply", "widely" and "connectively". Its mission is clearly set out: to ensure that students recognize the limitations of a monodisciplinary approach, especially in relation to today's highly complex global problems, and the need for a strategically integrated or holistic approach to such problems instead. A provisional HUIGS education programme was first offered to graduate students in 2007. The contents of the programme continue to be evaluated as well as strengthened by adding problem-based learning (PBL) courses, core lectures by IR3S and lectures by the Japan International Cooperation Agency (JICA) and the Global Land Project (GLP). In the spring of 2008, the newly established Center for Sustainability Science (CENSUS) took over the HUIGS programme. As a result, the programme has been officially recognized as an inter-graduate course of Hokkaido University (see Figure 6.8.2).

Currently, the HUIGS programme is supported by many of the university's faculty members from nine graduate schools (Letters, Law, Public Policy, Fisheries, Engineering, Agriculture, Economy, Environmental Science and Information Science). Students enrolled in the programme are from 13 graduate schools (Letters, Law, Public Policy, Fisheries, Engineering, Agriculture, Economy, Environmental Science, Medicine, Education, Information Science, International Media, and Communication & Tourism). The lecturers are from Japan, New Zealand and Nigeria, and the students are from Japan, Bangladesh, the Republic of Korea, China,

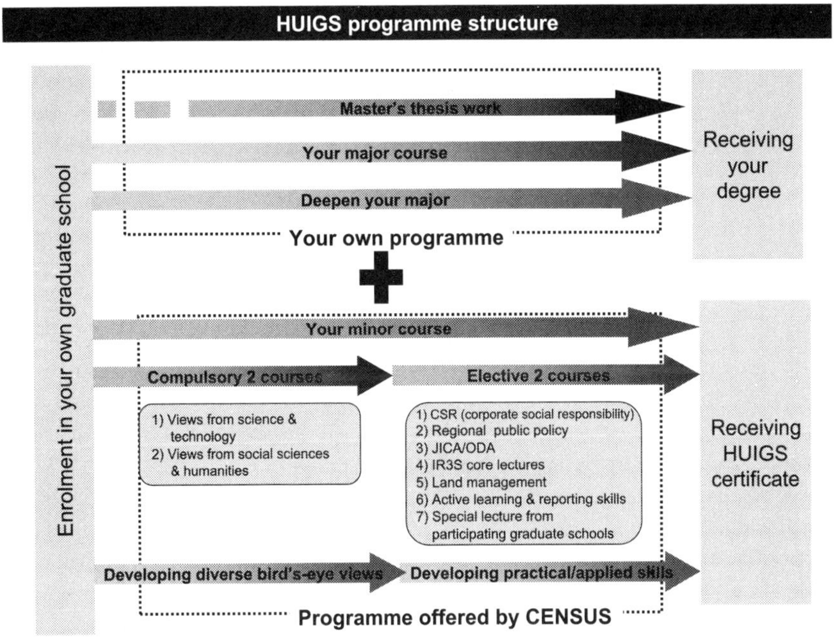

Figure 6.8.2 Structure of the HUIGS education programme.

Thailand, Indonesia, Malaysia, Tunisia and Nigeria. The programme courses are offered as compulsory and elective courses. These courses are mainly taught in English, although Japanese is sometimes used for the sake of the Japanese students.

Applicants to the HUIGS programme comprise students entering master's courses in any of Hokkaido University's graduate schools. Those recognized as possessing skills equivalent to or higher than a master's degree are also eligible to enrol. The programme requires students to select two compulsory "bird's-eye view" courses and at least two elective courses from a HUIGS minor list. Currently, 80 HUIGS minor courses are available from both participating graduate schools and CENSUS. Only 20 per cent of the minor courses from the graduate schools are currently taught in English. CENSUS is now offering five elective HUIGS minor courses in English and is strongly recommending these to the participating students.

Those students successfully completing their own degree programme and HUIGS courses are awarded a Certificate of Completion of the HUIGS programme by the director of CENSUS. At the end of the 2008 academic year, 30 students, half of whom were from overseas, had successfully completed the programme.

Establishment of the HUIGS programme is now complete. As a next step, CENSUS has launched an "intensive" training course for sustainability professionals based upon the foundation knowledge acquired during the HUIGS programme. This new programme aims to foster practitioners who will adopt a global perspective in developing regional sustainability, particularly in Asia and Africa.

6-8-6 Ibaraki University

Ibaraki University launched its Graduate Program on Sustainability Science in April 2009. Figure 6.8.3 shows an outline of this education programme. An interdisciplinary programme, it is designed for postgraduates and is composed of the Sustainability Science Course and the Sustainability Science Program. The Sustainability Science Course is one of the regular master's courses of the Urban System Planning Course of the Graduate School of Science and Engineering. The Sustainability Science Program is composed of minor courses offered by all graduate schools of Ibaraki University (Humanities, Education, Science and Engineering, and Agriculture).

These courses aim to develop not only advanced expertise but also three other competencies. The first competency is holistic knowledge of the broad range of issues associated with sustainability science, so as to enable students to adopt different viewpoints as well as position their own expertise in various fields. The second is the development of communication skills to enable students to understand others and form relationships, collaborating skills to encourage various stakeholders to address issues, and problem-solving skills to identify real issues and resolve conflicts. The third is "mind" or "heart", which includes the motivation to dedicate oneself to the public, having one's own beliefs to maintain that motivation, and synchronic/diachronic consciousness.

This education model can be referred to as "Education across Mind–Skill–Knowledge" (*Shin-Gi-Chi* in Japanese). This is an analogy derived from a proverb about traditional Japanese sports that defines *Shin-Gi-Tai* as the three elements essential to being a great athlete. *Shin-Gi-Tai* expresses the need in competitive sports for a comprehensive foundation combining physical ability (*Tai* = body), sophisticated athletic skills (*Gi* = skill) and a sound mind (*Shin* = mind). The replacement of *Tai* with *Chi* (knowledge) yields the expression *Shin-Gi-Chi*: "Mind–Skill–Knowledge".

In order to achieve this objective, the curriculum is composed of three main categories (Figure 6.8.3). First, basic subjects for holistic knowledge aim to help students understand the structures of the global system, the

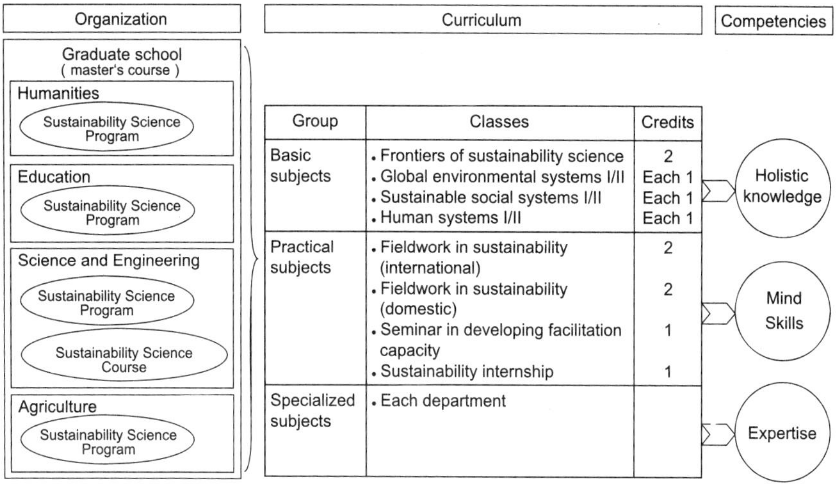

Figure 6.8.3 Graduate Program on Sustainability Science at Ibaraki University.

social system and the human system and to consider the interactions between these systems from an integrated viewpoint. Second, practical subjects aim to nurture the skills and mind needed to work in international or domestic fields where students will experience complex problems and must communicate and work with people living with those problems. Lastly, specialized subjects are offered with the expectation that recipients of sustainability education need expertise in a real field of specialization to which they can commit themselves.

The practical subjects in particular are one of the essential components of the education programme. In 2009, the programme initiated an international fieldwork seminar (Fieldwork in Sustainability) at Mai Khao village in Phuket, Thailand. To operate this fieldwork programme cooperatively, Ibaraki University established an academic exchange agreement with Phuket Rajabhat University in 2008. Additionally, a domestic fieldwork seminar deals with issues in Oarai City, in Ibaraki Prefecture.

Table 6.8.4 illustrates the difference between the Sustainability Science Course and the Sustainability Science Program. In the case of the Sustainability Science Course, subjects are included in the regular curriculum (a total of 30 credits must include 6 credits from basic subjects and 2 credits from practical subjects) and students earning the required units of credit will receive a Certificate of Sustainability Science Course in addition to their master's degree. In the case of the Sustainability Science Program, students can receive a Certificate of Sustainability Science Program once they have fulfilled the requirement of 6 credits from basic subjects or practical subjects and 4 credits from specialized subjects

Table 6.8.4 Differences between Ibaraki University's Sustainability Science Course and Sustainability Science Program

	Sustainability Science Course	Sustainability Science Program
Requirement	30 credits from the regular curriculum (includes 6 credits from basic subjects and 2 credits from practical subjects)	6 credits (from basic subjects or practical subjects) and completion of 4 credits in specialized subjects authorized by the respective graduate schools
Degree or Certificate	Certificate of Sustainability Science Course	Certificate of Sustainability Science Program
Students	Graduate students	Graduate students
Organization	Graduate School of Science and Engineering	All graduate schools at Ibaraki University (Humanities, Education, Science and Engineering, Agriculture)

authorized by the respective graduate schools (this requirement differs slightly depending on the graduate school).

Ibaraki University anticipates that graduates who finish this education programme will be equipped to contribute both locally and globally in such fields as government, business, education, international organizations, non-governmental organizations and non-profit organizations.

6-8-7 Joint Educational Program of IR3S

As the above descriptions make clear, each of the five participating IR3S universities runs its own distinctive sustainability science education programme, but all have cooperated in developing and running a single joint education programme as a common core. This is known as the Joint Educational Program of IR3S. Students who fulfil the requirements for completion of this Program receive a joint completion certificate under the name of IR3S.

The Program is composed of (1) a new course, Frontier of Sustainability Science, that is jointly offered by the five IR3S universities; (2) two subjects that are offered by individual universities and provide a holistic view of the diverse issues of sustainability; and (3) an additional two subjects that are locally available and defined by different universities. These subjects as a whole should provide 10 credits in total. Because the actual conditions for taking the Program may differ according to the particular system of the participating universities, each university issues its own guidance on the joint programme.

The Frontier of Sustainability Science course is jointly offered by the five universities and is delivered in a multi-point, interactive format using a remote lecture system. It is an omnibus course involving top-flight researchers in sustainability science from the five IR3S universities. With different lecturers at each university presenting their own unique courses, having a framework that allows the sharing of teaching materials among universities has proven particularly useful in such a broadly interdisciplinary subject as sustainability science. The first group of students completed the course in March 2009. This course has not only enabled a large number of students to experience lectures with a diversity of powerful messages from different universities, but also facilitated interaction between students from each university.

Note

1. See the website of GPSS at <http://www.sustainability.k.u-tokyo.ac.jp/intro/index_e.html> (accessed 21 June 2010).

REFERENCE

Komiyama, H. and K. Takeuchi (2006) "Sustainability Science: Building a New Discipline", *Sustainability Science* 1: 1–6.

6-9

Conclusion

Takashi Mino

This chapter was co-authored by the education officers of the five universities participating in the IR3S Joint Educational Program and is based on their discussions about sustainability science in the course of the programme's development. Thus far, the principles and background of sustainability science education have been explained, as well as its content, methods and objectives, and the educational initiatives of IR3S have also been introduced.

Since 2006, the five participating universities have each established postgraduate programmes in sustainability science. Upon implementing these programmes at the respective institutions, it was discovered that a significant number of students indicate a considerable level of interest in the study of sustainability science and its holistic approach to a transdisciplinary field using integrative methods. However, the fact remains that universities intending to offer this discipline to interested students are developing their own original curriculums owing to a lack of systematic organization of the educational content and pedagogy required for higher education. Based on the authors' experiences with sustainability science through the IR3S programme, a point that needs to be emphasized is that, as mentioned several times in the various sections of this chapter, simply providing knowledge of sustainability along with the skills for its application within society does not satisfy the aims of sustainability science education. Rather, the challenge lies in how to imbue students on the programme with the strong motivation necessary to move the world in a more sustainable direction.

There is one set of issues that has not been sufficiently addressed in this chapter, namely, the questions of what it means to contemplate sustainability in Asia and what type of sustainability science education Japan should be disseminating to the world. On the question of Asia – particularly when considering the social reforms taking place in China and India in regard to population issues and disparities of wealth, as well as Japan's response to past environmental problems and its shrinking population – it becomes apparent that sustainability education needs to be discussed in the context of Asia's regional and cultural idiosyncrasies. However, Asians can also contribute to global sustainability by introducing certain tenets of Asian philosophy to the world through education, particularly a worldview based on coexistence (that is, the perception that humans do not control nature but rather are a part of nature and are therefore linked in some way to all things within the natural environment). Although some may suggest that using international education as a forum for introducing traditional Asian or Japanese sensibilities is counterproductive in terms of both rationality and efficiency, it may in fact provide sustainability science, which is currently seeking a new paradigm, with the impetus it needs to create a new set of values for the next generation. These values have to be gleaned not only from Asia but from the diverse array of cultures that exist throughout the world.

7

Conclusion

7-1

Building a global meta-network for sustainability science

Kazuhiko Takeuchi

7-1-1 The G8 University Summit and a network of networks

Sustainability science has grown through the efforts, both individual and collective (via organizations such as the Alliance for Global Sustainability), of some of the top academic and research institutions in North America, Europe and Asia. The findings and proposals generated by these efforts often reflect the particular strengths of their institutions of origin. The Integrated Research System for Sustainability Science (IR3S), for example, tends to be stronger in the areas of engineering and the natural sciences, but less so in the humanities and the social sciences, an imbalance that has been frequently cited by the Organizational Evaluating Committee of the Program for Encouraging Development of Strategic Research Centers (Super COE) and that IR3S has sought to remedy. This committee, which consists of globally prominent experts in diverse fields who possess an objective viewpoint, ensures the transparency of evaluations and appraisals of the IR3S project. Despite its best efforts, however, even a research network such as IR3S, which brings together universities and other institutions throughout Japan, cannot by itself adequately address the multifarious problems associated with global sustainability.

IR3S has therefore committed itself to helping expand and strengthen the networks related to sustainability science now being set up in a number of countries, and to fostering international and interregional cooperation among these networks. The need for such an effort was a major topic of discussion at the G8 University Summit held in 2008 in Sapporo, Japan, concurrently with the G8 Summit in nearby Toyako. Attending the

University Summit were the heads of prominent universities from the G8 nations and outreach nations attending the Toyako Summit. Chaired by Hiroshi Komiyama, the G8 University Summit consisted of two days of active discussion of the role of universities in achieving global sustainability. The gathering concluded with the adoption of the Sapporo Sustainability Declaration, which cites the importance of developing knowledge innovation to support innovation in the sciences and society, and of constructing a "network of networks" that will unite research networks around the globe.

Integration and cooperation among diverse academic fields is a prerequisite for an effective response to the broad-ranging problems associated with sustainability. What sufficed in the past as collaboration on the individual or university level must now extend around the globe. The Sapporo Declaration therefore calls for the creation of a network of networks linking existing networks (such as IR3S) of universities and research institutions in a more extensive network to enable cooperation that will more effectively utilize the respective strengths of its members. By increasing opportunities for high-level joint research projects and student exchanges among members of existing networks, the network of networks will provide the framework for the development of a new, integrated base of scientific knowledge leading to solutions to the complex problems of sustainability.

Construction of this network of networks will facilitate the coordination and coexistence of global efforts to achieve a sustainable society on an international level and local efforts to preserve natural and cultural diversity on a regional level. The building of a sustainable society that embraces the scenarios for a low-carbon society, a resource-circulating society and a society in harmony with nature is an objective to be shared by people everywhere, but the specific means of achieving such a society will necessarily vary from nation to nation and region to region. Creation of a vital and diverse global society is predicated on the mutual fulfilment of both these objectives. To cite one example, renewable energy can take many forms – solar, wind, biomass, geothermal – and the choice should depend on which types of power maximize the potential of the region in question. Renewable energy is particularly suited to decentralized power generation and can therefore help promote the local production and consumption of energy.

7-1-2 A sustainability science meta-network

With additional funding received from the Special Coordination Funds for Promoting Science and Technology of Japan's Ministry of Education,

Culture, Sports, Science and Technology (MEXT), IR3S began constructing an international meta-network for sustainability research in 2008. Since then it has been actively engaged in outreach with research networks in North America and Europe. These efforts include an IR3S-hosted symposium at the annual meeting of the American Association for the Advancement of Science, whose members include Harvard University, the leading institution in sustainability science in North America. In Europe, IR3S has formed close ties with the Tyndall Centre for Climate Change Research in the United Kingdom, the Stockholm Resilience Centre in Sweden, and the Interuniversity Research Centre on Sustainable Development (CIRPS) in Italy. In February 2009, IR3S invited representatives of these networks to the International Conference on Sustainability Science (ICSS) 2009 held on The University of Tokyo's Hongo campus; this was the first such international gathering devoted to the formation of a research meta-network. The second ICSS conference was held in June 2010 at Sapienza University of Rome.

In Asia, a primary area of focus for IR3S, it is planned to initiate the formation of a sustainability science meta-network among universities in the region that have long been engaged in academic exchanges: Peking University, Tsinghua University and Zhejiang University in China, Seoul National University in the Republic of Korea, Vietnam National University, Hanoi, and the Asian Institute of Technology in Thailand. The hope is that this network will serve not only for research but also for educational exchanges as an Asian version of Europe's ERASMUS academic exchange programme. In the future, this linkage could facilitate the establishment of cooperative education programmes and dual or joint degrees among participating universities. In 2008, the author participated as a lecturer in one such joint education programme attended by students from Peking University and Seoul National University.

IR3S will also endeavour to establish a meta-network of universities and research institutes in Africa, utilizing the ties that United Nations University (UNU) has formed with institutions on that continent. In addition to joining IR3S as a cooperating institution, UNU has also partnered with IR3S to form the Joint Sustainability Initiative (IR3S/UNU-JSI), which will promote cooperative research and education in Africa through the Ghana-based UNU Institute for Natural Resources in Africa (UNU-INRA).

As an example of this commitment, UNU hosted a symposium in February 2009 on "The Role of Universities in the Promotion of Education for Sustainable Development" with support from MEXT. A second symposium, co-sponsored with IR3S, was held in Ghana in October 2009.

In a relatively short time, IR3S has thus evolved from its origins as a Japan-based sustainability science research network into a meta-network centre of global reach and perspective. To what extent IR3S can continue to meet the challenges of leadership in this meta-network building process remains to be seen. It is in the same spirit, however, that IR3S has published the international academic journal *Sustainability Science* since its inception. Through efforts such as these, IR3S hopes to contribute to the establishment of Asia as a locus of academic activity rivalling that of the West, and to the growth of academic activity in the developing nations of Asia and Africa.

7-1-3 The global contribution of IR3S

In terms of the scope of research undertaken, the activities of scholars and scientists in various parts of the world (including those with IR3S) associated with the creation of sustainability science represent a clear trend of expansion beyond an environmental studies framework towards the broader one of sustainability science. Environmental studies evolved by extracting the environment-related elements of various academic fields and melding them into a new, exclusively environment-oriented discipline. Naturally this process has involved some integration with other disciplines, but the prevailing tendency has been to seek to establish an independent field of purely environmental studies. By contrast, sustainability science aims to integrate entire existing disciplines into an all-inclusive area of endeavour.

The problem with the first approach is that the distillation of the environmental elements of various disciplines into a "pure" environmental studies does not lend itself to addressing the complex problems that must be solved to achieve a sustainable society. Sustainability science takes a more comprehensive, structural approach to these problems by treating the environment, economics and society as equally relevant components whose complex interactions give rise to the problems at hand. This approach makes feasible such concepts as devising a more sustainable society for the twenty-first century by linking the scenario of a low-carbon society to that of an ageing society. In fact, Hiroshi Komiyama, a co-editor of this volume, has proposed the implementation of just such a concept, which he calls the Platinum City Network, on an experimental basis in Japanese cities.

As noted in Chapter 1 of this volume, IR3S has, since its inception, treated sustainability science as a study of the interrelationships between global, social and human systems. It is extremely significant that this framework has been linked to the vision of a sustainable society that combines the attributes of a low-carbon society, a resource-circulating society and a society in harmony with nature.

The series of which this volume is a part overviews the concepts and methodology of sustainability (this volume), examines the three scenarios of a low-carbon society (Sumi, Mimura and Masui, eds, 2011, *Climate Change and Global Sustainability: A Holistic Approach*), a resource-circulating society (Morioka, Hanaki and Moriguchi, eds, 2011, *Establishing a Resource-Circulating Society in Asia: Challenges and Opportunities*) and a society in harmony with nature (Osaki, Braimoh and Nakagami, eds, 2011, *Designing Our Future: Local Perspectives on Bioproduction, Ecosystems and Humanity*), and discusses how to achieve global sustainability (Sawa, Iai and Ikkatai, eds, 2011, *Achieving Global Sustainability: Policy Recommendations*). The three visions of society are intimately linked through problems involving energy, resources and ecosystems. Merging these visions is a first step towards building a comprehensively sustainable society, and global-scale simulations have been conducted to this end.

A simulation project undertaken jointly by Japan's National Institute for Environmental Studies and The University of Tokyo studied a variety of problems under a scenario of reducing greenhouse gases to half their 1990 level by 2050, dramatically increasing the recycling and longevity of iron and other natural resources, and halting the degradation, both quantitative and qualitative, of the world's forests to achieve "no net loss" globally. The study concluded that, with a concerted effort by humanity, a sustainable society combining the aforementioned three societal visions could indeed be attained by 2050.

The integration of these three scenarios is closely linked to the coordination of efforts associated with several United Nations conventions that pertain directly to sustainability. The relationship between a low-carbon society and a society in harmony with nature has its direct corollary in the United Nations Framework Convention on Climate Change and the Convention on Biological Diversity. Meanwhile, the United Nations Convention to Combat Desertification is germane to the fact that climate and ecosystem changes contribute to the degradation of land and human livelihood in the arid regions of the world.

All three of these UN conventions emerged from the United Nations Conference on Environment and Development (better known as the Earth Summit) held in Rio de Janeiro in 1992. Since then, however, each convention has established its own office and held its own meetings of specialists and representatives from its signatory countries. This mutual isolation presents an obstacle to discussion of the interrelationships between these conventions and the issues they address, and many voices are now heard calling for efforts at greater synergy between the conventions.

Whatever mitigative measures are undertaken from this point on, the inevitability of climate change in this century makes the development of adaptive strategies a pressing issue. Because climate change exerts a powerful impact on ecosystems, it is particularly important that such strategies incorporate the concept of ecosystem resilience. For regions most vulnerable to climate change, such as the deltas and islands of Asia and the Pacific, or the arid lands of Asia and Africa, what is needed is more "adaptation research" on how to respond to the impact of climate and ecosystem changes.

To promote this kind of research, United Nations University[4] and IR3S are currently collaborating on the construction of a University Network for Climate and Ecosystems Change Adaptation Research (UN-CECAR) in Asia and Africa. In devising adaptation strategies through such research, it is crucial to give adequate weight to the natural, societal and cultural characteristics of each region studied. In the larger sense, this is simply a reflection of the fact that the response to global problems must always place priority on local solutions that accommodate local characteristics.

If one reviews the history of Japan's response to its own environmental problems, one sees that it began, and in many cases ended, with localized, "negative" solutions to air and water pollution. Although these initial measures did indeed improve local air and water quality, they were followed by calls for the creation of a more beautiful, more livable environment. Now, as environmental problems intensify on a global scale, the validity of this historically localized, negative-solution approach to such problems is called into question.

Global environmental problems cannot be solved merely through the application of technology-based strategies oriented to treating the environment, such as those employed to address pollution in the past. Fundamental changes in technology, economics, society and even values are required. At present, however, there is a lack of a clearly articulated, positive vision of the society that is desired for the future, as well as a road map for getting there from the stage of piecemeal removal of the problems immediately at hand. The disciplines known by such names as "hope studies" and "happiness theory" that support the cultivation of new value systems can and should contribute to this discussion.

The linkage of the low-carbon society scenario to that of an ageing society in debates over a future vision for nations and cities is a relevant example of how to design such visions. Both the low-carbon society and the ageing society need to be viewed as positive opportunities for restructuring nations and their cities so as to improve the welfare of local citizens. From the standpoint of sustainability, what is important here is that local revitalization be based on the real economy of the region in question.

This approach to discussion of the issues the Earth faces is made possible by the structuring of the problems and knowledge associated with sustainability science. In this regard, the concept of sustainability science as articulated by IR3S is both applicable to and significant for the world at large. As IR3S pursues the knowledge innovation and network-of-network (or meta-network) strategies described above, it is hoped to further solidify its collaboration with other groups around the world engaged in the development of sustainability science.

Index

"n" refers to notes

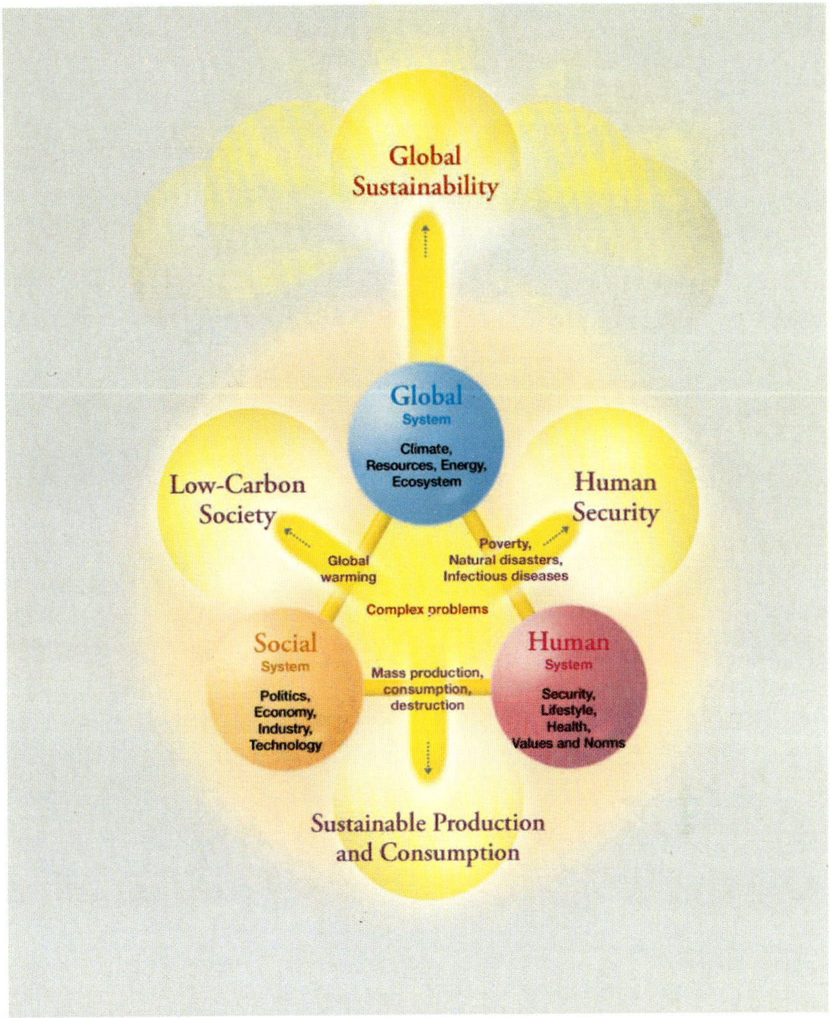

Figure 1.1.1 Addressing sustainability science through the lens of three systems, and the linkages among them.
Source: Komiyama and Takeuchi (2006).
Note: Please see page 8 for this figure's placement in the text.

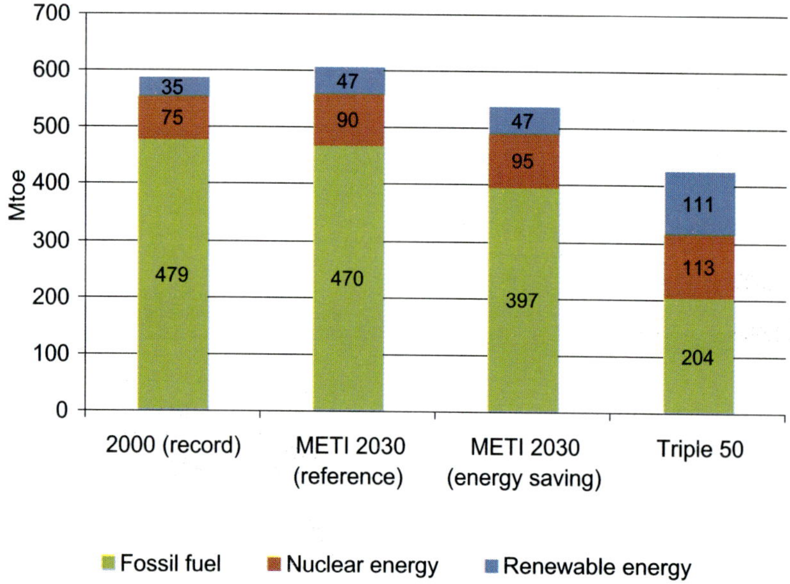

Figure 1.1.4 The "Triple 50" scenario for Japan: Forecasts of long-term energy demand in 2030 by the Agency for Natural Resources and Energy (Ministry of Economy, Trade and Industry) and Triple 50.
Source: Yuhara (2008: 4).
Note: Please see page 15 for this figure's placement in the text.

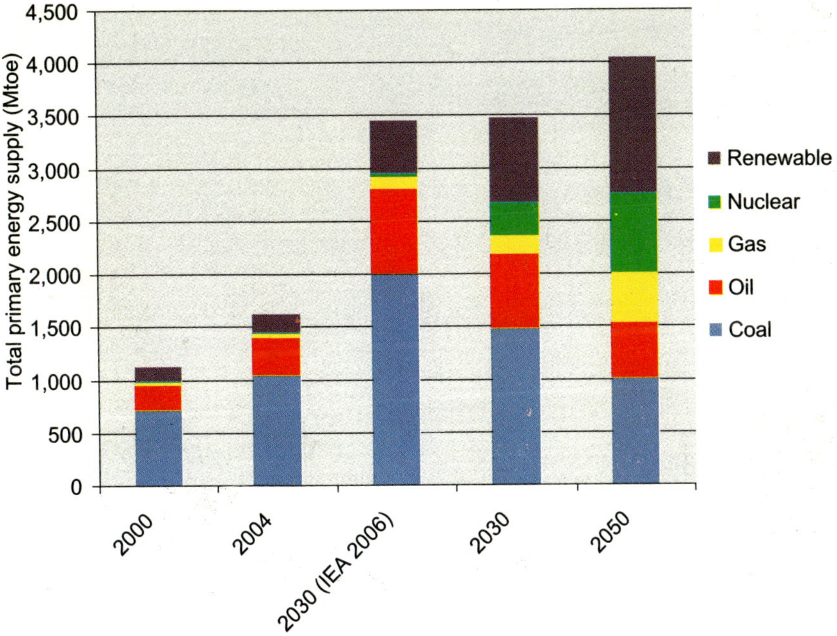

Figure 1.1.5 The "Triple 50" scenario for China: China's energy mix in 2000, 2030 and 2050.
Source: Yuhara (2008: 12).
Notes: 1. Given the current energy situation in China, the possibility of achieving the "Triple 50" scenario in China is not envisaged until 2050.
2. Please see page 16 for this figure's placement in the text.

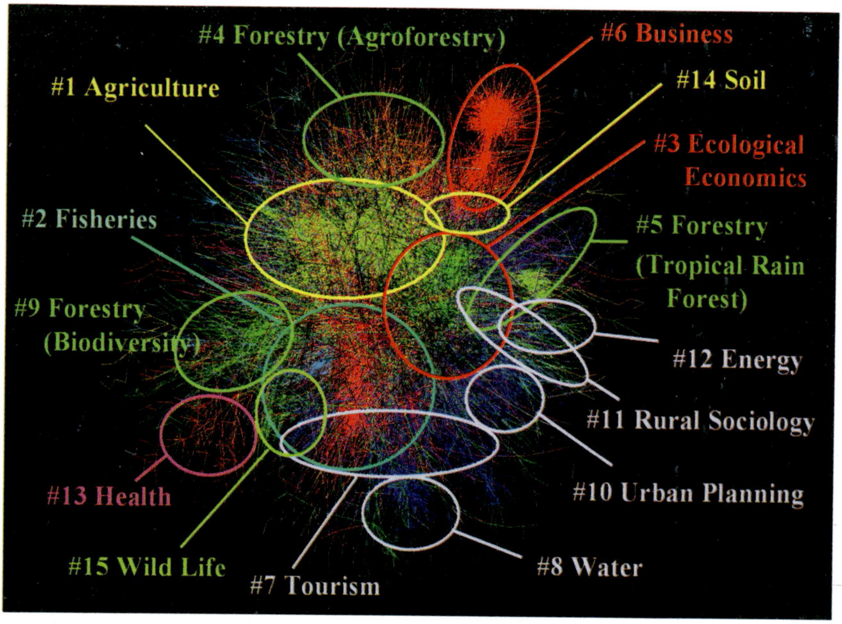

Figure 2.1.1 Visualization of the citation network of sustainability science.
Source: Kajikawa et al. (2007).
Note: Please see page 25 for this figure's placement in the text.

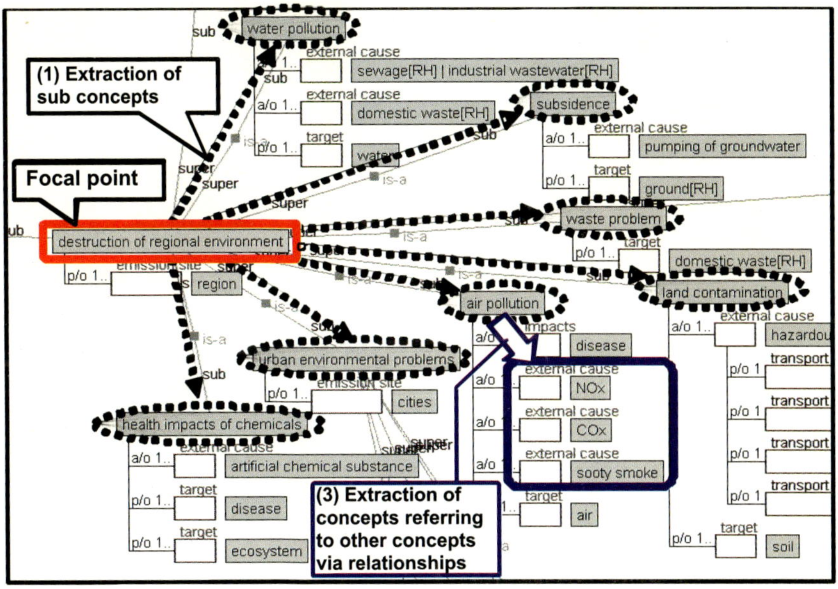

Figure 2.3.4 The mechanism of exploration.
Note: Please see page 60 for this figure's placement in the text.

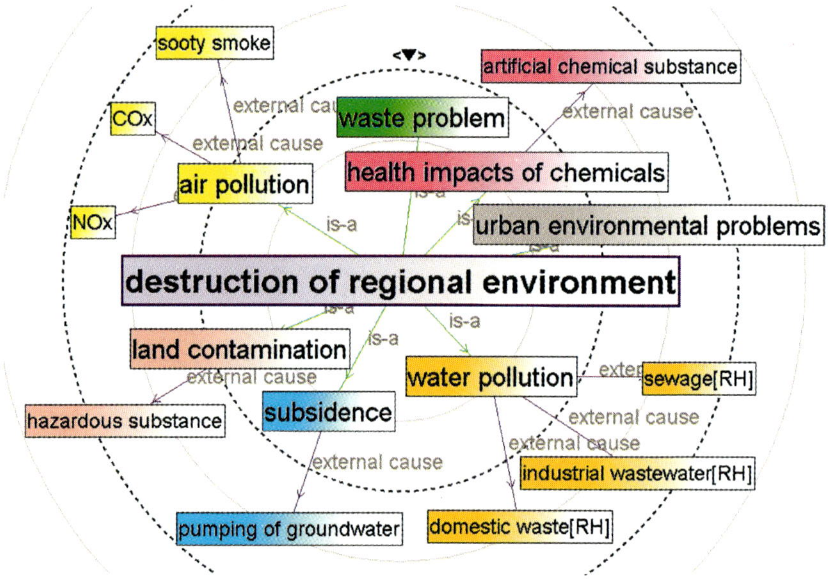

Figure 2.3.5 An example of a conceptual map.
Note: Please see page 61 for this figure's placement in the text.